THE COMPLETE U.S. ARMY SURVIVAL GUIDE TO

TROPICAL,

DESERT, COLD WEATHER, MOUNTAIN TERRAIN, SEA, AND NBC ENVIRONMENTS

U.S. Army Survival Series

THE COMPLETE U.S. ARMY SURVIVAL GUIDE TO TROPICAL, DESERT, COLD WEATHER, MOUNTAIN TERRAIN, SEA, AND NBC ENVIRONMENTS

Edited by Jay McCullough

Skyhorse Publishing

Skyhorse Publishing books may be purchased in bulk at special discounts for sales promotion, corporate gifts, fund-raising, or educational purposes. Special editions can also be created to specifications. For details, contact the Special Sales Department, Skyhorse Publishing, 307 West 36th Street, 11th Floor, New York, NY 10018 or info@skyhorsepublishing.com.

Visit our website at www.skyhorsepublishing.com.

10 9 8 7 6 5 4 3 2 1

Library of Congress Cataloging-in-Publication Data is available on file.

Cover design by Tom Lau

Print ISBN: 978-1-5107-0745-0
Ebook ISBN: 978-1-5107-0750-4

Printed in the United States of America

Table of Contents

Introduction

Although we are an incredibly adaptive and widespread species, all humans have certain criteria for survival, and the many habitats where we live—or may find ourselves—emphasize some criteria over others. Since the U.S. Army is global in nature, it has seen fit to train its soldiers in a variety of possible combat environments: The deserts of the Middle East; the jungles of Southeast Asia, Africa, or South America; the polar conditions of Alaska and Northern Europe; the mountain ranges of Afghanistan; as well as conditions at sea.

Here we find the tips, tactics, and techniques of basic survival in these environments, culled from the U.S. Army's own training manuals, and drawn from decades of hard-won practical experience. the best chance for survival anticipates the problems these environments present, and the primary way to achieve that is prior familiarity with those problems and how to solve them. In these pages we find out how to deal with sharks, how to begin rappelling, how and where to get water in the desert, how to treat fungal diseases, and over and above, how to face the many challenges to survival and comfort in those extreme conditions: nuclear, biological, and chemical environments.

There is no substitute for practical training. If we would be prepared for any eventuality, as the Army has seen to do with its training manuals, it would be wise to commit these techniques to memory, and to practice them as necessary.

—Jay McCullough
April 2016
North Haven, Connecticut

CHAPTER 1

TROPICAL SURVIVAL

Most people think of the tropics as a huge and forbidding tropical rain forest through which every step taken must be hacked out, and where every inch of the way is crawling with danger. Actually, over half of the land in the tropics is cultivated in some way.

A knowledge of field skills, the ability to improvise, and the application of the principles of survival will increase the prospects of survival. Do not be afraid of being alone in the jungle; fear will lead to panic. Panic will lead to exhaustion and decrease your chance of survival.

Everything in the jungle thrives, including disease germs and parasites that breed at an alarming rate. Nature will provide water, food, and plenty of materials to build shelters.

Indigenous peoples have lived for millennia by hunting and gathering. However, it will take an outsider some time to get used to the conditions and the nonstop activity of tropical survival.

TROPICAL WEATHER

High temperatures, heavy rainfall, and oppressive humidity characterize equatorial and subtropical regions, except at high altitudes. At low altitudes, temperature variation is seldom less than 10 degrees C and is often more than 35 degrees C. At altitudes over 1,500 meters, ice often forms at night. The rain has a cooling effect, but when it stops, the temperature soars.

Rainfall is heavy, often with thunder and lightning. Sudden rain beats on the tree canopy, turning trickles into raging torrents and causing rivers to rise. Just as suddenly, the rain stops. Violent storms may occur, usually toward the end of the summer months.

Hurricanes, cyclones, and typhoons develop over the sea and rush inland, causing tidal waves and devastation ashore. In choosing campsites, make sure you are above any potential flooding. Prevailing winds vary between winter and summer. The dry season has rain once a day and the monsoon has continuous rain. In Southeast Asia, winds from the Indian Ocean bring the monsoon, but it is dry when the wind blows from the landmass of China.

Tropical day and night are of equal length. Darkness falls quickly and daybreak is just as sudden.

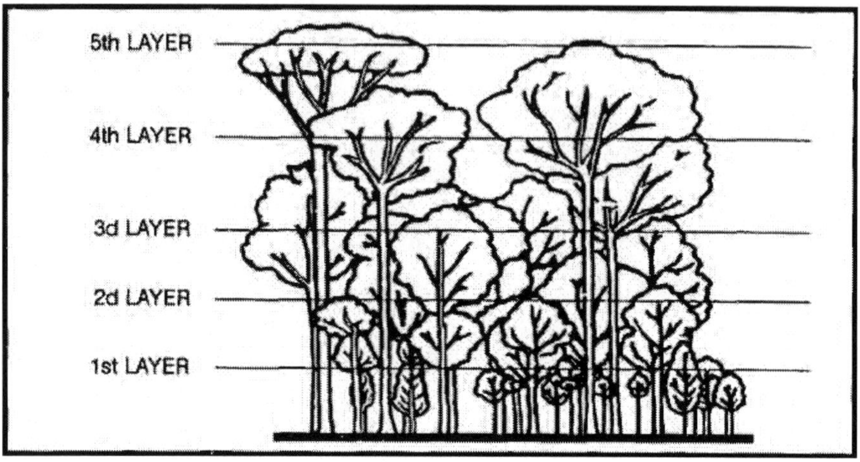

Figure 1-1: Five layers of tropical rain forest vegetation.

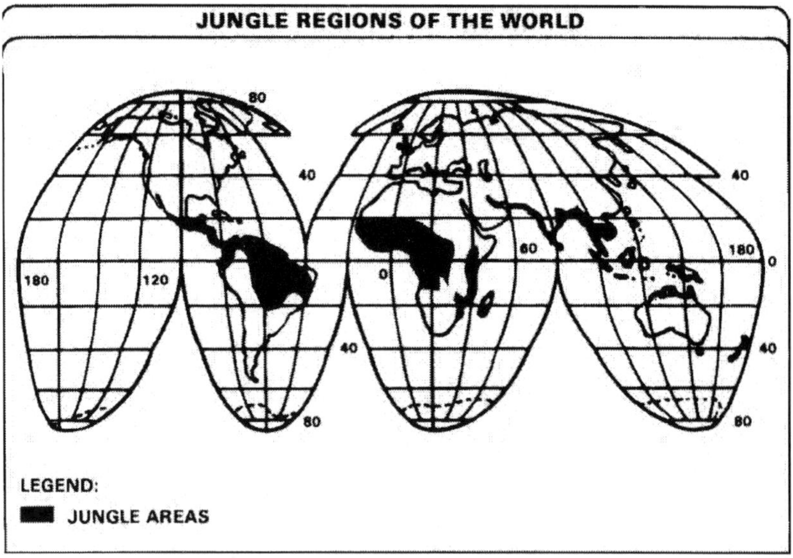

Figure 1-2: Jungle Regions of the World

TYPES OF JUNGLES

The jungle environment includes densely forested areas, grasslands, cultivated areas, and swamps. Jungles are classified as primary or secondary jungles based on the terrain and vegetation.

PRIMARY JUNGLES

These are tropical forests. Depending on the type of trees growing in these forests, primary jungles are classified either as tropical rain forests or as deciduous forests.

Tropical Rain Forests. The climate varies little in rain forests. You find these forests across the equator in the Amazon and Congo basins, parts of Indonesia, and several Pacific islands. Up to 3.5 meters of rain fall evenly throughout the year. Temperatures range from about 32 degrees C in the day to 21 degrees C at night.

Tropical Rain Forest

Deciduous Forest

There are five layers of vegetation in this jungle. Where untouched by man, jungle trees rise from buttress roots to heights of 60 meters. Below them, smaller trees produce a canopy so thick that little light reaches the jungle floor. Seedlings struggle beneath them to reach light, and masses of vines and lianas twine up to the sun. Ferns, mosses, and herbaceous plants push through a thick carpet of leaves, and a great variety of fungi grow on leaves and fallen tree trunks.

Because of the lack of light on the jungle floor, there is little undergrowth to hamper movement, but dense growth limits visibility to about 50 meters. A wet and soggy surface make vehicular traffic difficult. Foot movement is easier in tropical forests than in other types of jungle. You can easily lose your sense of direction in this jungle, and it is extremely hard for aircraft to see you.

Deciduous Forests. These are found in semitropical zones where there are both wet and dry seasons. In the wet season, trees are fully leaved; in the dry season, much of the foliage dies. Trees are generally less dense in deciduous forests than in rain forests. This allows more rain and sunlight to filter to the ground, producing thick undergrowth. In the wet season, with the trees in full leaf, observation both from the air and on the ground is limited. Movement is more difficult than in the rain forest. In the dry season, however, both observation and traffic ability improve.

The characteristics of the American and African semi-evergreen seasonal forests correspond with those of the Asian monsoon forests. These characteristics are—

- Their trees fall into two stories of tree strata. Those in the upper story average 18 to 24 meters; those in the lower story average 7 to 13 meters.
- The diameter of the trees averages 0.5 meter.
- Their leaves fall during a seasonal drought.

Except for the sage, nipa, and coconut palms, the same edible plants grow in these areas as in the tropical rain forests.

You find these forests in portions of Columbia and Venezuela and the Amazon basin in South America; in portions of southeast coastal Kenya, Tanzania, and Mozambique in Africa; in Northeastern India, much of Burma, Thailand, Indochina, Java, and parts of other Indonesian islands in Asia.

SECONDARY JUNGLES

These are found at the edge of the rain forest and the deciduous forest, and in areas where jungles have been cleared and abandoned.

Secondary Jungle

Secondary jungles appear when the ground has been repeatedly exposed to sunlight. These areas are typically overgrown with weeds, grasses, thorns, ferns, canes, and shrubs. Foot movement is extremely slow and difficult. Vegetation may reach to a height of 2 meters. This will limit observation to the front to only a few meters.

Such growth happens mainly along river banks, on jungle fringes, and where man has cleared rain forest. When abandoned, tangled masses of vegetation quickly reclaim these cultivated areas. You can often find cultivated food plants among this vegetation.

Tropical Scrub and Thorn Forests. The chief characteristics of tropical scrub and thorn forests are—

- There is a definite dry season.
- Trees are leafless during the dry season.
- The ground is bare except for a few tufted plants in bunches; grasses are uncommon.
- Plants with thorns predominate.
- Fires occur frequently.

You find tropical scrub and thorn forests on the west coast of Mexico, Yucatan peninsula, Venezuela, Brazil; on the northwest coast and central parts of Africa; and in Asia, in Turkestan and India.

Within the tropical scrub and thorn forest areas, you will find it hard to obtain food plants during the dry season. During the rainy season, plants are considerably more abundant.

COMMON JUNGLE FEATURES

SWAMPS

These are common to all low jungle areas where there is water and poor drainage. There are two basic types of swamps— mangrove and palm.

Mangrove Swamps. These are found in coastal areas wherever tides influence water flow. The mangrove is a shrub-like tree which grows 1 to 5 meters high. These trees have tangled root systems, both above and below the water level, which restrict movement to foot or small boats. Observation in mangrove swamps, both on the ground and from the air, is poor, and movement is extremely difficult. Sometimes, streams that you can raft form channels, but you usually must travel on foot through this swamp.

Mangrove Swamp

Palm Swamp

You find saltwater swamps in West Africa, Madagascar, Malaysia, the Pacific islands, Central and South America, and at the mouth of the Ganges River in India. The swamps at the mouths of the Orinoco and Amazon rivers and rivers of Guyana consist of mud and trees that offer little shade. Tides in saltwater swamps can vary as much as 12 meters.

Everything in a saltwater swamp may appear hostile to you, from leeches and insects to crocodiles and caimans. Avoid the dangerous animals in this swamp.

Avoid this swamp altogether if you can. If there are water channels through it, you may be able to use a raft to escape.

Palm Swamps. These exist in both salt and fresh water areas. Their characteristics are masses of thorny undergrowth, reeds, grasses, and occasional short palms that reduce visibility and make travel difficult. There are often islands that dot these swamps, allowing you to get out of the water. Wildlife is abundant in these swamps. Like movement in the mangrove swamps, movement through palm swamps is mostly restricted to foot (sometimes small boats). Vehicular traffic is nearly impossible except after extensive road construction by engineers. Observation and fields-of-fire are very limited. Concealment from both air and ground observation is excellent.

SAVANNA

This is a broad, open jungle grassland in which trees are scarce. The thick grass is broad-bladed and grows 1 to 5 meters high. It looks like a broad, grassy meadow, and frequently has red soil. It grows scattered trees that usually appear stunted and gnarled like apple trees. Palms also occur on savannas.

Movement in the savanna is generally easier than in other types of jungle areas, especially for vehicles. The sharp-edged, dense grass and extreme heat make foot movement a slow and tiring process. Depending on the height of the grass, ground observation may vary from poor to good. Concealment from air observation is poor for both troops and vehicles. You find savannas in parts of Venezuela, Brazil, and the Guianas in South America. In Africa, you find them in the southern Sahara (north-central Cameroon and Gabon and southern Sudan), Benin, Togo, most of Nigeria, northeastern Zaire, northern Uganda, western Kenya, part of Malawi, part of Tanzania, southern Zimbabwe, Mozambique, and western Madagascar.

Savanna

Bamboo

BAMBOO

This grows in clumps of varying size in jungles throughout the tropics. Large stands of bamboo are excellent obstacles for wheeled or tracked vehicles. Troop movement through bamboo is slow, exhausting, and noisy. Troops should bypass bamboo stands if possible.

CULTIVATED AREAS

These exist in jungles throughout the tropics and range from large, well-planned and well-managed farms and plantations to small tracts cultivated by individual farmers. There are three general types of cultivated areas—rice paddies, plantations, and small farms.

Rice Paddies. These are flat, flooded fields in which rice is grown. Flooding of the fields is controlled by a network of dikes and irrigation ditches which make movement by vehicles difficult even when the fields are dry. Concealment is poor in rice paddies. Cover is limited to the dikes, and then only from ground fire. Observation and fields of fire are excellent. Foot movement is poor when the fields are wet because soldiers must wade through water about 1/2 meter (2 feet)

Rice Paddies

deep and soft mud. When the fields are dry, foot movement becomes easier. The dikes, about 2 to 3 meters tall, are the only obstacles.

Plantations. These are large farms or estates where tree crops, such as rubber and coconut, are grown. They are usually carefully planned and free of undergrowth (like a well-tended park). Movement through plantations is generally easy. Observation along the rows of trees is generally good. Concealment and cover can be found behind the trees, but soldiers moving down the cultivated rows are exposed.

Plantation

Small Farms. These exist throughout the tropics. These small cultivated areas are usually hastily planned. After 1 or 2 years' use, they usually are abandoned, leaving behind a small open area which turns into secondary jungle. Movement through these areas may be difficult due to fallen trees and scrub brush.

Generally, observation and fields-of-fire are less restricted in cultivated areas than in uncultivated jungles. However, much of the natural cover and concealment are removed by cultivation, and troops will be more exposed in these areas.

Small Farm

LIFE IN THE JUNGLE

The jungle environment affects everyone. The degree to which you are trained to live and fight in harsh environments will determine your survival.

There is very little to fear from the jungle environment. Fear itself can be an enemy. Soldiers must be taught to control their fear of the jungle. A man overcome with fear is of little value in any situation. Soldiers in a jungle must learn that the most important thing is to keep their heads and calmly think out any situation.

Many of the stories written about out-of-the-way jungle places were written by writers who went there in search of adventure rather than facts. Practically without exception, these authors exaggerated or invented many of the thrilling experiences they relate. These thrillers are often a product of the author's imagination and are not facts.

Most Americans, especially those raised in cities, have lost the knack of taking care of themselves under all conditions. It would be foolish to say that, without proper training, they would be in no danger if lost in the jungles of Southeast Asia, South America, or some Pacific island. On the other hand, they would be in just as much danger if lost in the mountains of western Pennsylvania or in other undeveloped regions of our own country. The only difference would be that you are is less likely to panic when lost in your homeland than abroad.

IMMEDIATE CONSIDERATIONS

There is less likelihood of your rescue from beneath a dense jungle canopy than in other survival situations. You will probably have to travel to reach safety.

If you are the victim of an aircraft crash, the most important items to take with you from the crash site are a machete, a compass, a first aid kit, and a parachute or other material for use as mosquito netting and shelter.

Take shelter from tropical rain, sun, and insects. Malaria-carrying mosquitoes and other insects are immediate dangers, so protect yourself against bites.

Do not leave the crash area without carefully blazing or marking your route. Use your compass. Know what direction you are taking.

In the tropics, even the smallest scratch can quickly become dangerously infected. Promptly treat any wound, no matter how minor.

EFFECT OF CLIMATE

The discomforts of tropical climates are often exaggerated, but it is true that the heat is more persistent. In regions where the air contains a lot of moisture, the effect of the heat may seem worse than the same temperature in a dry climate. Many people experienced in jungle operations feel that the heat and discomfort in some US cities in the summertime are worse than the climate in the jungle.

Strange as it may seem, there may be more suffering from cold in the tropics than from the heat. Of course, very low temperatures do not occur, but chilly days and nights are common. In some jungles, in winter months, the nights are cold enough to require a wool blanket or poncho liner for sleeping.

Rainfall in many parts of the tropics is much greater than that in most areas of the temperate zones. Tropical downpours usually are followed by clear skies, and in most places the rains are predictable at certain times of the day. Except in those areas where rainfall may be continuous during the rainy season, there are not many days when the sun does not shine part of the time.

Precautions against malaria include:

- Taking Dapsone and chloroquine-primaquine
- Using insect repellent
- Wearing clothing that covers as much of the body as possible
- Using nets or screens at every opportunity
- Avoiding the worst-infested areas when possible

PROTECTION AGAINST MALARIA

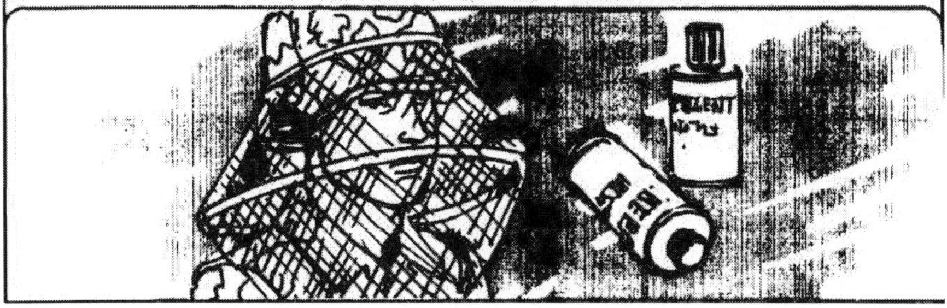

People who live in the tropics usually plan their activities so that they are able to stay under shelter during the rainy and hotter portions of the day. After becoming used to it, most tropical dwellers prefer the constant climate of the torrid zones to the frequent weather changes in colder climates.

INSECTS

Malaria-carrying mosquitoes are probably the most harmful of the tropical insects. Soldiers can contract malaria if proper precautions are not taken.

Mosquitoes are most prevalent early at night and just before dawn. Soldiers must be especially cautious at these times. Malaria is more common in populated areas than in uninhabited jungle, so soldiers must also be especially cautious when operating around villages. Mud packs applied to mosquito bites offer some relief from itching.

Wasps and bees may be common in some places, but they will rarely attack unless their nests are disturbed. When a nest is disturbed, the troops must leave the area and reassemble at the last rally point. In case of stings, mud packs are helpful. In some areas, there are tiny bees, called sweatbees, which may collect on exposed parts of the body during dry weather, especially if the body is sweating freely. They are annoying but stingless and will leave when sweating has completely stopped, or they may be scraped off with the hand.

The larger centipedes and scorpions can inflict stings which are painful but not fatal. They like dark places, so it is always advisable to shake out blankets before sleeping at night, and to make sure before dressing that they are not hidden in clothing or shoes. Spiders are commonly found in the jungle. Their bites may be painful, but are rarely serious. Ants can be dangerous to injured men lying on the ground and unable to move. Wounded soldiers should be placed in an area free of ants.

In Southeast Asian jungles, the rice-borer moth of the lowlands collects around lights in great numbers during certain seasons. It is a small, plain-colored moth with a pair of tiny black spots on the wings. It should never be brushed off roughly, as the small barbed hairs of its body may be ground into the skin. This causes a sore, much like a burn, that often takes weeks to heal.

LEECHES

Leeches are common in many jungle areas, particularly throughout most of the Southwest Pacific, Southeast Asia, and the Malay Peninsula. They are found in swampy areas, streams, and moist jungle country. They are not poisonous, but their bites may become infected if not cared for properly. The small wound that they cause may provide a point of entry for the germs which cause tropical ulcers or "jungle sores." Soldiers operating in the jungle should watch for leeches on the body and brush them off before they have had time to bite. When they have taken hold, they should not be pulled off forcibly because part of the leech may remain in the skin. Leeches will release themselves if touched with insect repellent, a moist piece of tobacco, the burning end of a cigarette, a coal from a fire, or a few drops of alcohol.

Straps wrapped around the lower part of the legs ("leech straps") will prevent leeches from crawling up the legs and into the crotch area. Trousers should be securely tucked into the boots.

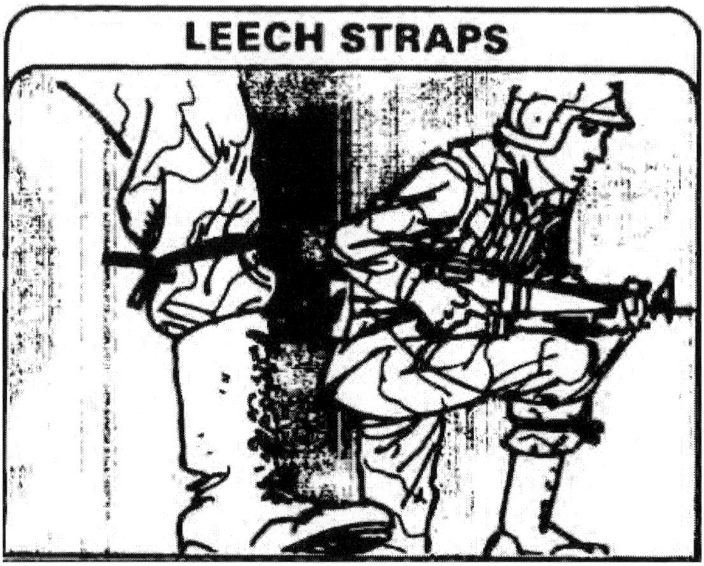

SNAKES

A soldier in the jungle probably will see very few snakes. When he does see one, the snake most likely will be making every effort to escape.

If a soldier should accidentally step on a snake or otherwise disturb a snake, it will probably attempt to bite. The chances of this happening to soldiers traveling along trails or waterways are remote if soldiers are alert and careful. Most jungle areas pose less of a snakebite danger than do the uninhabited areas of New Mexico, Florida, or Texas. This does not mean that soldiers should be careless about the possibility of snakebites, but ordinary precautions against them are enough. Soldiers should be particularly watchful when clearing ground. *Treat all snakebites as poisonous.*

VENOMOUS SNAKES

NOT TO SCALE

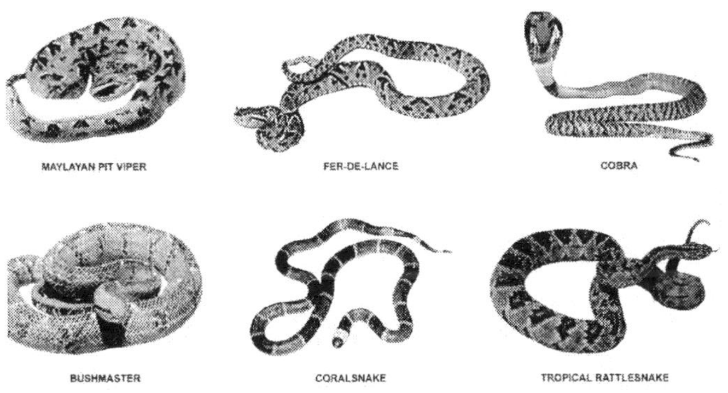

SNAKEBITE TREATMENT

Follow these steps if bitten:

- Remain calm, but act swiftly, and chances of survival are good. (Less than one percent of properly treated snakebites are fatal. Without treatment, the fatality rate is 10 to 15 percent.)
- Immobilize the affected part in a position below the level of the heart.
- Place a lightly constricting band 5 to 10 centimeters (2 to 4 inches) closer to the heart than the site of the bite. Reapply the constricting band ahead of the swelling if it moves up the arm or leg. The constricting band should be placed tightly enough to halt the flow of blood in surface vessels, but not so tight as to stop the pulse.
- Do not attempt to cut open the bite or suck out venom.
- Seek medical help. If possible, the snake's head with 5 to 10 centimeters (2 to 4 inches) of its body attached should be taken to the medics for identification. Identification insures use of the proper antivenom.

CROCODILES AND CAYMANS

Crocodiles and caymans are meat-eating reptiles which live in tropical areas. "Crocodile-infested rivers and swamps" is a catch-phrase often found in stories about the tropics. Asian jungles certainly have their share of crocodiles, but there are few authenticated cases of crocodiles actually attacking humans. Caymans, found in South and Central America, are not likely to attack unless provoked.

CROCODILES AND CAYMANS

CAYMAN

CROCODILE

ALLIGATOR

Before going into a jungle area, leaders must:

- **Make sure immunizations are current.**
- **Get soldiers in top physical shape.**
- **Instruct soldiers in personal hygiene.**

Upon arrival in the jungle area, leaders must:

- **Allow time to adjust (acclimate) to the new environment.**
- **Never limit the amount of water soldiers drink. (It is very important to replace the fluids lost through sweating.)**
- **Instruct soldiers on the sources of disease. Insects cause malaria, yellow fever, and scrub typhus. Typhoid, dysentery, cholera, and hepatitis are caused by dirty food and contaminated water.**

To prevent these diseases, soldiers should:

- Bathe often, and air- or sun-dry the body as often as possible.
- Wear clean, dry, loose-fitting clothing whenever possible.
- Not sleep in wet, dirty clothing. Soldiers should carry one dry set of clothes just for sleeping. Dirty clothing, even if wet, is put on again in the morning. This practice not only fights fungus, bacterial, and warm water immersion diseases but also prevents chills and allows soldiers to rest better.
- Not wear underwear during wet weather. Underwear dries slower than jungle fatigues, and causes severe chafing.
- Take off boots and massage feet as often as possible.
- Dust feet, socks, and boots with foot powder at every chance.
- Always carry several pairs of socks and change them frequently.
- Keep hair cut short.

WILD ANIMAL

In Africa, where lions, leopards, and other flesh-eating animals abound, they are protected from hunters by local laws and live on large preserves. In areas where the beasts are not protected, they are shy and seldom seen. When encountered, they will attempt to escape. All large animals can be dangerous if cornered or suddenly startled at close quarters. This is especially true of females with young. In the jungles of Sumatra, Bali, Borneo, Southeast Asia, and Burma there are tigers, leopards, elephants, and buffalo. Latin America's jungles have the jaguar. Ordinarily, these will not attack a man unless they are cornered or wounded.

Certain jungle animals, such as water buffalo and elephants, have been domesticated by the local people. Soldiers should also avoid these animals. They may appear tame, but this tameness extends only to people the animals are familiar with.

POISONOUS VEGETATION

Another area of danger is that of poisonous plants and trees. For example, nettles, particularly tree nettles, are one of the dangerous items of vegetation. These nettles have a severe stinging that will quickly educate the victim to recognize the plant. There are ringas trees in Malaysia which affect some people in much the same way as poison oak. The poison ivy and poison sumac of the continental US can cause many of the same type troubles that may be experienced in the jungle. The danger from poisonous plants in the woods of the US eastern seaboard is similar to that of the tropics. Thorny thickets, such as rattan, should be avoided as one would avoid a blackberry patch.

Some of the dangers associated with poisonous vegetation can be avoided by keeping sleeves down and wearing gloves when practical.

HEALTH AND HYGIENE

The climate in tropical areas and the absence of sanitation facilities increase the chance that soldiers may contract a disease. Disease is fought with good sanitation practices and preventive medicine. In past wars, diseases accounted for a significantly high percentage of casualties.

WATERBORNE DISEASES

Water is vital in the jungle and is usually easy to find. However, water from natural sources should be considered contaminated. Water purification procedures must be taught to all soldiers. Germs of serious diseases, like dysentery, are found in impure water. Other waterborne diseases, such as blood fluke, are caused by exposure of an open sore to impure water.

Soldiers can prevent waterborne diseases by:

- Obtaining drinking water from approved engineer water points.
- Using rainwater; however, rainwater should be collected after it has been
- Raining at least 15 to 30 minutes. This lessens the chances of impurity being washed from the jungle canopy into the water container. Even then the water should be purified.
- Insuring that all drinking water is purified.
- Not swimming or bathing in untreated water.
- Keeping the body fully clothed when crossing water obstacles.

HEAT INJURIES			
TYPE	CAUSE	SYMPTOMS	TREATMENT
Dehydration	Dehydration is caused by the loss of too much water. About two-thirds of the human body is water. When water is not replaced as it is lost, the body becomes dried out—dehydrated.	The symptoms are sluggishness and listlessness.	The treatment is to give the victim plenty of water.
Heat Exhaustion	Heat exhaustion is caused by the loss of too much water and salt.	The symptoms are: Dizziness. Nausea. Headache. Cramps. Rapid, weak pulse. Cool, wet skin.	The treatment consists of: Moving the victim to a cool, shaded place for rest. Loosening the clothing. Elevating the feet to improve circulation. Giving the victim cool salt water (two salt tablets dissolved in a canteen of water). Natural sea water should not be used.
Heat Cramps	Heat cramps are caused by the loss of too much salt.	The symptom is painful muscle cramps which are relieved as soon as salt is replaced.	The treatment is the same as for heat exhaustion.
Heatstrokes	Heatstroke (sunstroke) is caused by a breakdown in the body's heat control mechanism. The most likely victims are those who are not acclimated to the jungle, or those who have recently had bad cases of diarrhea. Heatstroke can kill if not treated quickly.	The symptoms are: Hot, red, dry skin (most important sign). No sweating (when sweating would be expected). Very high temperature (105 to 110 degrees). Rapid pulse. Spots before eyes. Headache, nausea, dizziness, mental confusion. Sudden collapse.	Treatment consists of: Cooling the victim immediately. This is achieved by putting him in a creek or stream; pouring canteens of water over him; fanning him; and using ice, if available. Giving him cool salt water (prepared as stated earlier) if he is conscious. Rubbing his arms and legs rapidly. Evacuating him to medical aid as soon as possible.

Heat injuries are prevented by:

- **Drinking plenty of water.**
- **Using extra salt with food and water.**
- **Slowing down movement.**

NOTE: For more details, see FM 21-10 for field hygiene and sanitation, and FM 21-11 for first aid for soldiers.

To prevent a conflict, leaders should insure that their soldiers:

- **Respect the natives' privacy and personal property**
- **Observe the local customs and taboos**
- **Do not enter a native house without being invited**
- **Do not pick fruits or cut trees without permission of their owners**
- **Treat the natives as friends**

Meats that can be found in most jungles include:	The following types of fruits and nuts are common in jungle areas:		Vegetables found in most jungles include:
Wild fowl	Bananas	Wild raspberries	Taro *
Wild cattle	Coconuts	Nakarika	Yam *
Wild pig	Oranges and lemons	Papaya	Yucca *
Freshwater fish *	Navele nuts	Mangoes	Hearts of palm trees
Saltwater fish	Breadfruit		
Fresh-water crawfish			

*These items must be cooked before eating.

FUNGUS DISEASES

These diseases are caused by poor personal health practices. The jungle environment promotes fungus and bacterial diseases of the skin and warm water immersion skin diseases. Bacteria and fungi are tiny plants which multiply fast under the hot, moist conditions of the jungle. Sweat-soaked skin invites fungus attack. The following are common skin diseases that are caused by long periods of wetness of the skin:

Warm Water Immersion Foot. This disease occurs usually where there are many creeks, streams, and canals to cross, with dry ground in between. The bottoms of the feet become white, wrinkled, and tender. Walking becomes painful.

Chafing. This disease occurs when soldiers must often wade through water up to their waists, and the trousers stay wet for hours. The crotch area becomes red and painful to even the lightest touch.

Most skin diseases are treated by letting the skin dry.

HEAT INJURIES

These result from high temperatures, high humidity, lack of air circulation, and physical exertion. All soldiers must be trained to prevent heat disorders.

NATIVES

Like all other regions of the world, the jungle also has its native inhabitants. Soldiers should be aware that some of these native tribes can be hostile if not treated properly.

There may be occasions, however, when hostile tribes attack without provocation.

FOOD

Food of some type is always available in the jungle—in fact, there is hardly a place in the world where food cannot be secured from plants and animals. All animals, birds, reptiles, and many kinds of insects of the jungle are edible. Some animals, such as toads and salamanders, have glands on the skin which should be removed before their meat is eaten. Fruits, flowers, buds, leaves, bark, and often tubers (fleshy plant roots) may be eaten. Fruits eaten by birds and monkeys usually may be eaten by man.

There are various means of preparing and preserving food found in the jungle. Fish, for example, can be cleaned and wrapped in wild banana leaves. This bundle is then tied with string made from bark, placed on a hastily constructed wood griddle, and roasted thoroughly until done. Another method is to roast the bundle of fish underneath a pile of red-hot stones.

Other meats can be roasted in a hollow section of bamboo, about 60 centimeters (2 feet) long. Meat cooked in this manner will not spoil for three or four days if left inside the bamboo stick and sealed.

Yams, taros, yuccas, and wild bananas can be cooked in coals. They taste somewhat like potatoes. Palm hearts can make a refreshing salad, and papaya a delicious dessert.

SHELTER

Jungle shelters are used to protect personnel and equipment from the harsh elements of the jungle. Shelters are necessary while sleeping, planning operations, and protecting sensitive equipment.

CLOTHING AND EQUIPMENT

Before deploying for jungle operations, troops are issued special uniforms and equipment. Some of these items are:

Jungle Fatigues

These fatigues are lighter and faster drying than standard fatigues. To provide the best ventilation, the uniform should fit loosely. It should never be starched.

Jungle Boots
These boots are lighter and faster drying than all-leather boots. Their cleated soles will maintain footing on steep, slippery slopes. The ventilating insoles should be washed in warm, soapy water when the situation allows.

Insect (Mosquito) Bar
The insect (mosquito) bar or net should be used any time soldiers sleep in the jungle. Even if conditions do not allow a shelter, the bar can be hung inside the fighting position or from trees or brush. No part of the body should touch the insect net when it is hung, because mosquitoes can bite through the netting. The bar should be tucked or laid loosely, not staked down. Although this piece of equipment is very light, it can be bulky if not folded properly. It should be folded inside the poncho as tightly as possible.

JUNGLE TRAVEL, NAVIGATION, AND TRACKING

TRAVEL THROUGH JUNGLE AREAS
With practice, movement through thick undergrowth and jungle can be done efficiently. Always wear long sleeves to avoid cuts and scratches.

To move easily, you must develop "jungle eye," that is, you should not concentrate on the pattern of bushes and trees to your immediate front. You must focus on the jungle further out and find natural breaks in the foliage. Look through the jungle, not at it. Stop and stoop down occasionally to look along the jungle floor. This action may reveal game trails that you can follow.

Stay alert and move slowly and steadily through dense forest or jungle. Stop periodically to listen and take your bearings. Use a machete to cut through dense vegetation, but do not cut unnecessarily or you will quickly wear yourself out. If using a machete, stroke upward when cutting vines to reduce noise because sound carries long distances in the jungle. Use a stick to part the vegetation. Using a stick will also help dislodge biting ants, spiders, or snakes. Do not grasp at brush or vines when climbing slopes; they may have irritating spines or sharp thorns.

Many jungle and forest animals follow game trails. These trails wind and cross, but frequently lead to water or clearings. Use these trails if they lead in your desired direction of travel.

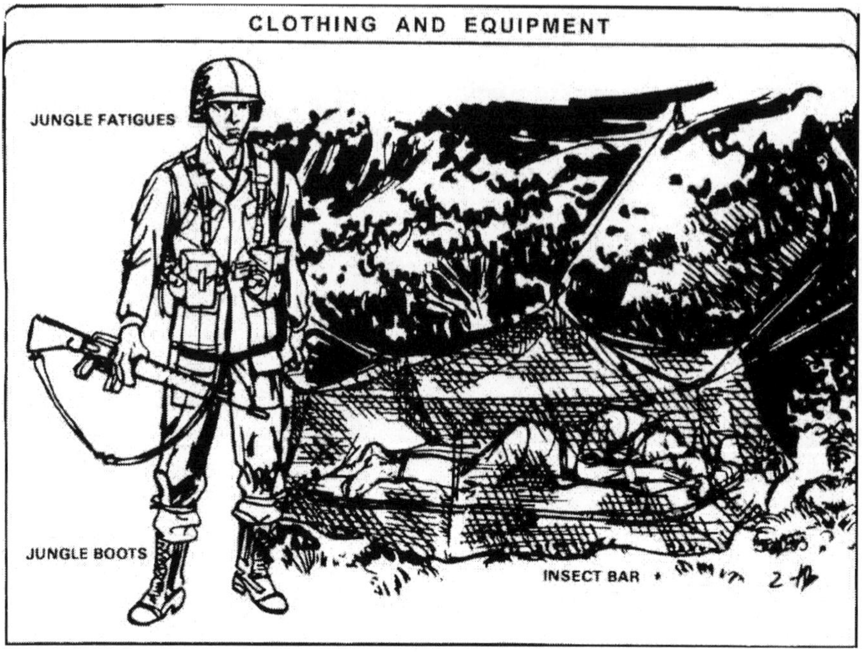

In many countries, electric and telephone lines run for miles through sparsely inhabited areas. Usually, the right-of-way is clear enough to allow easy travel. When traveling along these lines, be careful as you approach transformer and relay stations. In enemy territory, they maybe guarded.

JUNGLE NAVIGATION

Navigating in the jungle can be difficult for those troops not accustomed to it. This appendix outlines techniques which have been used successfully in jungle navigation. With training and practice, troops should be able to use these techniques to navigate in even the thickest jungle.

NAVIGATION TOOLS

Maps

Because of the isolation of many jungles, the rugged ground, and the presence of the canopy, topographic survey is difficult and is done mainly from the air. Therefore, although maps of jungle areas generally depict the larger features (hill, ridges, larger streams, etc.) fairly accurately, some smaller terrain features (gullies, small or intermittent streams, small swamps, etc.), which are actually on the ground, may not appear on the map. Also, many older maps are inaccurate. So, before going into the jungle, bring your maps up to date.

Compass

No one should move in the jungle without a compass. It should be tied to the clothing by a string or bootlace. The three most common methods used to follow the readings of a compass are:

Sighting along the desired azimuth. The compass man notes an object to the front (usually a tree or bush) that is on line with the proper azimuth and moves to that object.

This is not a good method in the jungle as trees and bushes tend to look very much alike.

Holding the compass at waist level and walking in the direction of a set azimuth. This is a good method for the jungle. The compass man sets the compass for night use with the long luminous line placed over the luminous north arrow and the desired azimuth under the black index line. There is a natural tendency to drift either left or right using this method. Jungle navigators must learn their own tendencies and allow for this drift.

Sighting along the desired azimuth and guiding a man forward until he is on line with the azimuth. The unit then moves to the man and repeats the process. This is the most accurate method to use in the jungle during daylight hours, but it is slow. In this method, the compass man cannot mistake the aiming point and is free to release the compass on its string and use both hands during movement to the next aiming point.

METHODS OF MAP UPDATING

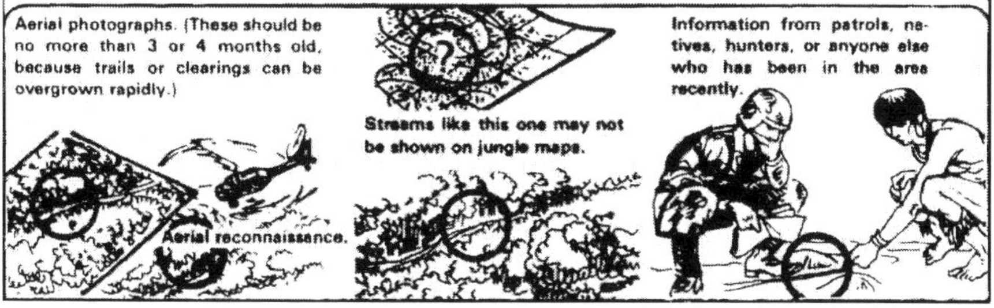
Aerial photographs. (These should be no more than 3 or 4 months old, because trails or clearings can be overgrown rapidly.)
Aerial reconnaissance.
?
Streams like this one may not be shown on jungle maps.
Information from patrols, natives, hunters, or anyone else who has been in the area recently.

The keys to navigation are maintaining the right direction and knowing the distance traveled. Skill with the compass (acquired through practice) takes care of the first requirement. Ways of knowing the distance traveled include checking natural features with the map, knowing the rate of movement, and pacing.

CHECKING FEATURES

Major recognizable features (hills, rivers, changes in the type of vegetation) should be noted as they are reached and then identified on the map. Jungle navigators must BE CAUTIOUS ABOUT TRAILS—the trail on the ground may not be the one on the map.

RATE OF MOVEMENT

Speed will vary with the physical condition of the troops, the load they carry, the danger of enemy contact, and the type of jungle growth. The normal error is to overestimate the distance traveled. The following can be used as a rough guide to the maximum distance covered in 1 hour during daylight.

PACING

In thick jungle, this is the best way of measuring distance. It is the only method which lets the soldier know how far he has traveled. With this information, he can estimate where he is at any given time. To be accurate, you must practice pacing over different types of terrain, and should make a PERSONAL PACE TABLE.

If possible, at least two men in each independent group should be compass men, and three or four should be keeping a pace count.

LOCATION OF AN OBJECTIVE

In open terrain, an error in navigation can be easily corrected by orienting on terrain features which are often visible from a long distance. In thick jungle, however, it is possible to be within 50 meters of a terrain feature and still not see it. Here are two methods which can aid in navigation.

DAYLIGHT MOVEMENT	
TYPE TERRAIN	MAXIMUM DISTANCE (in meters per hour)
TROPICAL RAIN FOREST	1,000
DECIDUOUS FOR-EST, SECONDARY JUNGLE, TALL GRASS	500
SWAMPS	100 TO 300
RICE PADDIES (WET)	800
RICE PADDIES (DRY)	2,000
PLANTATIONS	2,000
TRAILS	1,500

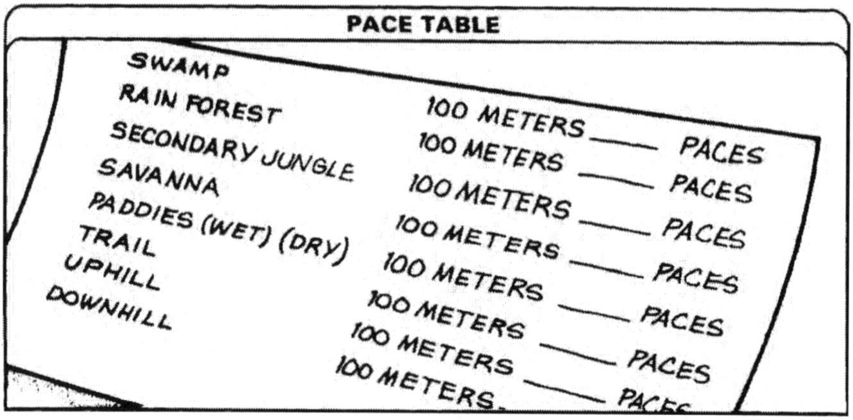
PACE TABLE
SWAMP
RAIN FOREST
SECONDARY JUNGLE
SAVANNA
PADDIES (WET) (DRY)
TRAIL
UPHILL
DOWNHILL
100 METERS ____ PACES
100 METERS ____ PACES
100 METERS ____ PACES
100 METERS ____ PACES
100 METERS ____ PACES
100 METERS ____ PACES
100 METERS ____ PACES
100 METERS

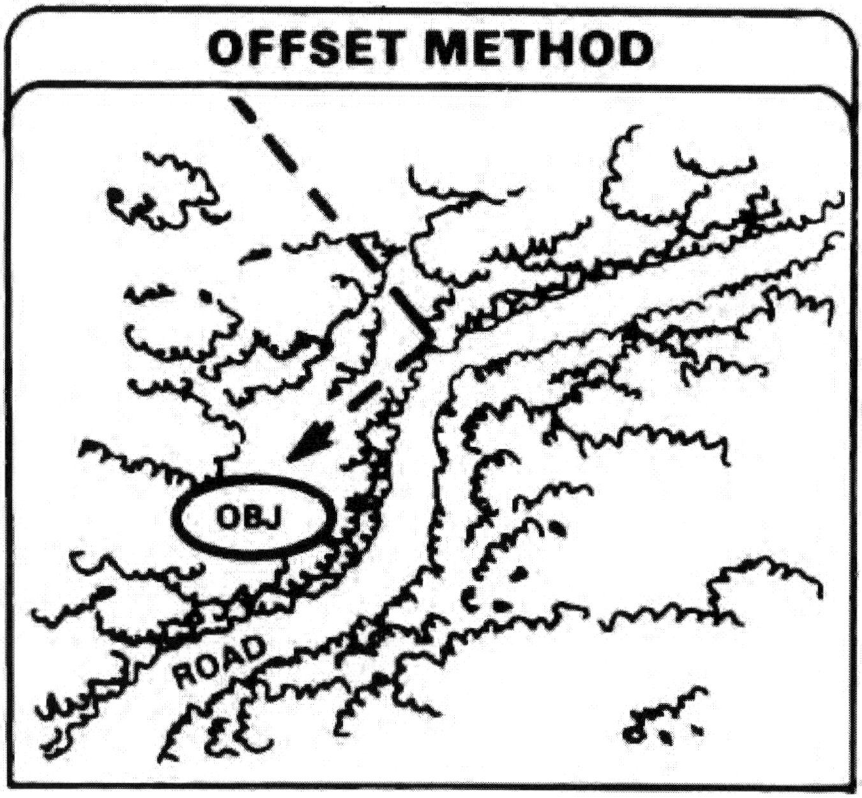
OFFSET METHOD
OBJ
ROAD

OFFSET METHOD

This method is useful in reaching an objective that is not large or not on readily identifiable terrain but is on a linear feature, such as a road, stream, or ridge. The unit plans a route following an azimuth which is a few degrees to the left or right of the objective. The unit then follows the azimuth to that terrain feature. Thus, when the unit reaches the terrain feature, the members know the objective is to their right or left, and the terrain feature provides a point of reference for movement to the objective.

ATTACK METHOD

This method is used when moving to an objective not on a linear feature. An easily recognizable terrain feature is chosen as close as possible to the objective. The unit then moves to that feature. Once there, the unit follows the proper azimuth and moves the estimated distance to get to the objective.

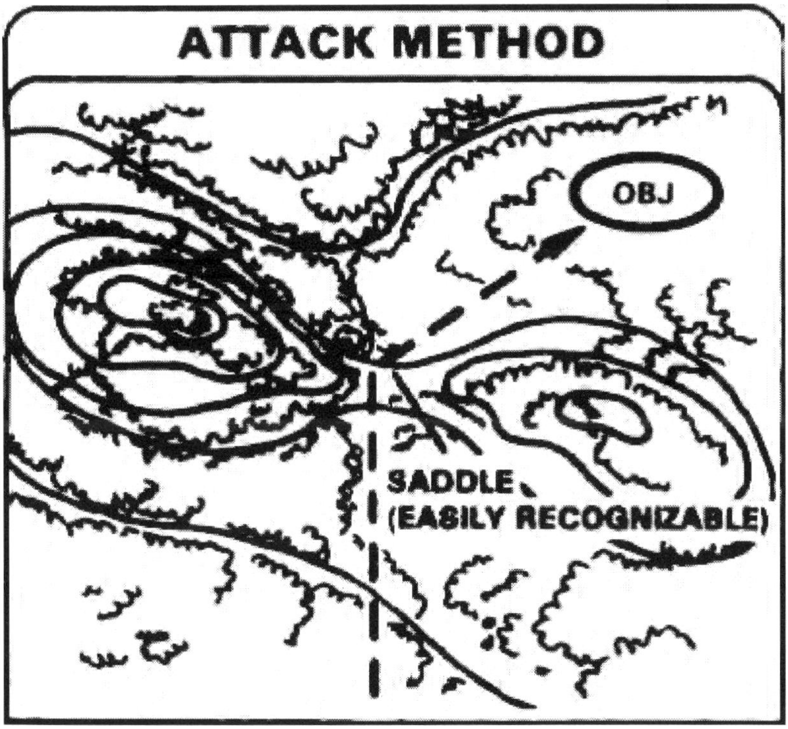

WHAT TO DO IF LOST

Do not panic. Few have ever been permanently lost in the jungle, although many have taken longer to reach their destination than they should.

Disoriented navigators should try to answer these questions. (If there are other navigators in the group, they all should talk it over.)

What was the last known location?

Did you go too far and pass the objective? (Compare estimates of time and distance traveled.)

Does the terrain look the way it should? (Compare the surroundings with the map.)

What features in the area will help to fix your location? (Try to find these features.)

NIGHT MOVEMENT

The principles for navigation at night are the same as those for day movement. The problem in night movement is one of control, not navigation. In clear weather, through sparse vegetation and under a bright moon, a unit can move almost as fast by night as by day. If the sky is overcast, vegetation is thick, or there is little or no moon, movement will be slow and hard to control. The following points can assist a unit during night movement.

Attach luminous tape to the back of each soldier's headgear. Two strips, side by side, each about the size of a lieutenant's bar, are recommended. The two strips aid depth perception and reduce the hypnotic effect that one strip can cause.

When there is no light at all, distance between soldiers should be reduced. When necessary to prevent breaks in contact, each soldier should hold on to the belt or the pack of the man in front of him.

The leader should carry a long stick to probe for sudden dropoffs or obstacles.

In limited visibility conditions, listening may become more important to security than observing. A unit which hears a strange noise should halt and listen for at least 1 minute. If the noise is repeated or cannot be identified, patrols should be sent out to investigate. Smell, likewise, can be an indication of enemy presence in an area.

All available night vision devices should be used.

NAVIGATIONAL TIPS

- Trust the map and compass, but understand the map's possible shortcomings. Use the compass bezel ring, especially during night navigation.
- Break brush. Do not move on trails or roads.
- Plan the move, and use the plan.
- Do not get frustrated. If in doubt, stop and think back over the route.
- Practice leads to confidence.

Visual tracking is following the paths of men or animals by the signs they leave, primarily on the ground or vegetation. Scent tracking is following men or animals by the odors they leave.

NAVIGATIONAL TIPS

- Trust the map and compass, but understand the map's possible shortcomings. Use the compass bezel ring, especially during night navigation.
- Break brush. Do not move on trails or roads.
- Plan the move, and use the plan.
- Do not get frustrated. If in doubt, stop and think back over the route.
- Practice leads to confidence.

Practice in tracking is required to achieve and maintain a high standard of skill. Because of the excellent natural concealment the jungle offers, all soldiers should be familiar with the general techniques of visual tracking to enable them to detect the presence of a concealed enemy, to follow the enemy, to locate and avoid mines or booby-traps, and to give early warning of ambush.

SIGNS

Men or animals moving through jungle areas leave signs of their passage. Some examples of these signs are listed below.

DECEPTION

The enemy may use any of the following methods to deceive or discourage trackers. They may, at times, mislead an experienced tracker.

TRACKING POINTS	
SAVANNA	**ROCKY GROUND**
NOTE:If the grass is high, above 3 feet, trails are easy to follow because the grass is knocked down and normally stays down for several days. If the grass is short, it springs back in a shorter length of time. ■ Grass that is tramped down will point in the direction that the person or animal is traveling. ■ Grass will show a contrast in color with the surrounding undergrowth when pressed down. ■ If the grass is wet with dew, the missing dew will show a trail where a person or an animal has traveled. ■ Mud or soil from boots may appear on some of the grass. ■ If new vegetation is showing through a track, the track is old. ■ In very short grass (12 inches or less) a boot will damage the grass near the ground and a footprint can be found.	■ Small stones and rocks are moved aside or rolled over when walked on. The soil is also disturbed, leaving a distinct variation in color and an impression. If the soil is wet, the underside of the stones will be much darker in color than the top when moved. ■ If the stone is brittle, it will chip and crumble when walked on. A light patch will appear where the stone is broken and the chips normally remain near the broken stone. ■ Stones on a loose or soft surface are pressed into the ground when walked upon. This leaves either a ridge around the edge of the stone where it has forced the dirt out, ora hole where the stone has been pushed below the surface of the ground. ■ Where moss is growing on rocks or stones, a boot or hand will scrape off some of the moss.

TRACKING POINTS CONTINUED

PRIMARY JUNGLES

NOTE: Within rain forests and deciduous forests, there are many ways to track. This terrain includes undergrowth, live and dead leaves and trees, streams with muddy or sandy banks, and moss on the forest floor and on rocks, which makes tracking easier.

- Disturbed leaves on the forest floor, when wet, show up a darker color when disturbed.
- Dead leaves are brittle and will crack or break under pressure of a person walking on them. The same is true of dry twigs.
- Where the undergrowth is thick, especially on the edges of the forest, green leaves of the bushes that have been pushed aside and twisted will show the underside of the leaf—this side is lighter in color than the upper surface. To find this sort of trail, the tracker must look through the jungle instead of directly at it.
- Boot impressions may be left on fallen and rotting trees.
- Marks may be left on the sides of logs lying across the path.
- Roots running across a path may show signs that something has moved through the area.
- Broken spiderwebs across a path indicate that something has moved through the area.

SECONDARY JUNGLE

- Broken branches and twigs.
- Leaves knocked off bushes and trees.
- Branches bent in the direction of travel.
- Footprints.
- Tunnels made through vegetation.
- Broken spiderwebs.
- Pieces of clothing caught on the sharp edges of bushes.

RIVERS, STREAMS, MARSHES, AND SWAMPS

- Footprints on the banks and in shallow water.
- Mud stirred up and discoloring the water.
- Rocks splashed with water in a quietly running stream.
- Water on the ground at a point of exit.
- Mud on grass or other vegetation near the edge of the water.

These deceptions include:

- Walking backwards. The heel mark tends to be deeper than that of the ball of the foot. The pace is shorter.
- More than one person stepping in the same tracks.
- Walking in streams.
- Splitting up into small groups.
- Walking along fallen trees or stepping from rock to rock.
- Covering tracks with leaves.

WARNING

A TRACKER SHOULD ALWAYS BE ALERT TO THE POSSIBILITY THAT THE ENEMY IS LEAVING FALSE SIGNS TO LEAD THE UNIT INTO AN AMBUSH.

CROSSING RIVERS AND STREAMS

There are several expedient ways to cross rivers and streams. The ways used in any situation depends on the width and depth of the water, the speed of the current, the time and equipment available, and the friendly and enemy situation.

There is always a possibility of equipment failure. For this reason, you should be able to swim. In all water crossings several strong swimmers should be stationed either at the water's edge or, if possible, in midstream to help anyone who gets into trouble.

If you accidentally fall into the water, swim with the current to the nearer bank. Swimming against the current is dangerous because the swimmer is quickly exhausted by the force of the current.

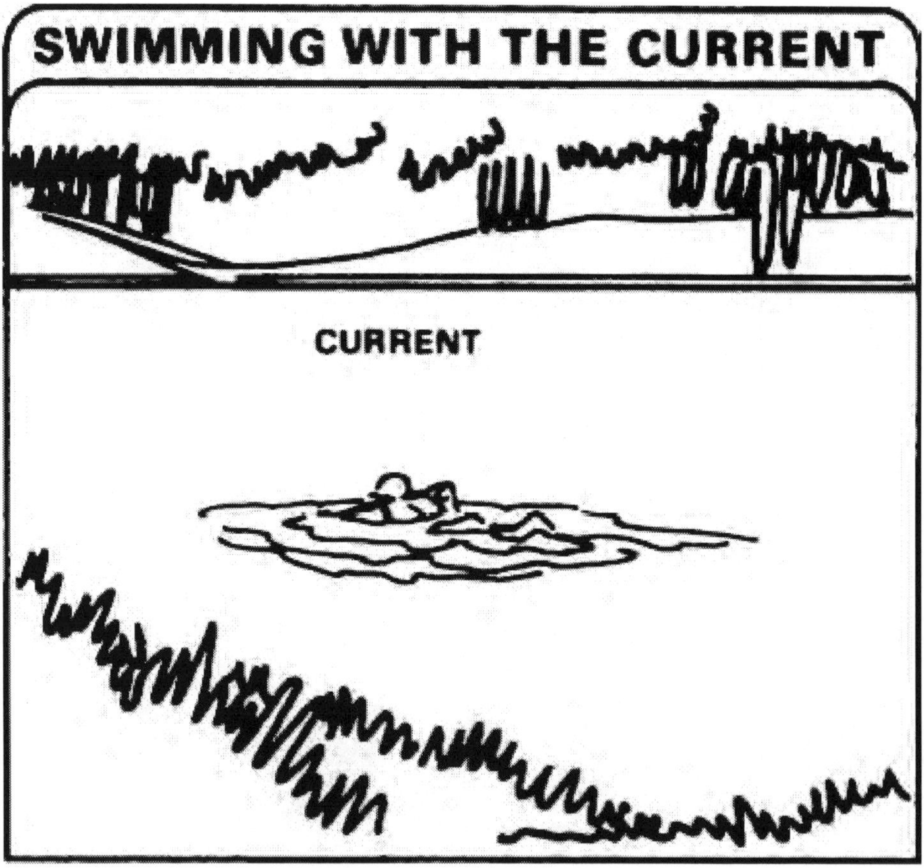

SOLDIER CROSSING A STREAM

- The standard air mattress
- Trousers

NOTE: Trousers must be soaked in water before using.

- Canteen safety belt
- Poncho life belt
- Water wings
- Poncho brush raft
- Australian poncho raft
- Log rafts

Fording

A good site to ford a stream has these characteristics:

- Good concealment on both banks.
- Few large rocks in the river bed. (Submerged large rocks are usually slippery and make it difficult to maintain footing.)
- Shallow water or a sandbar in the middle of the stream. Troops may rest or regain their footing on these sandbars.
- Low banks to make entry and exit easier. High banks normally mean deep water. Deep water near the far shore is especially dangerous as the soldiers may be tired and less able to get out.

You should cross at an angle against the current. Keep your feet wide apart and drag your legs through the water, not lift them, so that the current will not throw you off balance. Poles can be used to probe to help find deep holes and maintain footing.

FLOATING AIDS

For deeper streams which have little current, soldiers can use a number of floating aids such as the following:

When launching any poncho raft or leaving the water with it, take care not to drag it on the ground as this will cause punctures or tears.

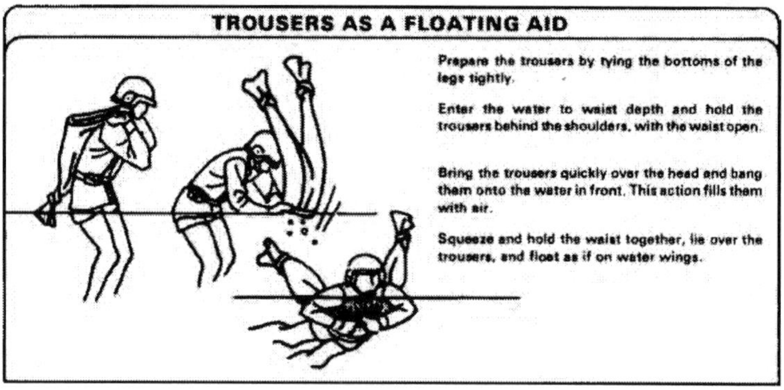

ROPE BRIDGES

For crossing streams and small rivers quickly, rope bridges offer a suitable temporary system, especially when there is a strong current. Because of the stretch factor of nylon ropes, they should not be used to cross gaps of more than 20 meters. For larger gaps, manila rope should be used.

In order to erect a rope bridge, the first thing to be done is to get one end of the rope across the stream. This task can be frustrating when there is a strong current. To get the rope across, anchor one end of a rope that is at least double the width of the stream at point A. Take the other end of the line upstream as far as it will go. Then, tie a sling rope around the waist of a strong swimmer and, using a snaplink, attach the line to him. He should swim diagonally downstream to the far bank, pulling the rope across.

One-Rope Bridge. A one-rope bridge can be constructed either above water level or at water level. The leader must decide which to construct. The bridge is constructed the same regardless of the level.

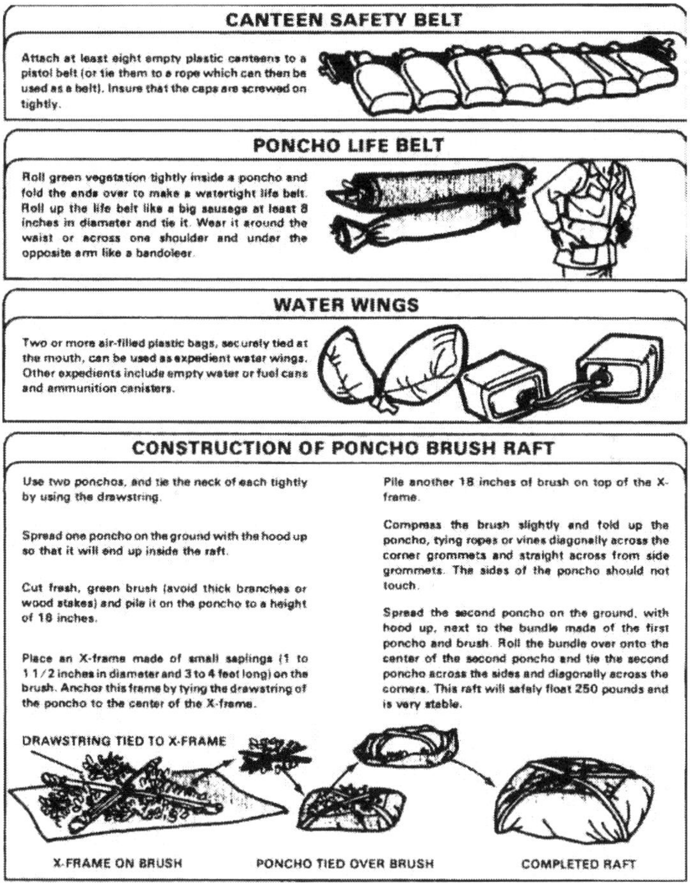

Crossing Method above Water Level. Use one of the following methods:

Commando crawl. Lie on the top of the rope with the instep of the right foot hooked on the rope. Let the left leg hang to maintain balance. Pull across with the hands and arms, at the same time pushing on the rope with the right foot. (For safety, tie a rappel seat and hooks the snaplink to the rope bridge.)

Monkey crawl. Hang suspended below the rope, holding the rope with the hands and crossing the knees over the top of the rope. Pull with the hands and push with the legs. (For safety, tie a rappel seat and hooks the snaplink to the rope bridge.) This is the safest and the best way to cross the one-rope bridge.

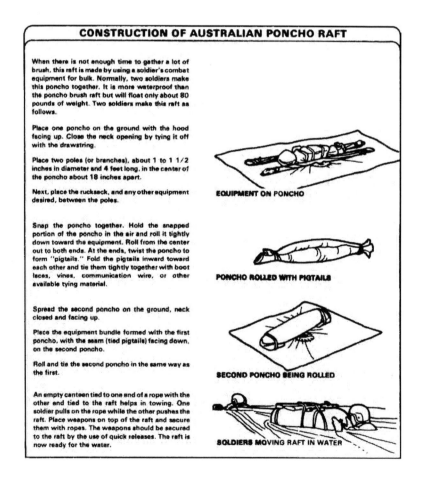

CONSTRUCTION OF AUSTRALIAN PONCHO RAFT

When there is not enough time to gather a lot of brush, this raft is made by using a soldier's combat equipment for bulk. Normally, two soldiers make this poncho together. It is more waterproof than the poncho brush raft but will float only about 80 pounds of weight. Two soldiers make this raft as follows.

Place one poncho on the ground with the hood facing up. Close the neck opening by tying it off with the drawstring.

Place two poles (or branches), about 1 to 1 1/2 inches in diameter and 4 feet long, in the center of the poncho about 18 inches apart.

Next, place the rucksack, and any other equipment desired, between the poles.

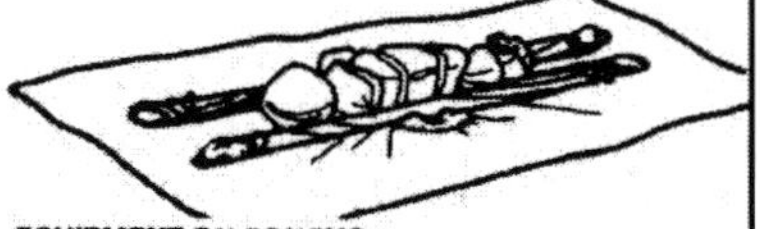

EQUIPMENT ON PONCHO

Snap the poncho together. Hold the snapped portion of the poncho in the air and roll it tightly down toward the equipment. Roll from the center out to both ends. At the ends, twist the poncho to form "pigtails." Fold the pigtails inward toward each other and tie them tightly together with boot laces, vines, communication wire, or other available tying material.

PONCHO ROLLED WITH PIGTAILS

Spread the second poncho on the ground, neck closed and facing up.

Place the equipment bundle formed with the first poncho, with the seam (tied pigtails) facing down, on the second poncho.

Roll and tie the second poncho in the same way as the first.

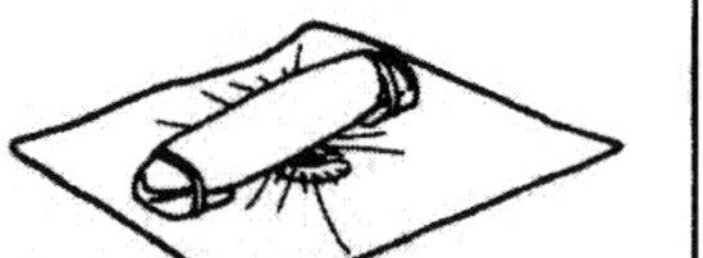

SECOND PONCHO BEING ROLLED

An empty canteen tied to one end of a rope with the other end tied to the raft helps in towing. One soldier pulls on the rope while the other pushes the raft. Place weapons on top of the raft and secure them with ropes. The weapons should be secured to the raft by the use of quick releases. The raft is now ready for the water.

SOLDIERS MOVING RAFT IN WATER

Crossing Method at Water Level. Hold onto the rope with both hands, face upstream, and walk into the water. Cross the bridge by sliding and pulling the hands along the rope. (For safety, tie a sling rope around your waist, leaving a working end of about 3 to 4 feet. Tie a bowline in the working end and attach a snaplink to the loop. Then, hook the snaplink to the rope bridge.)

CONSTRUCTION OF LOG RAFT

Logs, either singly or lashed together, can be used to float soldiers and equipment. Be careful when selecting logs for rafts. Some jungle trees will not float. To see whether certain wood is suitable, put a wood chip from a tree in the water. If the chip sinks, so will a raft made of that wood.

DETERMINATION OF RIVER WIDTH

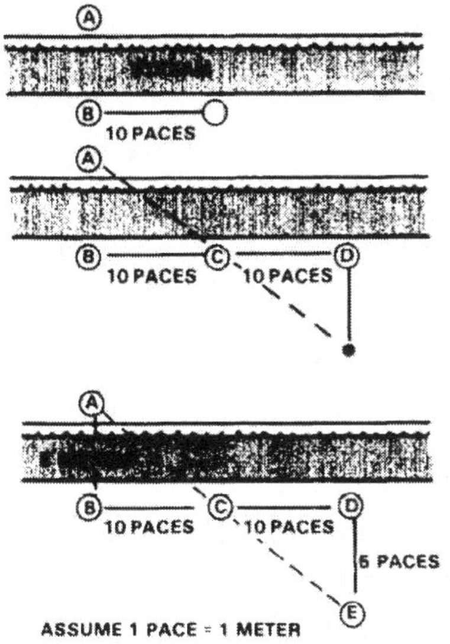

A method used to measure the width of a river or stream is described below.

Select a straight section of the stream.

Pick two points opposite each other (A and B).

Stand at B; turn in a direction parallel to the stream; walk off 10 paces. Mark that point as point C (B to C = 10 paces).

Continue walking in the same direction 10 more paces. Mark that point as point D (C to D = 10 paces).

Turn at a right angle away from the stream and walk until you are on line with points C and A. Mark this point as point E. Determine the distance between D and E by converting the pace count into meters. In this example, 1 pace is equal to 1 meter and the pace count is 5 paces. Therefore, the distance between D and E is 5 meters.

The distance from D to E is equal to the distance from A to B. Therefore, the width of the stream is also about 5 meters.

SWIMMER PULLING ROPE ACROSS

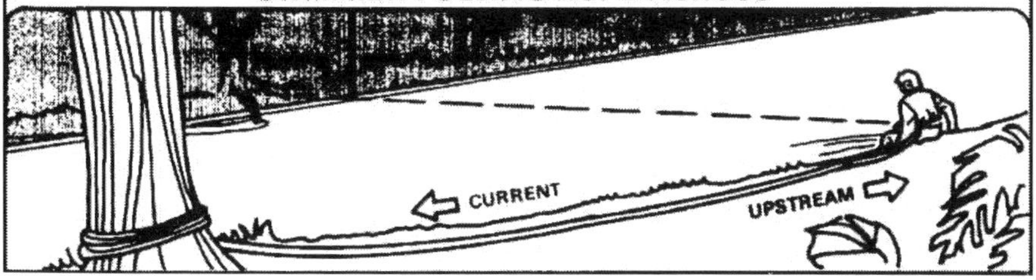

To recover the rope, the last person unties the rope, ties it around his waist and, after all slack is taken up, is pulled across.

Two-rope bridge. Construction of this bridge is similar to that of the one-rope bridge, except two ropes, a hand rope and a foot rope, are used. These ropes are spaced about 1.5 meters apart vertically at the anchor points. (For added safety, make snaplink attachments to the hand and foot ropes from a rope tied around the waist. Move across the bridge using the snaplink to allow the safety rope to slide.) To keep the ropes a uniform distance apart as men cross, spreader ropes should be tied between the two ropes every 15 feet. A sling rope is used and tied to each bridge rope with a round turn and two half-hitches.

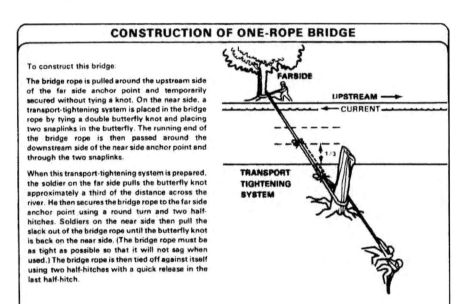

CONSTRUCTION OF ONE-ROPE BRIDGE

To construct this bridge:

The bridge rope is pulled around the upstream side of the far side anchor point and temporarily secured without tying a knot. On the near side, a transport-tightening system is placed in the bridge rope by tying a double butterfly knot and placing two snaplinks in the butterfly. The running end of the bridge rope is then passed around the downstream side of the near side anchor point and through the two snaplinks.

When this transport-tightening system is prepared, the soldier on the far side pulls the butterfly knot approximately a third of the distance across the river. He then secures the bridge rope to the far side anchor point using a round turn and two half-hitches. Soldiers on the near side then pull the slack out of the bridge rope until the butterfly knot is back on the near side. (The bridge rope must be as tight as possible so that it will not sag when used.) The bridge rope is then tied off against itself using two half-hitches with a quick release in the last half-hitch.

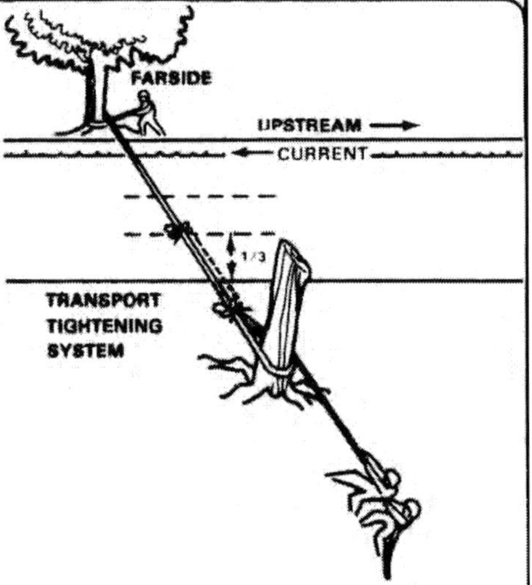

COMMANDO CRAWL

MONKEY CRAWL

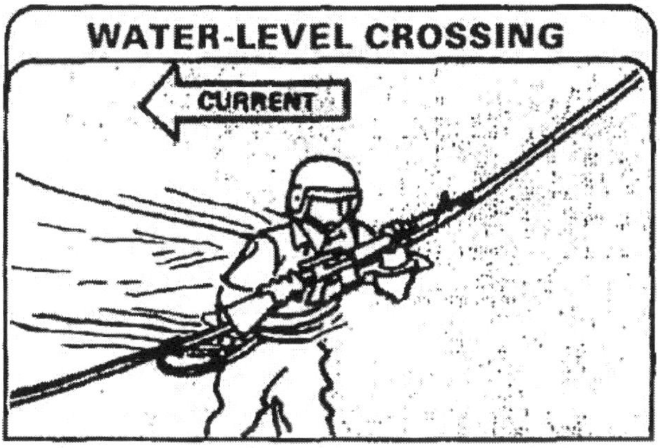
WATER-LEVEL CROSSING
CURRENT

TWO-ROPE BRIDGE CROSSING

CHAPTER 2

DESERT SURVIVAL

To survive and evade in arid or desert areas, you must understand and prepare for the environment you will face. You must determine your equipment needs, the tactics you will use, and how the environment will affect you and your tactics. Your survival will depend upon your knowledge of the terrain, basic climatic elements, your ability to cope with these elements, and your will to survive.

Desert terrain also varies considerably from place to place, the sole common denominator being lack of water with its consequent environmental effects, such as sparse, if any, vegetation. The basic land forms are similar to those in other parts of the world, but the topsoil has been eroded due to a combination of lack of water, heat, and wind to give deserts their characteristic barren appearance. The bedrock may be covered by a flat layer of sand, or gravel, or may have been exposed by erosion. Other common features are sand dunes, escarpments, wadis, and depressions.

It is important to realize that deserts are affected by seasons. Those in the Southern Hemisphere have summer between 21 December and 21 March. This 6-month difference from the United States makes acclimatization more difficult for persons coming from winter conditions.

TERRAIN

Key terrain in the desert is largely dependent on the restrictions to movement that are present. If the desert floor will not support wheeled vehicle traffic, the few roads and desert tracks become key terrain. Crossroads are vital as they control military operations in a large area. Desert warfare is often a battle for control of the lines of communication (LOC). The side that can protect its own LOC while interdicting those of the enemy will prevail. Water sources are vital, especially if a force is incapable of long distance resupply of its water requirements. Defiles play an important role, where they exist. In the Western Desert of Libya, an escarpment that paralleled the coast was a barrier to movement except through a few passes. Control of these passes was vital. Similar escarpments are found in Saudi Arabia and Kuwait.

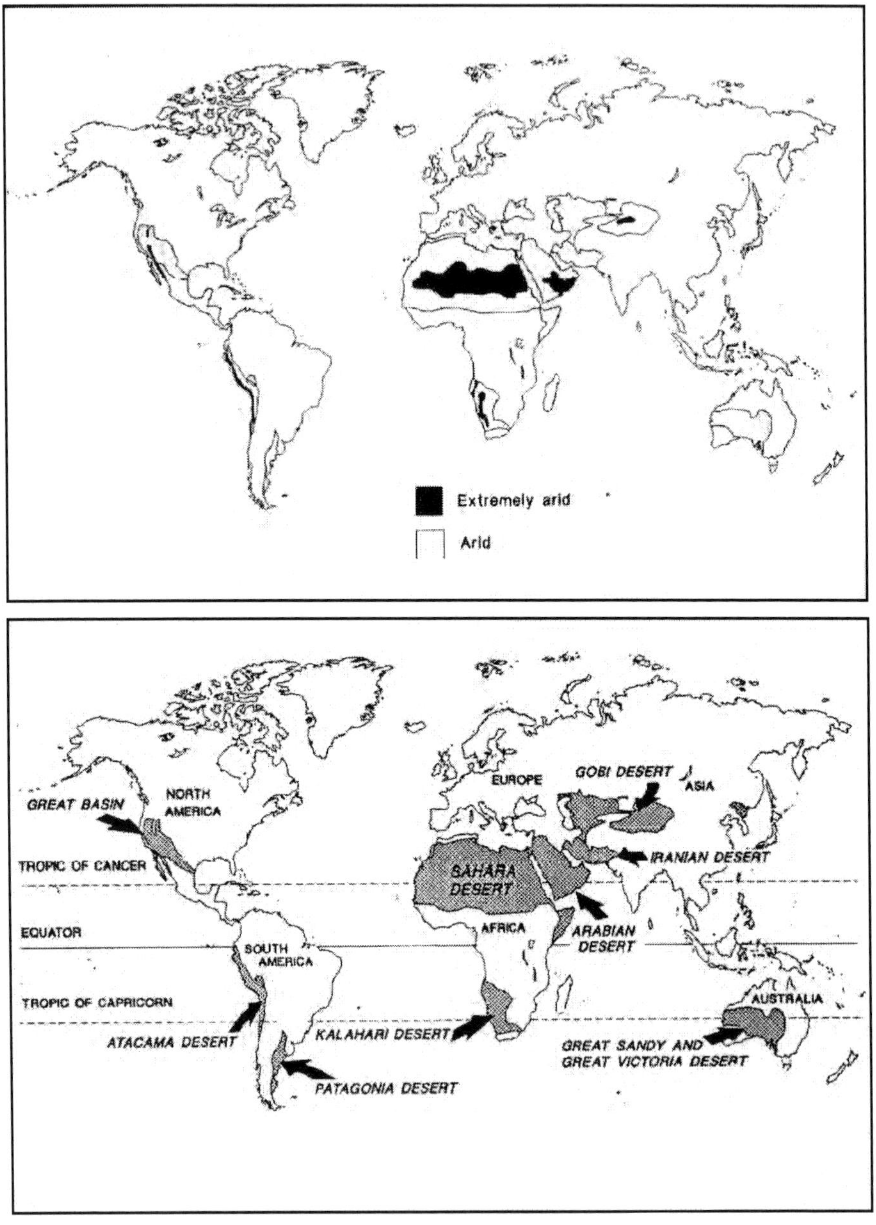

Deserts of the World

TYPES OF DESERT TERRAIN

There are three types of desert terrain: mountain, rocky plateau, and sandy or dune terrain. The following paragraphs discuss these types of terrain.

Mountain Deserts

Mountain deserts are characterized by scattered ranges or areas of barren hills or mountains, separated by dry, flat basins. See Figure below for an example of mountain desert terrain. High ground may rise gradually or abruptly from flat areas, to a height of several thousand feet above sea level. Most of the infrequent rainfall occurs on high ground and runs off in the form of flash floods, eroding deep gullies and ravines and depositing sand and gravel around the edges of the basins. Water evaporates rapidly, leaving the land as barren as before, although there may be short-lived vegetation. If sufficient water enters the basin to compensate for the rate of evaporation, shallow lakes may develop, such as the Great Salt Lake in Utah or the Dead Sea; most of these have a high salt content.

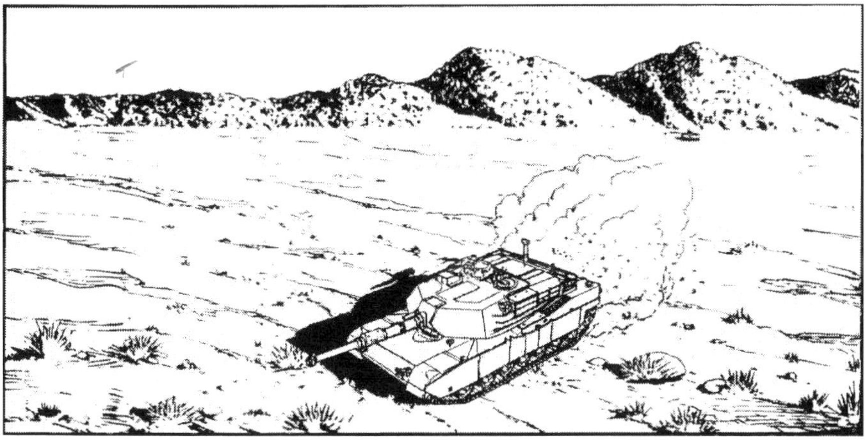

Mountain terrain.

Rocky Plateau Terrain

Rocky Plateau Deserts

Rocky plateau deserts are extensive flat areas with quantities of solid or broken rock at or near the surface. See Figure above for an example of a rocky plateau desert. They may be wet or dry, steep-walled eroded valleys, known as wadis, gulches, or canyons. Narrow valleys can be extremely dangerous to men and materiel due to flash flooding after rains; although their flat bottoms may be superficially attractive as assembly areas. The National Training Center and the Golan Heights are examples of rocky plateau deserts.

Sandy or Dune Deserts

Sandy or dune deserts are extensive flat areas covered with sand or gravel, the product of ancient deposits or modern wind erosion. "Flat" is relative in this case, as some areas may contain sand dunes that are over 1,000 feet high and 10-15 miles long; trafficability on this type of terrain will depend on windward/leeward gradients of the dunes and the texture of the sand. See Figure below for an example of a sandy desert. Other areas, however, may be totally flat for distances of 3,000 meters and beyond. Plant life may vary from none to scrub, reaching over 6 feet high. Examples of this type of desert include the ergs of the Sahara, the Empty Quarter of the Arabian desert, areas of California and New Mexico, and the Kalahari in South Africa. See Figure on next page for an example of a dune desert.

Sandy Desert Terrain

Dune Desert Terrain

Salt Marshes

Salt marshes are flat, desolate areas, sometimes studded with clumps of grass but devoid of other vegetation. They occur in arid areas where rainwater has collected, evaporated, and left large deposits of alkali salts and water with a high salt concentration. The water is so salty it is undrinkable. A crust that may be 2.5 to 30 centimeters thick forms over the saltwater.

In arid areas there are salt marshes hundreds of kilometers square. These areas usually support many insects, most of which bite. Avoid salt marshes. This type of terrain is highly corrosive to boots, clothing, and skin. A good example is the Shat-el-Arab waterway along the Iran-Iraq border.

Trafficability

Roads and trails are rare in the open desert. Complex road systems beyond simple commercial links are not needed. Road systems have been used for centuries to connect centers of commerce, or important religious shrines such as Mecca and Medina in Saudi Arabia. These road systems are supplemented by routes joining oil or other mineral deposits to collection outlet points. Some surfaces, such as lava beds or salt marshes, preclude any form of routine vehicular movement, but generally ground movement is possible in all directions. Speed of movement varies depending on surface texture. Rudimentary trails are used by minor caravans and nomadic tribesmen, with wells or oases approximately every 20 to 40 miles; although there are some waterless stretches which extend over 100 miles. Trails vary in width from a few meters to over 800 meters.

Vehicle travel in mountainous desert country may be severely restricted. Available mutes can be easily blinked by the enemy or by climatic conditions. Hairpin turns are common on the edges of precipitous mountain gorges, and the higher passes may be blocked by snow in the winter.

Natural Factors

The following terrain features require special considerations regarding trafficability.

Wadis or dried water courses, vary from wide, but barely perceptible depressions of soft sand, dotted with bushes, to deep, steep-sided ravines. There frequently is a passable route through the bottom of a dried wadi. Wadis can provide cover from ground observation and camouflage from visual air reconnaissance. The threat of flash floods after heavy rains poses a significant danger to troops and equipment downstream. Flooding may occur in these areas even if it is not

Wadi

raining in the immediate area. See Figure on previous page for an example of a wadi.

Salt marsh (sebkha) terrain is impassable to tracks and wheels when wet. When dry it has a brittle, crusty surface, negotiable by light wheel vehicles only. Salt marshes develop at points where the water in the subsoil of the desert rose to the surface. Because of the constant evaporation in the desert, the salts carried by the water are deposited, and results in a hard, brittle crust.

Salt marshes are normally impassable, the worst type being those with a dry crust of silt on top. Marsh mud used on desert sand will, however, produce an excellent temporary road. Many desert areas have salt marshes either in the center of a drainage basin or near the sea coast. Old trails or paths may cross the marsh, which are visible during the dry season but not in the wet season. In the wet season trails are indicated by standing water due to the crust being too hard or too thick for it to penetrate. However, such routes should not be tried by load-carrying vehicles without prior reconnaissance and marking. Vehicles may become mired so severely as to render equipment and units combat ineffective. Heavier track-laying vehicles, like tanks, are especially susceptible to these areas, therefore reconnaissance is critical.

Man-made Factors

The ruins of earlier civilizations, scattered across the deserts of the world, often are sited along important avenues of approach and frequently dominate the only available passes in difficult terrain. Control of these positions maybe imperative for any force intending

Example of Desert Nomads

Common man-made desert structures

to dominate the immediate area. Currently occupied dwellings have little impact on trafficability except that they are normally located near roads and trails. Apart from nomadic tribesmen who live in tents (see Figure on previous page for an example of desert nomads), the population lives in thick-walled structures with small windows, usually built of masonry or a mud and straw (adobe) mixture. Figure 2-9 shows common man-made desert structures.

Because of exploration for and production of oil and other resources, wells, pipelines, refineries, quarries, and crushing plants may be of strategic importance in the desert. Pipelines are often raised 1 meter off the ground—where this is the case, pipelines will inhibit movement. Subsurface pipelines can also be an obstacle. In Southwest Asia, the subsurface pipelines were indicated on maps. Often they were buried at such a shallow depth that they could be damaged by heavy vehicles traversing them. Furthermore, if a pipeline is ruptured, not only is the spill of oil a consideration, but the fumes maybe hazardous as well.

Agriculture in desert areas has little effect on trafficability except that canals limit surface mobility. Destruction of an irrigation system, which may be a result of military operations, could have a devastating effect on the local population and should be an important consideration in operational estimates. Figure 2-10 shows an irrigation ditch.

Irrigation ditch

TEMPERATURE

The highest known ambient temperature recorded in a desert was 136 degrees Fahrenheit (58 degrees Celsius). Lower temperatures than this produced internal tank temperatures approaching 160 degrees Fahrenheit (71 degrees Celsius) in the Sahara Desert during the Second World War. Winter temperatures in Siberian deserts and in the Gobi reach minus 50 degrees Fahrenheit (minus 45 degrees Celsius). Low temperatures are aggravated by very strong winds producing high windchill factors. The cloudless sky of the desert permits the earth to heat during sunlit hours, yet cool to near freezing at night. In the inland Sinai, for example, day-to-night temperature fluctuations are as much as 72 degrees Fahrenheit.

WINDS

Example of Wind Erosion

Desert winds can achieve velocities of near hurricane force; dust and sand suspended within them make life intolerable, maintenance very difficult, and restrict visibility to a few meters. The Sahara "Khamseen", for example, lasts for days at a time; although it normally only occurs in the spring and summer. The deserts of Iran are equally well known for the "wind of 120 days," with sand blowing almost constantly from the north at wind velocities of up to 75 miles per hour.

Although there is no danger of a man being buried alive by a sandstorm, individuals can become separated from their units. In all deserts, rapid temperature changes invariably follow strong winds. Even without wind, the telltale clouds raised by wheels, tracks, and marching troops give away movement. Wind aggravates the problem. As the day gets warmer the wind increases and the dust signatures of vehicles may drift downwind for several hundred meters.

In the evening the wind normally settles down. In many deserts a prevailing wind blows steadily from one cardinal direction for most of the year, and eventually switches to another direction for the remaining months. The equinoctial gales raise huge sandstorms that rise to several thousand feet and may last for several days. Gales and sandstorms in the winter months can be bitterly cold. See Figure 2-11 for an example of wind erosion.

MIRAGES

Mirages are optical phenomena caused by the refraction of light through heated air rising from a sandy or stony surface. They occur in the interior of the desert about 10 kilometers from the coast. They make objects that are 1.5 kilometers or more away appear to move.

This mirage effect makes it difficult for you to identify an object from a distance. It also blurs distant range contours so much that you feel surrounded by a sheet of water from which elevations stand out as "islands."

The mirage effect makes it hard for a person to identify targets, estimate range, and see objects clearly. However, if you can get to high ground (3 meters or more above the desert floor), you can get above the superheated air close to the ground and overcome the mirage effect. Mirages make land navigation difficult because they obscure natural features. You can survey the area at dawn, dusk, or by moonlight when there is little likelihood of mirage.

Example of a typical oasis

Light levels in desert areas are more intense than in other geographic areas. Moonlit nights are usually crystal clear, winds die down, haze and glare disappear, and visibility is excellent. You can see lights, red flashlights, and blackout lights at great distances. Sound carries very far.

Conversely, during nights with little moonlight, visibility is extremely poor. Traveling is extremely hazardous. You must avoid getting lost, falling into ravines, or stumbling into enemy positions. Movement during such a night is practical only if you have a compass and have spent the day in a shelter, resting, observing and memorizing the terrain, and selecting your route.

Sandstorms are likely to form suddenly and stop just as suddenly. In a severe sandstorm, sand permeates everything making movement nearly impossible, not only because of limited visibility, but also because blowing sand damages moving parts of machinery.

WATER

The lack of water is the most important single characteristic of the desert. The population, if any, varies directly with local water supply.

A Sahara oasis may, for its size, be one of the most densely occupied places on earth (see Figure 2-12 for a typical oasis).

Desert rainfall varies from one day in the year to intermittent showers throughout the winter. Severe thunderstorms bring heavy rain, and usually far too much rain falls far too quickly to organize collection on a systematic basis. The water soon soaks into the ground and may result in flash floods. In some cases the rain binds the sand much like a beach after the tide ebbs allowing easy maneuver however, it also turns loam into an impassable quagmire obstacle. Rainstorms tend to be localized, affecting only a few square kilometers at a time. Whenever possible, as storms approach, vehicles should move to rocky areas or high ground to avoid flash floods and becoming mired.

Permanent rivers such as the Nile, the Colorado, or the Kuiseb in the Namib Desert of Southwest Africa are fed by heavy precipitation outside the desert so the river survives despite a high evaporation rate.

Subsurface water may be so far below the surface, or so limited, that wells are normally inadequate to support any great number of people. Because potable water is absolutely vital, a large natural supply may be both tactically and strategically important. Destruction of a water supply system may become a political rather than military decision, because of its lasting effects on the resident civilian population.

Finding Water

When there is no surface water, tap into the earth's water table for ground water. Access to this table and its supply of generally pure water depends on the contour of the land and the type of soil. See Figure 2-13 for water tables.

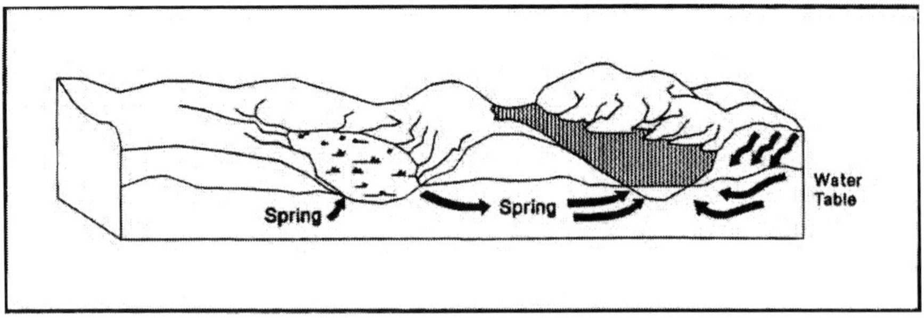

Water Tables

From Rocky Soil

Look for springs and seepages. Limestone has more and larger springs than any other type rock. Because limestone is easily dissolved, caverns are readily etched in it by ground water. Look in these caverns for springs. Lava rock is a good source of seeping ground water because it is porous. Look for springs along the walls of valleys that cross the lava flow. Look for seepage where a dry canyon cuts through a layer of porous sandstone.

Watch for water indicators in desert environments. Some signs to look for are the direction in which certain birds fly, the location of plants, and the convergence of game trails. Asian sand grouse, crested larks, and zebra birds visit water holes at least once a day. Parrots and pigeons must live within reach of water. Cattails, greasewoods, willows, elderberry, rushes, and salt grass grow only where ground water is near the surface. Look for these signs and dig. If you do not have a bayonet or entrenching tool, dig with a flat rock or sharp stick.

Desert natives often know of lingering surface pools in low places. They cover their surface pools, so look under brush heaps or in sheltered nooks, especially in semiarid and brush country.

Places that are visibly damp, where animals have scratched, or where flies hover, indicate recent surface water. Dig in such places for water. Collect dew on clear nights by sponging it up with a handkerchief. During a heavy dew you should be able to collect about a pint an hour.

Dig in dry stream beds because water may be found under the gravel. When in snow fields, put in a water container and place it in the sun out of the wind.

From Plants

If unsuccessful in your search for ground or runoff water, or if you do not have time to purify the questionable water, a water-yielding plant may be the best source. Clear sap from many plants is easily obtained. This sap is pure and is mostly water.

Plant tissues. Many plants with fleshy leaves or stems store drinkable water. Try them wherever you find them. The barrel cactus of the southwestern United States is a possible source of water (see Figure 2-14). Use it only as a last resort and only if you have the energy to cut through the tough, spine-studded outer rind. Cut off the top of the cactus and smash the pulp within the plant. Catch the liquid in a container. Chunks may be carried as an emergency water source. A barrel cactus 3-1/2 feet high will yield about a quart of milky juice and is an exception to the rule that milky or colored sap-bearing plants should not be eaten.

Barrel Cactus as a Source of Water

Roots of desert plants. Desert plants often have their roots near the surface. The Australian water tree, desert oak, and bloodwood are some examples. Pry these roots out of the ground, cut them into 24-36 inch lengths, remove the bark, and suck the water.

Vines. Not all vines yield palatable water, but try any vine found. Use the following method for tapping a vine--it will work on any species:

Step 1. Cut a deep notch in the vine as high up as you can reach.
Step 2. Cut the vine off close to the ground and let the water drip into your mouth or into a container.
Step 3. When the water ceases to drip, cut another section off the top. Repeat this until the supply of fluid is exhausted.

Palms. Burl, coconut, sugar and nipa palms contain a drinkable sugary fluid. To start the flow in coconut palm, bend the flower stalk downward and cut off the top. If a thin slice is cut off the stalk every 12 hours, you can renew the flow and collect up to a quart a day.

Coconut. Select green coconuts. They can be opened easily with a knife and they have more milk than ripe coconuts. The juice of a ripe coconut is extremely laxative; therefore, do not drink more than three or four cups a day.

The milk of a coconut can be obtained by piercing two eyes of the coconut with a sharp object such as a stick or a nail. To break off the outer fibrous covering of the coconut without a knife, slam the coconut forcefully on the point of a rock or protruding stump.

Survival Water Still

You can build a cheap and simple survival still that will produce drinking water in a dry desert. Basic materials for setting up this still are—

- 6-foot square sheet of clean plastic.
- A 2- to 4-quart capacity container.
- A 5-foot piece of flexible plastic tubing.

Pick an unshaded spot for the still, and dig a hole. If no shovel is available, use a stick or even your hands. The hole should be about 3 feet across for a few inches down, then slope the hole toward the bottom as shown in Figure 2-15 which depicts a cross section of a survival still. The hole should be deep enough so the point of the plastic cone will be about 18 inches below ground and will still clear the top of the container. Once the hole is properly dug, tape one end of the plastic drinking tube inside the container and center the container in

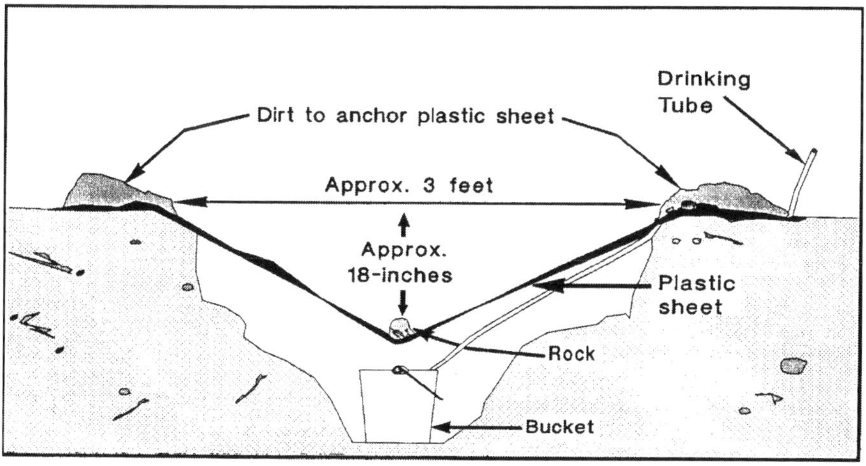

Cross-Section of a Survival Still

the bottom of the hole. Leave the top end of the drinking tube free, lay the plastic sheet over the hole, and pile enough dirt around the edge of the plastic to hold it securely. Use a fist-size rock to weight down the center of the plastic; adjust the plastic as necessary to bring it within a couple of inches of the top of the container. Heat from the sun vaporizes the ground water. This vapor condenses under the plastic, trickles down, and drops into the container.

VEGETATION

The indigenous vegetation and wildlife of a desert have physiologically adapted to the conditions of the desert environment. For example, the cacti of the American desert store moisture in enlarged stems. Some plants have drought-resistant seeds that may lie dormant for years, followed by a brief, but colorful display of growth after a rainstorm. The available vegetation is usually inadequate to provide much shade, shelter, or concealment, especially from the air. Some plants, like the desert gourd, have vines which grow to 4.5 meters (15 feet). Others have wide lateral roots just below the surface to take advantage of rain and dew, while still others grow deep roots to tap subsurface water. Presence of palm trees usually indicates water within a meter of the surface, salt grass within 2 meters, cottonwood and willows up to 4 meters. In addition to indicating the presence of water, some plants are edible.

WILDLIFE

Invertebrates such as ground-dwelling spiders, scorpions, and centipedes, together with insects of almost every type, are in the desert. Drawn to man as a source of moisture or food, lice, mites, and flies can be extremely unpleasant and carry diseases such as scrub typhus and dysentery. The stings of scorpions and the bites of centipedes and spiders are extremely painful, though seldom fatal. Some species of scorpion, as well as black widow and recluse spiders, can cause death. The following paragraphs describe some of the wildlife that are encountered in desert areas and the hazards they may pose to man.

Scorpions

Scorpions are prevalent in desert regions. They prefer damp locations and are particularly active at night. Scorpions are easily recognizable by their crab-like appearance, and by their long tail which ends in a sharp stinger. Adult scorpions vary from less than an inch to almost 8 inches in length. Colors range from nearly black to straw to striped. Scorpions hide in clothing, boots, or bedding, so troops should routinely shake these items before using. Although scorpion stings are rarely fatal, they can be painful.

Flies

Flies are abundant throughout desert environments. Filth-borne disease is a major health problem posed by flies. Dirt or insects in the desert can cause infection in minor cuts and scratches.

Fleas

Avoid all dogs and rats which are the major carriers of fleas. Fleas are the primary carriers of plague and murine typhus.

Reptiles

Reptiles are perhaps the most characteristic group of desert animals. Lizards and snakes occur in quantity, and crocodiles are common in some desert rivers.Lizards are normally harmless and can be ignored; although exceptions occur in North America and Saudi Arabia.

Snakes, ranging from the totally harmless to the lethal, abound in the desert. A bite from a poisonous snake under two feet long can easily become infected. Snakes seek shade (cool areas) under bushes, rocks, trees, and shrubs. These areas should be checked before sitting or resting. Troops should always check clothing and boots before putting them on. Vehicle operators should look for snakes when initially conducting before-operations maintenance. Look for snakes in and around suspension components and engine compartments as snakes may seek the warm areas on recently parked vehicles to avoid the cool night temperatures,

Sand vipers have two long and distinctive fangs that may be covered with a curtain of flesh or folded back into the mouth. Sand vipers usually are aggressive and dangerous in spite of their size. A sand viper usually buries itself in the sand and may strike at a passing man; its presence is alerted by a characteristic coiling pattern left on the sand.

The Egyptian cobra can be identified by its characteristic cobra combative posture. In this posture, the upper portion of the body is raised vertically and the head tilted sharply forward. The neck is usually flattened to form a hood. The Egyptian cobra is often found around rocky places and ruins and is fairly common. The distance the cobra can strike in a forward direction is equal to the distance the head is raised above the ground. Poking around in holes and rock piles is particularly dangerous because of the likelihood of encountering a cobra. See Figure 2-16 for an example of a viper and cobra. For more information on dangerous snakes, see Part IV, Chapter 4, "Poisonous Snakes and Lizards."

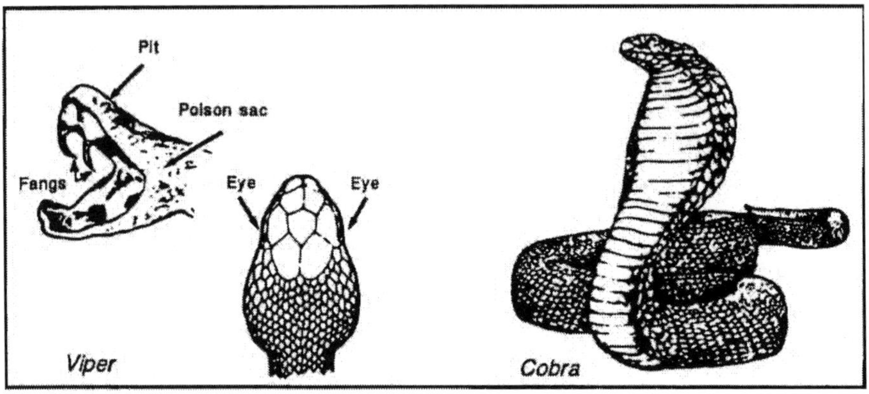

Sand Viper and Cobra

Mammals

The camel is the best known desert mammal. The urine of the camel is very concentrated to reduce water loss, allowing it to lose 30 percent of its body weight without undue distress. A proportionate loss would be fatal to man. The camel regains this weight by drinking up to 27 gallons (120 liters) of water at a time. It cannot, however, live indefinitely without water and will die of dehydration as readily as man in equivalent circumstances. Other mammals, such as gazelles, obtain most of their required water supply from the vegetation they eat and live in areas where there is no open water. Smaller animals, including rodents, conserve their moisture by burrowing underground away from the direct heat of the sun, only emerging for foraging at night. All these living things have adapted to the environment over a period of thousands of years; however, man has not made this adaptation and must carry his food and water with him and must also adapt to this severe environment.

Dogs are often found near mess facilities and tend to be in packs of 8 or 10. Dogs are carriers of rabies and should be avoided.

Commanders must decide how to deal with packs of dogs; extermination and avoidance are two options. Dogs also carry fleas which may be transferred upon bodily contact. Rabies is present in most desert mammal populations. Do not take any chances of contracting fleas or rabies from any animal by adopting pets.

Rats are carriers of various parasites and gastrointestinal diseases due to their presence in unsanitary locations.

There is no reason to fear the desert environment, and it should not adversely affect the morale of a soldier / marine who is prepared for it. Lack of natural concealment has been known to induce temporary agoraphobia (fear of open spaces) in some troops new to desert conditions, but this fear normally disappears with acclimatization. Remember that there is nothing unique about either living or fighting in deserts; native tribesmen have lived in the Sahara for thousands of years. The British maintained a field army and won a campaign in the Western Desert in World War II at the far end of a 12,000-mile sea line of communication with equipment considerably inferior to that in service now. The desert is neutral, and affects both sides equally; the side whose personnel are best prepared for desert operations has a distinct advantage.

The desert is fatiguing, both physically and mentally. A high standard of discipline is essential, as a single individual's lapse may cause serious damage to his unit or to himself. Commanders must exercise a high level of leadership and train their subordinate leaders to assume greater responsibilities required by the wide dispersion of units common in desert warfare. Soldiers / marines with good leaders are more apt to accept heavy physical exertion and uncomfortable conditions. Every soldier / marine must clearly understand why he is fighting in such harsh conditions and should be kept informed of the operational situation. Ultimately, however, the maintenance of discipline will depend on individual training.

Commanders must pay special attention to the welfare of troops operating in the desert, as troops are unable to find any "comforts" except those provided by the command. Welfare is an essential factor in the maintenance of morale in a harsh environment, especially to the inexperienced. There is more to welfare than the provision of mail and clean clothing. Troops must be kept healthy and physically fit; they must have adequate, palatable, regular food, and be allowed periods of rest and sleep. These things will not always be possible and discomfort is inevitable, but if troops know that their commanders are doing everything they can to make life tolerable, they will more readily accept the extremes brought on by the environment.

HEAT

The extreme heat of the desert can cause heat exhaustion and heatstroke and puts troops at risk of degraded performance. For optimum mental and physical performance, body temperatures must be maintained within narrow limits. Thus, it is important that the body lose the heat it gains during work. The amount of heat accumulation in the human body depends upon the amount of physical activity, level of hydration, and the state of personal heat acclimatization. Unit leaders must monitor their troops carefully for signs of heat distress and adjust schedules, work rates, rest, and water consumption according to conditions.

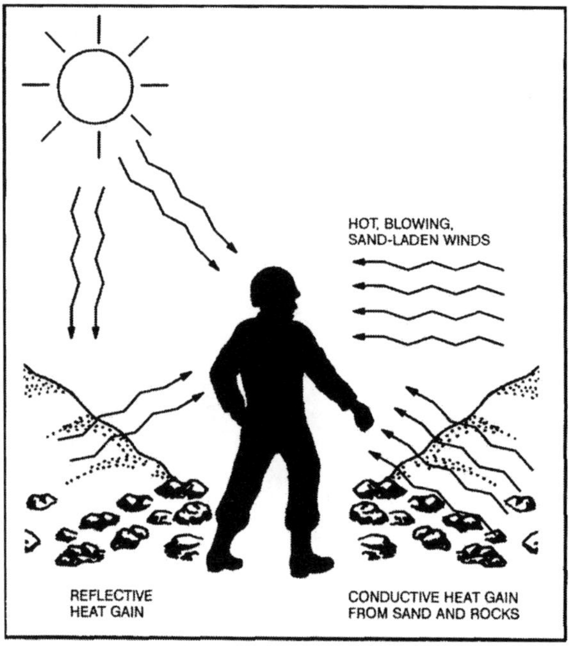

Types of Heat Gain

Normally, several physical and physiological mechanisms (e.g., convection and evaporation) assure transfer of excess body heat to the air. But when air temperature is above skin temperature (around 92 degrees Fahrenheit) the evaporation of sweat is the only operative mechanism. Following the loss of sweat, water must be consumed to replace the body's lost fluids. If the body fluid lost through sweating is not replaced, dehydration will follow. This will hamper heat dissipation and can lead to heat illness. When humidity is high, evaporation of sweat is inhibited and there is a greater risk of dehydration or heat stress. Consider the following to help prevent dehydration:

- Heat, wind, and dry air combine to produce a higher individual water requirement, primarily through loss of body water as sweat. Sweat rates can be high even when the skin looks and feels dry.
- Dehydration nullifies the benefits of heat acclimatization and physical fitness, it increases the susceptibility to heat injury, reduces the capacity to work, and decreases appetite and alertness. A lack of alertness can indicate early stages of dehydration.
- Thirst is not an adequate indicator of dehydration. The soldier/marine will not sense when he is dehydrated and will fail to replace body water losses, even when drinking water is available. The universal experience in the desert is that troops exhibit "voluntary dehydration" that is, they maintain their hydration status at about 2 percent of body weight (1.5 quarts) below their ideal hydration status without any sense of thirst.

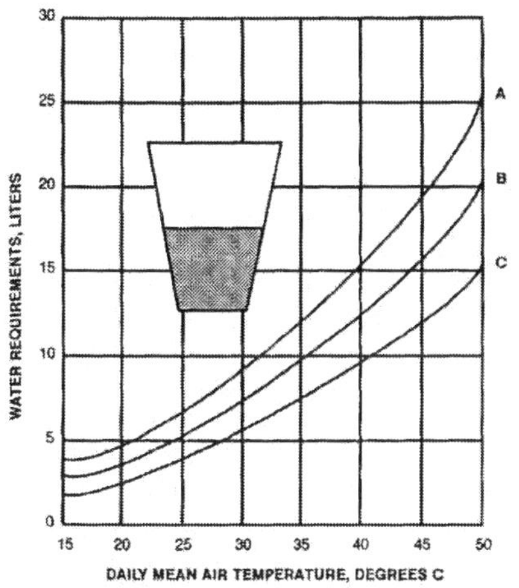

A: Hard work in sun (creeping and crawling with equipment on).
B: Moderate work in sun (cleaning weapons and equipment).
C: Rest in shade.

This graph shows water needs, in liters per day, for men at three activity levels in relation to the daily mean air temperature. For example, if one is doing 8 hours of hard work in the sun (curve A) when the average temperature for the day is 50 degrees C (horizontal scale), one's water requirement for the day will be approximately 25 liters (vertical scale).

From Technical Report E-P118. Southwest Asia: Environment and Its Relationship to Military Activities. July 1959.

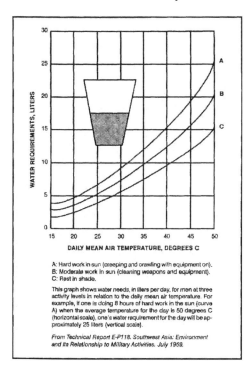

Daily water requirements for three different levels of activity

Chronic dehydration increases the incidence of several medical problems: constipation (already an issue in any field situation), piles (hemorrhoids), kidney stones, and urinary infections. The likelihood of these problems occurring can be reduced by enforcing mandatory drinking schedules.

Resting on hot sand will increase heat stress--the more a body surface is in contact with the sand, the greater the heat stress. Ground or sand in full sun is hot, usually 30-45 degrees hotter than the air, and may reach 150 degrees Fahrenheit when the air temperature is 120 degrees Fahrenheit. Cooler sand is just inches below the surface; a shaded trench will provide a cool resting spot.

At the first evidence of heat illness, have the troops stop work, get into shade, and rehydrate. Early intervention is important. Soldiers/ marines who are not taken care of can become more serious casualties.

WARNING

One heat casualty is usually followed by others and is a warning that the entire unit may be at risk. This is the "Weak Link Rule." The status of the whole unit is assessed at this point.

ACCLIMATIZATION

Acclimatization to heat is necessary to permit the body to reach and maintain efficiency in its cooling process. A period of approximately 2 weeks should be allowed for acclimatization, with progressive increases in heat exposure and physical exertion. Significant acclimatization can be attained in 4-5 days, but full acclimatization takes 7-14 days, with 2-3 hours per day of exercise in the heat. Gradually increase physical activity until full acclimatization is achieved.

Acclimatization does not reduce, and may increase, water requirements. Although this strengthens heat resistance, there is no such thing as total protection against the debilitating effects of heat. Situations may arise where it is not possible for men to become fully acclimatized before being required to do heavy labor. When this happens heavy activity should be limited to cooler hours and troops should be allowed to rest frequently. Check the weather daily. Day-to-day and region-to-region variations in temperatures, wind, and humidity can be substantial.

CLIMATIC STRESS

Climatic stress on the human body in hot deserts can be caused by any combination of air temperature, humidity, air movement, and radiant heat. The body is also adversely affected by such factors as lack of acclimatization, being overweight, dehydration, alcohol consumption, lack of sleep, old age, and poor health.

The body maintains its optimum temperature of 98.6 degrees Fahrenheit by conduction/convection, radiation, and evaporation (sweat). The most important of these in the daytime desert is evaporation, as air temperature alone is probably already above skin temperature. If, however, relative humidity is high, air will not easily evaporate sweat and the cooling effect is reduced. The following paragraphs describe the effects of radiant light, wind, and sand on personnel in desert areas.

RADIANT LIGHT

Radiant light comes from all directions. The sun's rays, either direct or reflected off the ground, affect the skin and can also produce eyestrain and temporarily impaired vision. Not only does glare damage the eyes but it is very tiring; therefore, dark glasses or goggles should be worn.

Overexposure to the sun can cause sunburn. Persons with fair skin, freckledskin, ruddy complexions, or red hair are more susceptible to sunburn than others, but all personnel are susceptible to some degree. Personnel with darker complexions can also sunburn. This is difficult to monitor due to skin pigmentation, so leaders must be ever vigilant to watch for possible sunburn victims. Sunburn is characterized by painful reddened skin, and can result in blistering and lead to other forms of heat illness.

Soldier/marines should acquire a suntan in gradual stages (preferably in the early morning or late afternoon) to gain some protection against sunburn. They should not be permitted to expose bare skin to the sun for longer than five minutes on the first day, increasing exposure gradually at the rate of five minutes per day. They should be fully clothed in loose garments in all operational situations. This will also reduce sweat loss. It is important to remember that—

- The sun is as dangerous on cloudy days as it is on sunny days.
- Sunburn ointment is not designed to give complete protection against excessive exposure.
- Sunbathing or dozing in the desert sun can be fatal.

Wind

The wind can be as physically demanding as the heat, burning the face, arms, and any exposed skin with blown sand. Sand gets into eyes, nose, mouth, throat, lungs, ears, and hair, and reaches every part of the body. Even speaking and listening can be difficult. Continual exposure to blown sand is exhausting and demoralizing. Technical work spaces that are protected from dust and sand are likely to be very hot. Work/rest cycles and enforced water consumption will be required.

The combination of wind and dust or sand can cause extreme irritation to mucous membranes, chap the lips and other exposed skin surfaces, and can cause nosebleed. Cracked, chapped lips make eating difficult and cause communication problems. Irritative conjunctivitis, caused when fine particles enter the eyes, is a frequent complaint of vehicle crews, even those wearing goggles. Lip balm and skin and eye ointments must be used by all personnel. Constant wind noise is tiresome and increases soldier/marine fatigue, thus affecting alertness.

When visibility is reduced by sandstorms to the extent that military operations are impossible, soldiers/marines should not be allowed to leave their group for any purpose unless secured by lines for recovery.

The following are special considerations when performing operations in dust or sand:

- Contact lenses are very difficult to maintain in the dry dusty environment of the desert and should not be worn except by military personnel operating in air conditioned environments, under command guidance.
- Mucous membranes can be protected by breathing through a wet face cloth, snuffing small amounts of water into nostrils (native water is not safe for this purpose) or coating the

nostrils with a small amount of petroleum jelly. Lips should be protected by lip balm.

- Moving vehicles create their own sandstorms and troops traveling in open vehicles should be protected.
- Scarves and bandannas can be used to protect the head and face.
- The face should be washed as often as possible. The eyelids should be cleaned daily.

High Mineral Content

All arid regions have areas where the surface soil has a high mineral content (borax, salt, alkali, and lime). Material in contact with this soil wears out quickly, and water in these areas is extremely hard and undrinkable. Wetting your uniform in such water to cool off may cause a skin rash. The Great Salt Lake area in Utah is an example of this type of mineral-laden water and soil. There is little or no plant life; therefore, shelter is hard to find. Avoid these areas if possible.

BASIC HEAT INJURY PREVENTION

The temperature of the body is regulated within very narrow limits. Too little salt causes heat cramps; too little salt and insufficient water causes heat exhaustion. Heat exhaustion will cause a general collapse of the body's cooling mechanism. This condition is heatstroke, and is potentially fatal. To avoid these illnesses, troops should maintain their physical fitness by eating adequately, drinking sufficient water, and consuming adequate salt. If soldiers/marines expend more calories than they take in, they will be more prone to heat illnesses. Since troops may lose their desire for food in hot climates, they must be encouraged to eat, with the heavier meal of the day scheduled during the cooler hours.

It is necessary to recognize heat stress symptoms quickly. When suffering from heatstroke, the most dangerous condition, there is a tendency for a soldier/marine to creep away from his comrades and attempt to hide in a shady and secluded spot; if not found and treated, he will die. When shade is required during the day, it can best be provided by tarpaulins or camouflage nets, preferably doubled to allow air circulation between layers and dampened with any surplus water.

Approximately 75 percent of the human body is fluid. All chemical activities in the body occur in a water solution, which assists in the removal of toxic body wastes and plays a vital part in the maintenance of an even body temperature. A loss of 2 quarts of body fluid (2.5 percent of body weight) decreases efficiency by 25 percent and a loss of fluid equal to 15 percent of body weight is usually fatal. The following are some considerations when operating in a desert environment:

- Consider water a tactical weapon. Reduce heat injury by forcing water consumption. Soldiers/marines in armored vehicles, MOPP gear, and in body armor need to increase their water intake.
- When possible, drink cool (50-55 degrees Fahrenheit) water.
- Drink one quart of water in the morning, at each meal, and before strenuous work. In hot climates drink at least one quart of water each hour. At higher temperatures hourly water requirements increase to over two quarts.
- Take frequent drinks since they are more often effective than drinking the same amount all at once. Larger soldiers/marines need more water.
- Replace salt loss through eating meals.
- When possible, work loads and/or duration of physical activity should be less during the first days of exposure to heat, and then should gradually be increased to follow acclimatization.
- Modify activities when conditions that increase the risk of heat injury (fatigue/loss of sleep, previous heat exhaustion, taking medication) are present.
- Take frequent rest periods in the shade, if possible. Lower the work rate and work loads as the heat condition increases.
- Perform heavy work in the cooler hours of the day such as early morning or late evening, if possible.

A description of the symptoms and treatment for heat illnesses follows:

- Heat cramps.
    - Symptoms: Muscle cramps of arms, legs, and/or stomach. Heavy sweating (wet skin) and extreme thirst.
    - First aid: Move soldier/marine to a shady area and loosen clothing. Slowly give large amounts of cool water. Watch

the soldier/marine and continue to give him water, if he accepts it. Get medical help if cramps continue.

- Heat exhaustion.
  - Symptoms: Heavy sweating with pale, moist, cool skin; headache, weakness, dizziness, and/or loss of appetite; heat cramps, nausea (with or without vomiting), rapid breathing, confusion, and tingling of the hands and/or feet.
  - First aid: Move the soldier/marine to a cool, shady area and loosen/remove clothing. Pour water on the soldier/marine and fan him to increase the cooling effect. Have the soldier/ marine slowly drink at least one full canteen of water. Elevate the soldier's/marine's legs. Get medical help if symptoms continue; watch the soldier/marine until the symptoms are gone or medical aid arrives.
- Heatstroke.
  - Symptoms: Sweating stops (red, flushed, hot dry skin).
  - First aid: Evacuate to a medical facility immediately. Move the soldier/marine to a cool, shady area and loosen or remove clothing if the situation permits. Start cooling him immediately. Immerse him in water and fan him. Massage his extremities and skin and elevate his legs. If conscious, have the soldier/marine slowly drink one full canteen of water.

WATER SUPPLY

Maintaining safe, clean, water supplies is critical. The best containers for small quantities of water (5 gallons) are plastic water cans or coolers. Water in plastic cans will be good for up to 72 hours; storage in metal containers is safe only for 24 hours. Water trailers, if kept cool, will keep water fresh up to five days. If the air temperature exceeds 100 degrees Fahrenheit, the water temperature must be monitored. When the temperature exceeds 92 degrees Fahrenheit, the water should be changed, as bacteria will multiply. If the water is not changed the water can become a source of sickness, such as diarrhea. Ice in containers keeps water cool. If ice is put in water trailers, the ice must be removed prior to moving the trailer to prevent damage to the inner lining of the trailer.

Potable drinking water is the single most important need in the desert. Ensure nonpotable water is never mistaken for drinking water. Water that is not fit to drink but is not otherwise dangerous (it may be merely oversalinated) may be used to aid cooling. It can be used to wet clothing, for example, so the body does not use too much of its internal store of water.

Use only government-issued water containers for drinking water. Carry enough water on a vehicle to last the crew until the next planned resupply. It is wise to provide a small reserve. Carry water containers in positions that—

- Prevent vibration by clamping them firmly to the vehicle body.
- Are in the shade and benefit from an air draft.
- Are protected from puncture by shell splinters.
- Are easily dismounted in case of vehicle evacuation.

Troops must be trained not to waste water. Water that has been used for washing socks, for example, is perfectly adequate for a vehicle cooling system.

Obtain drinking water only from approved sources to avoid disease or water that may have been deliberately polluted. Be careful to guard against pollution of water sources. If rationing is in effect, water should be issued under the close supervision of officers and noncommissioned officers.

Humans cannot perform to maximum efficiency on a decreased water intake. An acclimatized soldier/marine will need as much (if not more) water as the nonacclimatized soldier/marine, as he sweats more readily. If the ration water is not sufficient, there is no alternative but to reduce physical activity or restrict it to the cooler parts of the day.

In very hot conditions it is better to drink smaller quantities of water often rather than large quantities occasionally. Drinking large quantities causes excessive sweating and may induce heat cramps. Use of alcohol lessens resistance to heat due to its dehydrating effect. As activities increase or conditions become more severe, increase water intake accordingly.

The optimum water drinking temperature is between 10 degrees Celsius and 15.5 degrees Celsius (50-60 degrees Fahrenheit). Use lister bags or even wet cloth around metal containers to help cool water.

Units performing heavy activities on a sustained basis, such as a forced march or digging in, at 80 degrees wet bulb globe temperature index, may require more than 3 gallons of drinking water per man.

Any increase in the heat stress will increase this need. In high temperatures, the average soldier/marine will require 9 quarts of water per day to survive, but 5 gallons are recommended. Details on water consumption and planning factors are contained in Appendix G.

While working in high desert temperatures, a man at rest may lose as much as a pint of water per hour from sweating. In very high temperatures and low humidity, sweating is not noticeable as it evaporates so fast the skin will appear dry. Whenever possible, sweat should be retained on the skin to improve the cooling process; however, the only way to do this is to avoid direct sun on the skin. This is the most important reason why desert troops must remain fully clothed. If a soldier/marine is working, his water loss through sweating (and subsequent requirement for replenishment) increases in proportion to the amount of work done (movement). Troops will not always drink their required amount of liquid readily and will need to be encouraged or ordered to drink more than they think is necessary as the sensation of thirst is not felt until there is a body deficit of 1 to 2 quarts of water. This is particularly true during the period of acclimatization. Packets of artificial fruit flavoring encourages consumption due to the variety of pleasant tastes.

All unit leaders must understand the critical importance of maintaining the proper hydration status. Almost any contingency of military operations will act to interfere with the maintenance of hydration. Urine provides the best indicator of proper hydration. The following are considerations for proper hydration during desert operations:

- Water is the key to your health and survival. Drink before you become thirsty and drink often, When you become thirsty you will be about a "quart and a half low."
- Carry as much water as possible when away from approved sources of drinking water. Man can live longer without food than without water. Drink before you work; carry water in your belly, do not "save" it in your canteen. Learn to drink a quart or more of water at one time and drink frequently to replace sweat losses.
- Ensure troops have at least one canteen of water in reserve, and know where and when water resupply will be available.
- Carbohydrate/electrolyte beverages (e.g., Gatorade) are not required, and if used, should not be the only source of water. They are too concentrated to be used alone. Many athletes prefer to dilute these 1:1 with water. Gaseous drinks, sodas,

beer, and milk are not good substitutes for water because of their dehydrating effects.

- If urine is more colored than diluted lemonade, or the last urination cannot be remembered, there is probably insufficient water intake. Collect urine samples in field expedient containers and spot check the color as a guide to ensuring proper hydration. Very dark urine warns of dehydration. Soldiers/marines should observe their own urine, and use the buddy system to watch for signs of dehydration in others.
- Diseases, especially diarrheal diseases, will complicate and often prevent maintenance of proper hydration.

Salt, in correct proportions, is vital to the human body; however, the more a man sweats, the more salt he loses. The issue ration has enough salt for a soldier/marine drinking up to 4 quarts of water per day. Unacclimatized troops need additional salt during their first few days of exposure and all soldiers/marines need additional salt when sweating heavily. If the water demand to balance sweat loss rises, extra salt must be taken under medical direction. Salt, in excess of body requirements, may cause increased thirst and a feeling of sickness, and can be dangerous. Water must be tested before adding salt as some sources are already saline, especially those close to the sea.

What can individuals or small groups do when they are totally cut off from normal water supply? If you are totally cut off from the normal water supply, the first question you must consider is whether you should try to walk to safety or stay put and hope for rescue. Walking requires 1 gallon of water for every 20 miles covered at night, and 2 gallons for every 20 miles covered during the day. Without any water and walking only at night, you may be able to cover 20 to 25 miles before you collapse. If your chance of being rescued is not increased by walking 20 miles, you may be better off staying put and surviving one to three days longer. If you do not know where you are going, do not try to walk with a limited supply of water.

If you decide to walk to safety, follow the following guidelines in addition to the general conservation practices listed in the next section:

- Take as much water as you have and can carry, and carry little or no food.
- Drink as much as you can comfortably hold before you set out.
- Walk only at night.

Whether you decide to walk or not, you should follow the principles listed below to conserve water in emergency situations:

- Avoid the sun. Stay in shade as much as possible. If you are walking, rest in shade during the day. This may require some ingenuity. You may want to use standard or improvised tents, lie under vehicles, or dig holes in the ground.
- Cease activity. Do not perform any work that you do not have to for survival.
- Remain clothed. It will reduce the water lost to evaporation.
- Shield yourself from excessive winds. Winds, though they feel good also increase the evaporation rate.
- Drink any potable water you have as you feel the urge. Saving it will not reduce your body's need for it or the rate at which you use it.
- Do not drink contaminated water from such sources as car radiators or urine. It will actually require more water to remove the waste material. Instead, in emergencies, use such water to soak your clothing as this reduces sweating.
- Do not eat unless you have plenty of water.

Do not count on finding water if you are stranded in the desert. Still, in certain cases, some water can be found. It does rain sometimes in the desert (although it may be 20 years between showers) and some water will remain under the surface. Signs of possible water are green plants or dry lake beds. Sometimes water can be obtained in these places by digging down until the soil becomes moist and then waiting for water to seep into the hole. Desert trails lead from one water point to another, but they may be further apart than you can travel without water. Almost all soils contain some moisture.

COLD

The desert can be dangerously cold. The dry air, wind, and clear sky can combine to produce bone-chilling discomfort and even injury. The ability of the body to maintain body temperature within a narrow range is as important in the cold as in the heat. Loss of body heat to the environment can lead to cold injury; a general lowering of the body temperature can result in hypothermia, and local freezing of body tissues can lead to frostbite. Hypothermia is the major threat from the cold in the desert, but frostbite also occurs.

Troops must have enough clothing and shelter to keep warm. Remember, wood is difficult to find; any that is available is probably already in use. Troops may be tempted to leave clothing and equipment behind that seems unnecessary (and burdensome) during the heat of the day. Cold-wet injuries (immersion foot or trench foot) may be a problem for dismounted troops operating in the coastal marshes of the Persian Gulf during the winter. Some guidelines to follow when operating in the cold are

- Anticipate an increased risk of cold-wet injuries if a proposed operation includes lowland or marshes. Prolonged exposure of the feet in cold water causes immersion foot injury, which is completely disabling.
- Check the weather—know what conditions you will be confronting. The daytime temperature is no guide to the nighttime temperature; 90-degree-Fahrenheit days can turn into 30-degree-Fahrenheit nights.
- The effects of the wind on the perception of cold is well known. Windchill charts contained in FM 21-10 allow estimation of the combined cooling power of air temperature and wind speed compared to the effects of an equally cooling still-air temperature.

CLOTHING

Uniforms should be worn to protect against sunlight and wind. Wear the uniform loosely. Use hats, goggles, and sunscreen. Standard lightweight clothing is suitable for desert operations but should be camouflaged in desert colors, not green. Wear nonstarched long-sleeved shirts, and full-length trousers tucked into combat boots. Wear a scarf or triangular bandanna loosely around the neck (as a sweat rag) to protect the face and neck during sandstorms against the sand and the sun. In extremely hot and dry conditions a wet sweat rag worn loosely around the neck will assist in body cooling.

Combat boots wear out quickly in desert terrain, especially if the terrain is rocky. The leather dries out and cracks unless a nongreasy mixture such as saddle soap is applied. Covering the ventilation holes on jungle boots with glue or epoxies prevents excessive sand from entering the boots. Although difficult to do, keep clothing relatively clean by washing in any surplus water that is available. When water is not available, air and sun clothing to help kill bacteria.

Change socks when they become wet. Prolonged wear of wet socks can lead to foot injury. Although dry desert air promotes evaporation of water from exposed clothing and may actually promote cooling, sweat tends to accumulate in boots.

Soldier/marines may tend to stay in thin clothing until too late in the desert day and become susceptible to chills--so respiratory infections may be common. Personnel should gradually add layers of clothing at night (such as sweaters), and gradually remove them in the morning. Where the danger of cold weather injury exists in the desert, commanders must guard against attempts by inexperienced troops to discard cold weather clothing during the heat of the day.

Compared to the desert battle dress uniform (DBDU) the relative impermeability of the battle dress overgarment (BDO) reduces evaporative cooling capacity. Wearing underwear and the complete DBDU, with sleeves rolled down and under the chemical protective garment, provides additional protection against chemical poisoning. However, this also increases the likelihood of heat stress casualties.

HYGIENE AND SANITATION

Personal hygiene is absolutely critical to sustaining physical fitness. Take every opportunity to wash. Poor personal hygiene and lack of attention to siting of latrines cause more casualties than actual combat. Field Marshal Rommel lost over 28,400 soldiers of his Afrika Corps to disease in 1942. During the desert campaigns of 1942, for every one combat injury, there were three hospitalized for disease.

Proper standards of personal hygiene must be maintained not only as a deterrent to disease but as a reinforcement to discipline and morale. If water is available, shave and bathe daily. Cleaning the areas of the body that sweat heavily is especially important; change underwear and socks frequently, and use foot powder often. Units deployed in remote desert areas must have a means of cutting hair therefore, barber kits should be maintained and inventoried prior to any deployment. If sufficient water is not available, troops should clean themselves with sponge baths, solution-impregnated pads, a damp rag, or even a dry, clean cloth. Ensure that waste water is disposed of in an approved area to prevent insect infestation. If sufficient water is not available for washing, a field expedient alternative is powder baths, that is, using talcum or baby powder to dry bathe.

Check troops for any sign of injury, no matter how slight, as dirt or insects can cause infection in minor cuts and scratches. Small quantities of disinfectant in washing water reduces the chance of infection. Minor sickness can have serious effects in the desert. Prickly heat for example, upsets the sweating mechanism and diarrhea increases water loss, making the soldier/marine more prone to heat illnesses. The buddy system helps to ensure that prompt attention is given to these problems before they incapacitate individuals.

Intestinal diseases can easily increase in the desert. Proper mess sanitation is essential. Site latrines well away and downwind of troop areas and lagers. Trench-type latrines should be used where the soil is suitable but must be dug deeply, as shallow latrines become exposed in areas of shifting sand. Funnels dug into a sump work well as urinals. Layer the bottom of slit trenches with lime and cover the top prior to being filled in. Ensure lime is available after each use of the latrine. Flies area perpetual source of irritation and carry infections. Only good sanitation can keep the fly problem to a minimum. Avoid all local tribe camps since they are frequently a source of disease and vermin.

DESERT SICKNESS

Diseases common to the desert include plague, typhus, malaria, dengue fever, dysentery, cholera, and typhoid. Diseases which adversely impact hydration, such as those which include nausea, vomiting, and diarrhea among their symptoms, can act to dramatically increase the risk of heat (and cold) illness or injury. Infectious diseases can result in a fever; this may make it difficult to diagnose heat illness. Occurrences of heat illness in troops suffering from other diseases complicate recovery from both ailments.

Many native desert animals and plants are hazardous. In addition to injuries as a result of bites, these natural inhabitants of the desert can be a source of infectious diseases.

Many desert plants and shrubs have a toxic resin that can cause blisters, or spines that can cause infection. Consider milky sap, all red beans, and smoke from burning oleander shrubs, poisonous. Poisonous snakes, scorpions, and spiders are common in all deserts. Coastal waters of the Persian Gulf contain hazardous marine animals including sea snakes, poisonous jellyfish, and sea urchins.

Skin diseases can result from polluted water so untreated water should not be used for washing clothes; although it can be used for vehicle cooling systems or vehicle decontamination.

The excessive sweating common in hot climates brings on prickly heat and some forms of fungus infections of the skin. The higher the humidity, the greater the possibility of their occurrence. Although many deserts are not humid, there are exceptions, and these ailments are common to humid conditions.

The following are additional health-related considerations when operating in a desert environment:

- The most common and significant diseases in deserts include diarrheal and insect borne febrile (i.e., fever causing) illnesses-both types of these diseases are preventable.
- Most diarrheal diseases result from ingestion of water or food contaminated with feces. Flies, mosquitoes, and other insects carry fever-causing illnesses such as malaria, sand fly fever, dengue (fever with severe pain in the joints), typhus, and tick fevers.
- There are no safe natural water sources in the desert. Standing water is usually infectious or too brackish to be safe for consumption. Units and troops must always know where and how to get safe drinking water.
- Avoid brackish water (i.e., salty). It, like sea water, increases thirst; it also dehydrates the soldier/marine faster than were no water consumed. Brackish water is common even in public water supplies, iodine tablets only kill germs and do not reduce brackishness.
- Water supplies with insufficient chlorine residuals, native food and drink, and ice from all sources are common sources of infective organisms.

PREVENTIVE MEASURES

Both diarrheal and insect borne diseases can be prevented through a strategy which breaks the chain of transmission from infected sources to susceptible soldiers/marines by effectively applying preventive measures. Additional preventive measures are described below:

- Careful storage, handling and purification/preparation of water and food are the keys to prevention of diarrheal disease. Procure all food, water, ice, and beverages from US military approved sources and inspect them routinely.

- Well-cooked foods that are "steaming hot" when eaten are generally safe, as are peeled fruits and vegetables.
- Local dairy products and raw leafy vegetables are generally unsafe.
- Consider the food in native markets hazardous. Avoid local food unless approved by medical personnel officials.
- Assume raw ice and native water to be contaminated--raw ice cannot be properly disinfected. Ice has been a major source of illness in all prior conflicts; therefore, use ice only from approved sources.
- If any uncertainty exists concerning the quality of drinking water, troops should disinfect their supplies using approved field-expedient methods (e.g., hypochlorite for lister bags, iodine tablets for canteens, boiling).
- Untreated water used for washing or bathing risks infection.
- Hand washing facilities should be established at both latrines and mess facilities. Particular attention should be given to the cleanliness of hands and fingernails. Dirty hands are the primary means of transmitting disease.
- Dispose of human waste and garbage. Additional considerations regarding human waste and garbage are—
    - Sanitary disposal is important in preventing the spread of disease from insects, animals, and infected individuals, to healthy soldiers/marines.
    - Construction and maintenance of sanitary latrines are essential.
    - Burning is the best solution for waste.
    - Trench latrines can be used if the ground is suitable, but they must be dug deeply enough so that they are not exposed to shifting sand, and they must have protection against flies and other insects that can use them as breeding places.
    - Food and garbage attract animals--do not sleep where you eat and keep refuse areas away from living areas.
    - Survey the unit area for potential animal hazards.

Shake out boots, clothing, and bedding before using them.

CHAPTER 3:

COLD WEATHER SURVIVAL

One of the most difficult survival situations is a cold weather scenario. Remember, cold weather is an adversary that can be as dangerous as an enemy soldier. Every time you venture into the cold, you are pitting yourself against the elements. With a little knowledge of the environment, proper plans, and appropriate equipment, you can overcome the elements. As you remove one or more of these factors, survival becomes increasingly difficult. Remember, winter weather is highly variable. Prepare yourself to adapt to blizzard conditions even during sunny and clear weather.

Cold is a far greater threat to survival than it appears. It decreases your ability to think and weakens your will to do anything except to get warm. Cold is an insidious enemy; as it numbs the mind and body, it subdues the will to survive. Cold makes it very easy to forget your ultimate goal—to survive.

COLD REGIONS AND LOCATIONS

Cold regions include arctic and subarctic areas and areas immediately adjoining them. You can classify about 48 percent of the northern hemisphere's total landmass as a cold region due to the influence and extent of air temperatures. Ocean currents affect cold weather and cause large areas normally included in the temperate zone to fall within the cold regions during winter periods. Elevation also has a marked effect on defining cold regions.

Within the cold weather regions, you may face two types of cold weather environments—wet or dry. Knowing in which environment your area of operations falls will affect planning and execution of a cold weather operation.

Wet Cold Weather Environments

Wet cold weather conditions exist when the average temperature in a 24-hour period is -10 degrees C or above. Characteristics of this condition are freezing during the colder night hours and thawing during the day. Even though the temperatures are warmer during this condition, the terrain is usually very sloppy due to slush and mud. You must concentrate on protecting yourself from the wet ground and from freezing rain or wet snow.

Dry Cold Weather Environments

Dry cold weather conditions exist when the average temperature in a 24-hour period remains below -10 degrees C. Even though the temperatures in this condition are much lower than normal, you do not have to contend with the freezing and thawing. In these conditions, you need more layers of inner clothing to protect you from temperatures as low as -60 degrees C. Extremely hazardous conditions exist when wind and low temperature combine.

WINDCHILL

Windchill increases the hazards in cold regions. Windchill is the effect of moving air on exposed flesh. For instance, with a 27.8-kph (15-knot) wind and a temperature of -10 degrees C, the equivalent windchill temperature is -23 degrees C. Table 3-1 gives the windchill factors for various temperatures and wind speeds.

Remember, even when there is no wind, you will create the equivalent wind by skiing, running, being towed on skis behind a vehicle, working around aircraft that produce wind blasts.

COOLING POWER OF WIND EXPRESSED AS "EQUIVALENT CHILL TEMPERATURE"																						
WIND SPEED		TEMPERATURE (DEGREES C)																				
CALM	CALM	4	2	-1	-4	-7	-9	-12	-15	-18	-21	-23	-26	-29	-32	-34	-37	-40	-43	-46	-48	-51
KNOTS	*KPH*	EQUIVALENT CHILL TEMPERATURE																				
4	8	2	-1	-4	-7	-9	-12	-15	-18	-21	-23	-26	-29	-32	-34	-37	-40	-43	-46	-48	-54	-57
9	16	-1	-7	-9	-12	-15	-18	-23	-26	-29	-32	-37	-40	-43	-46	-51	-54	-57	-59	-62	-68	-71
13	24	-4	-9	-12	-18	-21	-23	-29	-32	-34	-40	-43	-46	-51	-54	-57	-62	-65	-68	-73	-76	-79
17	32	-7	-12	-15	-18	-23	-26	-32	-34	-37	-43	-46	-51	-54	-59	-62	-65	-71	-73	-79	-82	-84
22	40	-9	-12	-18	-21	-26	-29	-34	-37	-43	-46	-51	-54	-59	-62	-68	-71	-76	-79	-84	-87	-93
26	48	-12	-15	-18	-23	-29	-32	-34	-40	-46	-48	-54	-57	-62	-65	-71	-73	-79	-82	-87	-90	-96
30	56	-12	-15	-21	-23	-29	-34	-37	-40	-46	-51	-54	-59	-62	-68	-73	-76	-82	-84	-90	-93	-98
35	64	-12	-18	-21	-26	-29	-34	-37	-43	-48	-51	-57	-59	-65	-71	-73	-79	-82	-87	-90	-96	-101
(Higher winds have little additional effects)		LITTLE DANGER					INCREASING DANGER (Flesh may freeze within 1 minute)						GREAT DANGER (Flesh may freeze within 30 seconds)									
		DANGER OF FREEZING EXPOSED FLESH FOR PROPERLY CLOTHED PERSONS																				

Windchill Table

BASIC PRINCIPLES OF COLD WEATHER SURVIVAL

It is more difficult for you to satisfy your basic water, food, and shelter needs in a cold environment than in a warm environment. Even if you have the basic requirements, you must also have adequate protective clothing and the will to survive. The will to survive is as important as the basic needs. There have been incidents when trained and well-equipped individuals have not survived cold weather situations because they lacked the will to live. Conversely, this will has sustained individuals less well-trained and equipped.

There are many different items of cold weather equipment and clothing issued by the U.S. Army today. Specialized units may have access to newer, lightweight underwear, outerwear and boots, and other special equipment. Remember, however, the older gear will keep you warm as long as you apply a few cold weather principles. If the newer types of clothing are available, use them. If not, then your clothing should be entirely wool, with the possible exception of a windbreaker.

You must not only have enough clothing to protect you from the cold, you must also know how to maximize the warmth you get from it. For example, always keep your head covered. You can lose 40 to 45 percent of body heat from an unprotected head and even more from the unprotected neck, wrist, and ankles. These areas of the body are good radiators of heat and have very little insulating fat. The brain is very susceptible to cold and can stand the least amount of cooling. Because there is much blood circulation in the head, most of which is on the surface, you can lose heat quickly if you do not cover your head.

There are four basic principles to follow to keep warm. An easy way to remember these basic principles is to use the word COLD—

C - Keep clothing *clean*.
O - Avoid *overheating*.
L - Wear clothes loose and in *layers*.
D - Keep clothing *dry*.

C - *Keep clothing clean*. This principle is always important for sanitation and comfort. In winter, it is also important from the standpoint of warmth. Clothes matted with dirt and grease lose much of their insulation value. Heat can escape more easily from the body through the clothing's crushed or filled up air pockets.

O - *Avoid overheating*. When you get too hot, you sweat and your clothing absorbs the moisture. This affects your warmth in two ways: dampness decreases the insulation quality of clothing, and as sweat evaporates, your body cools. Adjust your clothing so that you do not sweat. Do this by partially opening your parka or jacket, by removing an inner layer of clothing, by removing heavy outer mittens, or by throwing back your parka hood or changing to lighter headgear. The head and hands act as efficient heat dissipaters when overheated.

L - *Wear your clothing loose and in layers*. Wearing tight clothing and footgear restricts blood circulation and invites cold injury. It also decreases the volume of air trapped between the layers, reducing its insulating value. Several layers of lightweight clothing are better than one equally thick layer of clothing, because the layers have dead-air space between them. The dead-air space provides extra insulation. Also, layers of clothing allow you to take off or add clothing layers to prevent excessive sweating or to increase warmth.

D - *Keep clothing dry*. In cold temperatures, your inner layers of clothing can become wet from sweat and your outer layer, if not water repellent, can become wet from snow and frost melted by body heat. Wear water repellent outer clothing, if available. It will shed most of the water collected from melting snow and frost. Before entering a heated shelter, brush off the snow and frost. Despite the precautions you take, there will be times when you cannot keep from getting wet. At such times, drying your clothing may become a major problem. On the march, hang your damp mittens and socks on your rucksack. Sometimes in freezing temperatures, the wind and sun will dry this clothing. You can also place damp socks or mittens, unfolded, near your body so that your body heat can dry them. In a campsite, hang damp clothing inside the shelter near the top, using drying lines or improvised racks. You may even be able to dry each item by holding it before an open fire. Dry leather items slowly. If no other means are available for drying your boots, put them between your sleeping bag shell and liner. Your body heat will help to dry the leather.

A heavy, down-lined sleeping bag is a valuable piece of survival gear in cold weather. Ensure the down remains dry. If wet, it loses a lot of its insulation value. If you do not have a sleeping bag, you can make one out of parachute cloth or similar material and natural dry material, such as leaves, pine needles, or moss. Place the dry material between two layers of the material.

Other important survival items are a knife; waterproof matches in a waterproof container, preferably one with a flint attached; a durable compass; map; watch; waterproof ground cloth and cover; flashlight; binoculars; dark glasses; fatty emergency foods; food gathering gear; and signaling items.

Remember, a cold weather environment can be very harsh. Give a good deal of thought to selecting the right equipment for survival in the cold. If unsure of an item you have never used, test it in an "overnight backyard" environment before venturing further. Once you have selected items that are essential for your survival, do not lose them after you enter a cold weather environment.

HYGIENE

Although washing yourself may be impractical and uncomfortable in a cold environment, you must do so. Washing helps prevent skin rashes that can develop into more serious problems.

In some situations, you may be able to take a snow bath. Take a handful of snow and wash your body where sweat and moisture accumulate, such as under the arms and between the legs, and then wipe yourself dry. If possible, wash your feet daily and put on clean, dry socks. Change your underwear at least twice a week. If you are unable to wash your underwear, take it off, shake it, and let it air out for an hour or two.

If you are using a previously used shelter, check your body and clothing for lice each night. If your clothing has become infested, use insecticide powder if you have any. Otherwise, hang your clothes in the cold, then beat and brush them. This will help get rid of the lice, but not the eggs.

If you shave, try to do so before going to bed. This will give your skin a chance to recover before exposing it to the elements.

MEDICAL ASPECTS

When you are healthy, your inner core temperature (torso temperature) remains almost constant at 37 degrees C (98.6 degrees F). Since your limbs and head have less protective body tissue than your torso, their temperatures vary and may not reach core temperature.

Your body has a control system that lets it react to temperature extremes to maintain a temperature balance. There are three main factors that affect this temperature balance—heat production, heat loss, and evaporation. The difference between the body's core temperature and the environment's temperature governs the heat production rate.

Your body can get rid of heat better than it can produce it. Sweating helps to control the heat balance. Maximum sweating will get rid of heat about as fast as maximum exertion produces it.

Shivering causes the body to produce heat. It also causes fatigue that, in turn, leads to a drop in body temperature. Air movement around your body affects heat loss. It has been calculated that a naked man exposed to still air at or about 0 degrees C can maintain a heat balance if he shivers as hard as he can. However, he can't shiver forever.

It has also been calculated that a man at rest wearing the maximum arctic clothing in a cold environment can keep his internal heat balance during temperatures well below freezing. To withstand really cold conditions for any length of time, however, he will have to become active or shiver.

COLD INJURIES

The best way to deal with injuries and sicknesses is to take measures to prevent them from happening in the first place. Treat any injury or sickness that occurs as soon as possible to prevent it from worsening.

The knowledge of signs and symptoms and the use of the buddy system are critical in maintaining health. Following are cold injuries that can occur.

Hypothermia

Hypothermia is the lowering of the body temperature at a rate faster than the body can produce heat. Causes of hypothermia may be general exposure or the sudden wetting of the body by falling into a lake or spraying with fuel or other liquids.

The initial symptom is shivering. This shivering may progress to the point that it is uncontrollable and interferes with an individual's ability to care for himself. This begins when the body's core (rectal) temperature falls to about 35.5 degrees C (96 degrees F). When the core temperature reaches 35 to 32 degrees C (95 to 90 degrees F), sluggish thinking, irrational reasoning, and a false feeling of warmth may occur. Core temperatures of 32 to 30 degrees C (90 to 86 degrees F) and below result in muscle rigidity, unconsciousness, and barely detectable signs of life. If the victim's core temperature falls below 25 degrees C (77 degrees F), death is almost certain.

To treat hypothermia, rewarm the entire body. If there are means available, rewarm the person by first immersing the trunk area only in warm water of 37.7 to 43.3 degrees C (100 to 110 degrees F).

> ⚠ **CAUTION**
> Rewarming the total body in a warm water bath should be done only in a hospital environment because of the increased risk of cardiac arrest and rewarming shock.

One of the quickest ways to get heat to the inner core is to give warm water enemas. Such an action, however, may not be possible in a survival situation. Another method is to wrap the victim in a warmed sleeping bag with another person who is already warm; both should be naked.

> ⚠ **CAUTION**
> The individual placed in the sleeping bag with victim could also become a hypothermia victim if left in the bag too long.

If the person is conscious, give him hot, sweetened fluids. One of the best sources of calories is honey or dextrose; if unavailable, use sugar, cocoa, or a similar soluble sweetener.

> ⚠ **CAUTION**
> Do not force an unconscious person to drink.

There are two dangers in treating hypothermia-rewarming too rapidly and "after drop." Rewarming too rapidly can cause the victim to have circulatory problems, resulting in heart failure. After drop is the sharp body core temperature drop that occurs when taking the victim from the warm water. Its probable muse is the return of previously stagnant limb blood to the core (inner torso) area as recirculation occurs. Concentrating on warming the core area and stimulating peripheral circulation will lessen the effects of after drop. Immersing the torso in a warm bath, if possible, is the best treatment.

Frostbite

This injury is the result of frozen tissues. Light frostbite involves only the skin that takes on a dull whitish pallor. Deep frostbite extends to a depth below the skin. The tissues become solid and immovable. Your feet, hands, and exposed facial areas are particularly vulnerable to frostbite.

Do	Don't
• Periodically check for frostbite.	• Rub injury with snow.
• Rewarm light frostbite.	• Drink alcoholic beverages.
• Keep injured areas from refreezing.	• Smoke.
	• Try to thaw out a deep frostbite injury if you are away from definitive medical care.

Frostbite Do's and Don'ts

The best frostbite prevention, when you are with others, is to use the buddy system. Check your buddy's face often and make sure that he checks yours. If you are alone, periodically cover your nose and lower part of your face with your mittened hand.

The following pointers will aid you in keeping warm and preventing frostbite when it is extremely cold or when you have less than adequate clothing:

- *Face.* Maintain circulation by twitching and wrinkling the skin on your face making faces. Warm with your hands.
- *Ears.* Wiggle and move your ears. Warm with your hands.
- *Hands.* Move your hands inside your gloves. Warm by placing your hands close to your body.
- *Feet.* Move your feet and wiggle your toes inside your boots.

A loss of feeling in your hands and feet is a sign of frostbite. If you have lost feeling for only a short time, the frostbite is probably light. Otherwise, assume the frostbite is deep. To rewarm a light frostbite, use your hands or mittens to warm your face and ears. Place your hands under your armpits. Place your feet next to your buddy's stomach. A deep frostbite injury, if thawed and refrozen, will cause more damage than a nonmedically trained person can handle. Table above lists some do's and don'ts regarding frostbite.

Trench Foot and Immersion Foot

These conditions result from many hours or days of exposure to wet or damp conditions at a temperature just above freezing. The symptoms are a sensation of pins and needles, tingling, numbness, and then pain. The skin will initially appear wet, soggy, white, and shriveled. As it progresses and damage appears, the skin will take on a red and then a bluish or black discoloration. The feet become cold, swollen, and have a waxy appearance. Walking becomes difficult and the feet feel heavy and numb. The nerves and muscles sustain the main damage, but gangrene can occur. In extreme cases, the flesh dies and it may become necessary to have the foot or leg amputated. The best prevention is to keep your feet dry. Carry extra socks with you in a waterproof packet. You can dry wet socks against your torso (back or chest). Wash your feet and put on dry socks daily.

Dehydration

When bundled up in many layers of clothing during cold weather, you may be unaware that you are losing body moisture. Your heavy clothing absorbs the moisture that evaporates in the air. You must drink water to replace this loss of fluid. Your need for water is as great in a cold environment as it is in a warm environment (see Chapter 2). One way to tell if you are becoming dehydrated is to check the color of your urine on snow. If your urine makes the snow dark yellow, you are becoming dehydrated and need to replace body fluids. If it makes the snow light yellow to no color, your body fluids have a more normal balance.

Cold Diuresis

Exposure to cold increases urine output. It also decreases body fluids that you must replace.

Sunburn

Exposed skin can become sunburned even when the air temperature is below freezing. The sun's rays reflect at all angles from snow, ice, and water, hitting sensitive areas of skin—lips, nostrils, and eyelids. Exposure to the sun results in sunburn more quickly at high altitudes than at low altitudes. Apply sunburn cream or lip salve to your face when in the sun.

Snow Blindness

The reflection of the sun's ultraviolet rays off a snow-covered area causes this condition. The symptoms of snow blindness are a sensation of grit in the eyes, pain in and over the eyes that increases with eyeball movement, red and teary eyes, and a headache that intensifies with continued exposure to light. Prolonged exposure to these rays can result in permanent eye damage. To treat snow blindness, bandage your eyes until the symptoms disappear.

You can prevent snow blindness by wearing sunglasses. If you don't have sunglasses, improvise. Cut slits in a piece of cardboard, thin wood, tree bark, or other available material (Figure 15-3). Putting soot under your eyes will help reduce shine and glare.

Constipation

It is very important to relieve yourself when needed. Do not delay because of the cold condition. Delaying relieving yourself because of the cold, eating dehydrated foods, drinking too little liquid, and irregular eating habits can cause you to become constipated. Although not disabling, constipation can cause some discomfort. Increase your fluid intake to at least 2 liters above your normal 2 to 3 liters daily intake and, if available, eat fruit and other foods that will loosen the stool.

Insect Bites

Insect bites can become infected through constant scratching. Flies can carry various disease-producing germs. To prevent insect bites, use insect repellent, netting, and wear proper clothing.

SHELTERS

Your environment and the equipment you carry with you will determine the type of shelter you can build. You can build shelters in wooded areas, open country, and barren areas. Wooded areas usually provide the best location, while barren areas have only snow as building-material. Wooded areas provide timber for shelter construction, wood for fire, concealment from observation, and protection from the wind.

> ✍ **NOTE**
>
> In extreme cold, do not use metal, such as an aircraft fuselage, for shelter. The metal will conduct away from the shelter what little heat you can generate.

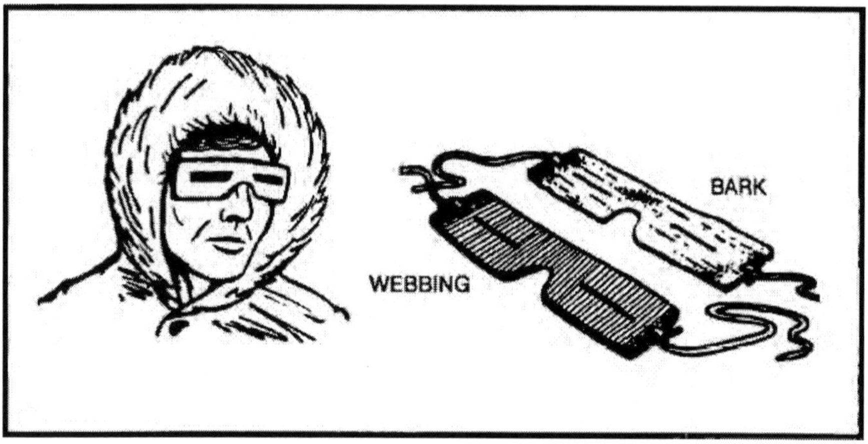

FIGURE 3-1: IMPROVISED SUNGLASSES

Shelters made from ice or snow usually require tools such as ice axes or saws. You must also expend much time and energy to build such a shelter. Be sure to ventilate an enclosed shelter, especially if you intend to build a fire in it. Always block a shelter's entrance, if possible, to keep the heat in and the wind out. Use a rucksack or snow block. Construct a shelter no larger than needed. This will reduce the amount of space to heat. A fatal error in cold weather shelter construction is making the shelter so large that it steals body heat rather than saving it. Keep shelter space small.

Never sleep directly on the ground. Lay down some pine boughs, grass, or other insulating material to keep the ground from absorbing your body heat.

Never fall asleep without turning out your stove or lamp. Carbon monoxide poisoning can result from a fire burning in an unventilated shelter. Carbon monoxide is a great danger. It is colorless and odorless. Anytime you have an open flame, it may generate carbon monoxide. Always check your ventilation. Even in a ventilated shelter, incomplete combustion can cause carbon monoxide poisoning. Usually, there are no symptoms. Unconsciousness and death can occur without warning. Sometimes, however, pressure at the temples, burning of the eyes, headache, pounding pulse, drowsiness, or nausea may occur. The one characteristic, visible sign of carbon monoxide poisoning is a cherry red coloring in the tissues of the lips, mouth, and inside of the eyelids. Get into fresh air at once if you have any of these symptoms.

There are several types of field-expedient shelters you can quickly build or employ. Many use snow for insulation.

FIRE

Fire is especially important in cold weather. It not only provides a means to prepare food, but also to get warm and to melt snow or ice for water. It also provides you with a significant psychological boost by making you feel a little more secure in your situation.

Use the techniques described in *The Complete U.S. Army Survival Guide to Firecraft Skills, Tactics, and Techniques* to build and light your fire. If you are in enemy territory, remember that the smoke, smell, and light from your fire may reveal your location. Light reflects from surrounding trees or rocks, making even indirect light a source of danger. Smoke tends to go straight up in cold, calm weather, making it a beacon during the day, but helping to conceal the smell at night. In warmer weather, especially in a wooded area, smoke tends to hug the ground, making it less visible in the day, but making its odor spread.

If you are in enemy territory, cut low tree boughs rather than the entire tree for firewood. Fallen trees are easily seen from the air.

All wood will burn, but some types of wood create more smoke than others. For instance, coniferous trees that contain resin and tar create more and darker smoke than deciduous trees.

There are few materials to use for fuel in the high mountainous regions of the arctic. You may find some grasses and moss, but very little. The lower the elevation, the more fuel available. You may find some scrub willow and small, stunted spruce trees above the tree line. On sea ice, fuels are seemingly nonexistent. Driftwood or fats may be the only fuels available to a survivor on the barren coastlines in the arctic and subarctic regions.

Abundant fuels within the tree line are—

- Spruce trees are common in the interior regions. As a conifer, spruce makes a lot of smoke when burned in the spring and summer months. However, it burns almost smoke-free in late fall and winter.
- The tamarack tree is also a conifer. It is the only tree of the pine family that loses its needles in the fall. Without its needles, it looks like a dead spruce, but it has many knobby buds and cones on its bare branches. When burning, tamarack wood makes a lot of smoke and is excellent for signaling purposes.

- Birch trees are deciduous and the wood burns hot and fast, as if soaked with oil or kerosene. Most birches grow near streams and lakes, but occasionally you will find a few on higher ground and away from water.
- Willow and alder grow in arctic regions, normally in marsh areas or near lakes and streams. These woods burn hot and fast without much smoke.

Dried moss, grass, and scrub willow are other materials you can use for fuel. These are usually plentiful near streams in tundras (open, treeless plains). By bundling or twisting grasses or other scrub vegetation to form a large, solid mass, you will have a slower burning, more productive fuel.

If fuel or oil is available from a wrecked vehicle or downed aircraft, use it for fuel. Leave the fuel in the tank for storage, drawing on the supply only as you need it. Oil congeals in extremely cold temperatures, therefore, drain it from the vehicle or aircraft while still warm if there is no danger of explosion or fire. If you have no container, let the oil drain onto the snow or ice. Scoop up the fuel as you need it.

⚠ CAUTION

Do not expose flesh to petroleum, oil, and lubricants in extremely cold temperatures. The liquid state of these products is deceptive in that it can cause frostbite.

Some plastic products, such as MRE spoons, helmet visors, visor housings, aid foam rubber will ignite quickly from a burning match. They will also burn long enough to help start a fire. For example, a plastic spoon will burn for about 10 minutes.

In cold weather regions, there are some hazards in using fires, whether to keep warm or to cook. For example—

- Fires have been known to burn underground, resurfacing nearby. Therefore, do not build a fire too close to a shelter.
- In snow shelters, excessive heat will melt the insulating layer of snow that may also be your camouflage.
- A fire inside a shelter lacking adequate ventilation can result in carbon monoxide poisoning.

- A person trying to get warm or to dry clothes may become careless and burn or scorch his clothing and equipment.
- Melting overhead snow may get you wet, bury you and your equipment, and possibly extinguish your fire.

In general, a small fire and some type of stove is the best combination for cooking purposes. A hobo stove is particularly suitable to the arctic. It is easy to make out of a tin can, and it conserves fuel. A bed of hot coals provides the best cooking heat. Coals from a crisscross fire will settle uniformly. Make this type of fire by crisscrossing the firewood. A simple crane propped on a forked stick will hold a cooking container over a fire.

For heating purposes, a single candle provides enough heat to warm an enclosed shelter. A small fire about the size of a man's hand is ideal for use in enemy territory. It requires very little fuel, yet it generates considerable warmth and is hot enough to warm liquids.

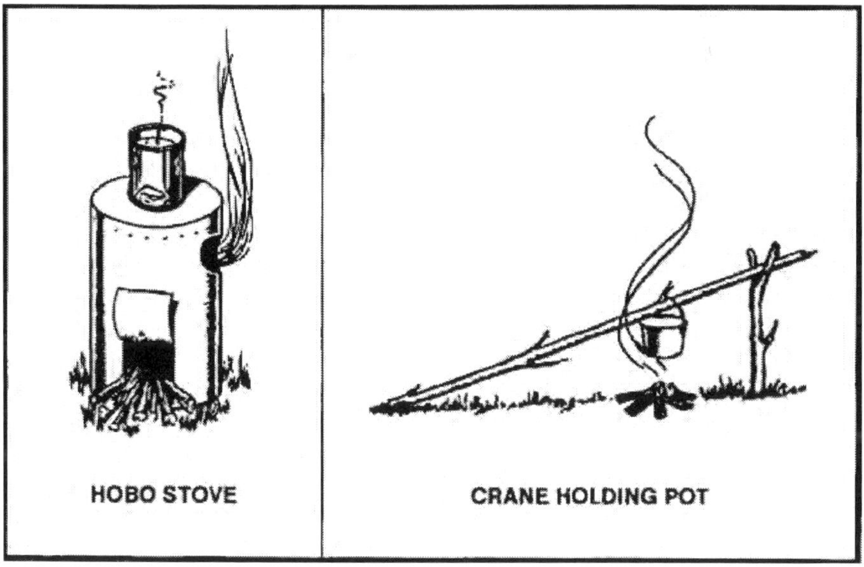

COOKING FIRE/STOVE

WATER

There are many sources of water in the arctic and subarctic. Your location and the season of the year will determine where and how you obtain water.

Water sources in arctic and subarctic regions are more sanitary than in other regions due to the climatic and environmental conditions. However, always purify the water before drinking it. During the summer months, the best natural sources of water are freshwater lakes, streams, ponds, rivers, and springs. Water from ponds or lakes may be slightly stagnant, but still usable. Running water in streams, rivers, and bubbling springs is usually fresh and suitable for drinking.

The brownish surface water found in a tundra during the summer is a good source of water. However, you may have to filter the water before purifying it.

You can melt freshwater ice and snow for water. Completely melt both before putting them in your mouth. Trying to melt ice or snow in your mouth takes away body heat and may cause internal cold injuries. If on or near pack ice in the sea, you can use old sea ice to melt for water. In time, sea ice loses its salinity. You can identify this ice by its rounded corners and bluish color.

You can use body heat to melt snow. Place the snow in a water bag and place the bag between your layers of clothing. This is a slow process, but you can use it on the move or when you have no fire.

> ✍ **NOTE**
>
> Do not waste fuel to melt ice or snow when drinkable water is available from other sources.

When ice is available, melt it, rather than snow. One cup of ice yields more water than one cup of snow. Ice also takes less time to melt. You can melt ice or snow in a water bag, MRE ration bag, tin can, or improvised container by placing the container near a fire. Begin with a small amount of ice or snow in the container and, as it turns to water, add more ice or snow.

Another way to melt ice or snow is by putting it in a bag made from porous material and suspending the bag near the fire. Place a container under the bag to catch the water.

During cold weather, avoid drinking a lot of liquid before going to bed.

Crawling out of a warm sleeping bag at night to relieve yourself means less rest and more exposure to the cold.

Once you have water, keep it next to you to prevent refreezing. Also, do not fill your canteen completely. Allowing the water to slosh around will help keep it from freezing.

FOOD

There are several sources of food in the arctic and subarctic regions. The type of food—fish, animal, fowl, or plant—and the ease in obtaining it depend on the time of the year and your location.

Fish

During the summer months, you can easily get fish and other water life from coastal waters, streams, rivers, and lakes.

The North Atlantic and North Pacific coastal waters are rich in seafood. You can easily find crawfish, snails, clams, oysters, and king crab. In areas where there is a great difference between the high and low tide water levels, you can easily find shellfish at low tide. Dig in the sand on the tidal flats. Look in tidal pools and on offshore reefs. In areas where there is a small difference between the high- and low-tide water levels, storm waves often wash shellfish onto the beaches.

The eggs of the spiny sea urchin that lives in the waters around the Aleutian Islands and southern Alaska are excellent food. Look for the sea urchins in tidal pools. Break the shell by placing it between two stones. The eggs are bright yellow in color.

Most northern fish and fish eggs are edible. Exceptions are the meat of the arctic shark and the eggs of the sculpins.

The bivalves, such as clams and mussels, are usually more palatable than spiral-shelled seafood, such as snails.

> **WARNING**
>
> The black mussel, a common mollusk of the far north, may be poisonous in any season. Toxins sometimes found in the mussel's tissue are as dangerous as strychnine.The sea cucumber is another edible sea animal. Inside its body are five long white muscles that taste much like clam meat.

In early summer, smelt spawn in the beach surf. Sometimes you can scoop them up with your hands.

You can often find herring eggs on the seaweed in midsummer. Kelp, the long ribbonlike seaweed, and other smaller seaweed that grow among offshore rocks are also edible.

Sea Ice Animals

You find polar bears in practically all arctic coastal regions, but rarely inland. Avoid them if possible. They are the most dangerous of all bears. They are tireless, clever hunters with good sight and an extraordinary sense of smell. If you must kill one for food, approach it cautiously. Aim for the brain; a bullet elsewhere will rarely kill one. Always cook polar bear meat before eating it.

> ⚠ **CAUTION**
>
> Do not eat polar bear liver as it contains a toxic concentration of vitamin A.

Earless seal meat is some of the best meat available. You need considerable skill, however, to get close enough to an earless seal to kill it. In spring, seals often bask on the ice beside their breathing holes. They raise their heads about every 30 seconds, however, to look for their enemy, the polar bear.

To approach a seal, do as the Eskimos do—stay downwind from it, cautiously moving closer while it sleeps. If it moves, stop and imitate its movements by lying flat on the ice, raising your head up and down, and wriggling your body slightly. Approach the seal with your body sideways to it and your arms close to your body so that you look as much like another seal as possible. The ice at the edge of the breathing hole is usually smooth and at an incline, so the least movement of the seal may cause it to slide into the water. Therefore, try to get within 22 to 45 meters of the seal and kill it instantly (aim for the brain). Try to reach the seal before it slips into the water. In winter, a dead seal will usually float, but it is difficult to retrieve from the water.

Keep the seal blubber and skin from coming into contact with any scratch or broken skin you may have. You could get "spekk-finger," that is, a reaction that causes the hands to become badly swollen.

Keep in mind that where there are seals, there are usually polar bears, and polar bears have stalked and killed seal hunters.

You can find porcupines in southern subarctic regions where there are trees. Porcupines feed on bark; if you find tree limbs stripped bare, you are likely to find porcupines in the area.

Ptarmigans, owls, Canadian jays, grouse, and ravens are the only birds that remain in the arctic during the winter. They are scarce north of the tree line. Ptarmigans and owls are as good for food as any game bird. Ravens are too thin to be worth the effort it takes to catch them. Ptarmigans, which change color to blend with their surroundings, are hard to spot. Rock ptarmigans travel in pairs and you can easily approach them. Willow ptarmigans live among willow clumps in bottomlands. They gather in large flocks and you can easily snare them. During the summer months all arctic birds have a 2- to 3-week molting period during which they cannot fly and are easy to catch.

Skin and butcher game while it is still warm. If you do not have time to skin the game, at least remove its entrails, musk glands, and genitals before storing. If time allows, cut the meat into usable pieces and freeze each separately so that you can use the pieces as needed. Leave the fat on all animals except seals. During the winter, game freezes quickly if left in the open. During the summer, you can store it in underground ice holes.

Plants

Although tundras support a variety of plants during the warm months, all are small, however, when compared to plants in warmer climates. For instance, the arctic willow and birch are shrubs rather than trees. The following is a list of some plant foods found in arctic and subarctic regions.

ARCTIC FOOD PLANTS

- Arctic raspberry and blueberry
- Arctic willow
- Bearberry
- Cranberry
- Crowberry
- Dandelion
- Eskimo potato
- Fireweed
- Iceland moss
- Marsh marigold
- Reindeer moss
- Rock tripe
- Spatterdock

There are some plants growing in arctic and subarctic regions that are poisonous if eaten. Use the plants that you know are edible. When in doubt, *The Complete U.S. Army Survival Guide to Foraging Skills, Tactics, and Techniques.*

TRAVEL

As a survivor or an evader in an arctic or subarctic region, you will face many obstacles. Your location and the time of the year will determine the types of obstacles and the inherent dangers. You should—

- Avoid traveling during a blizzard.
- Take care when crossing thin ice. Distribute your weight by lying flat and crawling.
- Cross streams when the water level is lowest. Normal freezing and thawing action may cause a stream level to vary as much as 2 to 2.5 meters per day. This variance may occur any time during the day, depending on the distance from a glacier, the temperature, and the terrain. Consider this variation in water level when selecting a campsite near a stream.
- Consider the clear arctic air. It makes estimating distance difficult. You more frequently underestimate than overestimate distances.
- Do not travel in "whiteout" conditions. The lack of contrasting colors makes it impossible to judge the nature of the terrain.
- Always cross a snow bridge at right angles to the obstacle it crosses. Find the strongest part of the bridge by poking ahead of you with a pole or ice axe. Distribute your weight by crawling or by wearing snowshoes or skis.
- Make camp early so that you have plenty of time to build a shelter.
- Consider frozen or unfrozen rivers as avenues of travel. However, some rivers that appear frozen may have soft, open areas that make travel very difficult or may not allow walking, skiing, or sledding.
- Use snowshoes if you are traveling over snow-covered terrain. Snow 30 or more centimeters deep makes traveling difficult. If you do not have snowshoes, make a pair using willow, strips of cloth, leather, or other suitable material.

It is almost impossible to travel in deep snow without snowshoes or skis. Traveling by foot leaves a well-marked trail for any pursuers to follow. If you must travel in deep snow, avoid snow-covered streams.

The snow, which acts as an insulator, may have prevented ice from forming over the water. In hilly terrain, avoid areas where avalanches appear possible. Travel in the early morning in areas where there is danger of avalanches. On ridges, snow gathers on the lee side in overhanging piles called cornices. These often extend far out from the ridge and may break loose if stepped on.

WEATHER SIGNS

There are several good indicators of climatic changes.

Wind

You can determine wind direction by dropping a few leaves or grass or by watching the treetops. Once you determine the wind direction, you can predict the type of weather that is imminent. Rapidly shifting winds indicate an unsettled atmosphere and a likely change in the weather.

Clouds

Clouds come in a variety of shapes and patterns. A general knowledge of clouds and the atmospheric conditions they indicate can help you predict the weather.

Smoke

Smoke rising in a thin vertical column indicates fair weather. Low rising or "flattened out" smoke indicates stormy weather.

Birds and Insects

Birds and insects fly lower to the ground than normal in heavy, moisture-laden air. Such flight indicates that rain is likely. Most insect activity increases before a storm, but bee activity increases before fair weather.

Low-Pressure Front

Slow-moving or imperceptible winds and heavy, humid air often indicate a low-pressure front. Such a front promises bad weather that will probably linger for several days. You can "smell" and "hear" this front. The sluggish, humid air makes wilderness odors more pronounced than during high-pressure conditions. In addition, sounds are sharper and carry farther in low-pressure than high-pressure conditions.

CHAPTER 4:

SURVIVAL IN MOUNTAIN TERRAIN

SECTION 1: MOUNTAIN TERRAIN AND WEATHER

Operations in the mountains require soldiers to be physically fit and leaders to be experienced in operations in this terrain. Problems arise in moving men and transporting loads up and down steep and varied terrain in order to accomplish the mission. Chances for success in this environment are greater when a leader has experience operating under the same conditions as his men. Acclimatization, conditioning, and training are important factors in successful military mountaineering.

4-1. DEFINITION

Mountains are land forms that rise more than 500 meters above the surrounding plain and are characterized by steep slopes. Slopes commonly range from 4 to 45 degrees. Cliffs and precipices may be vertical or overhanging. Mountains may consist of an isolated peak, single ridges, glaciers, snowfields, compartments, or complex ranges extending for long distances and obstructing movement. Mountains usually favor the defense; however, attacks can succeed by using detailed planning, rehearsals, surprise, and well-led troops.

4-2. COMPOSITION

All mountains are made up of rocks and all rocks of minerals (compounds that cannot be broken down except by chemical action). Of the approximately 2,000 known minerals, seven rock-forming minerals make up most of the earth's crust: quartz and feldspar make up granite and sandstone; olivene and pyroxene give basalt its dark color; and amphibole and biotite (mica) are the black crystalline specks in granitic rocks. Except for calcite, found in limestone, they all contain silicon and are often referred to as silicates.

4-3. ROCK AND SLOPE TYPES

Different types of rock and different slopes present different hazards. The following paragraphs discuss the characteristics and hazards of the different rocks and slopes.

a. Granite. Granite produces fewer rockfalls, but jagged edges make pulling rope and raising equipment more difficult. Granite is abrasive and increases the danger of ropes or accessory cords being cut. Climbers must beware of large loose boulders. After a rain, granite dries quickly. Most climbing holds are found in cracks. Face climbing can be found, however, it cannot be protected.

b. Chalk and Limestone. Chalk and limestone are slippery when wet. Limestone is usually solid; however, conglomerate type stones may be loose. Limestone has pockets, face climbing, and cracks.

c. Slate and Gneiss. Slate and gneiss can be firm and or brittle in the same area (red coloring indicates brittle areas). Rockfall danger is high, and small rocks may break off when pulled or when pitons are emplaced.

d. Sandstone. Sandstone is usually soft causing handholds and footholds to break away under pressure. Chocks placed in sandstone may or may not hold. Sandstone should be allowed to dry for a couple of days after a rain before climbing on it wet sandstone is extremely soft. Most climbs follow a crack. Face climbing is possible, but any outward pull will break off handholds and foot holds, and it is usually difficult to protect.

e. Grassy Slopes. Penetrating roots and increased frost cracking cause a continuous loosening process. Grassy slopes are slippery after rain, new snow, and dew. After long, dry spells clumps of the slope tend to break away. Weight should be distributed evenly; for example, use flat hand push holds instead of finger pull holds.

f. Firm Spring Snow (Firm Snow). Stopping a slide on small, leftover snow patches in late spring can be difficult. Routes should be planned to avoid these dangers. Self-arrest should be practiced before encountering this situation. Beginning climbers should be secured with rope when climbing on this type surface. Climbers can glissade down firm snow if necessary. Firm snow is easier to ascend than walking up scree or talus.

g. Talus. Talus is rocks that are larger than a dinner plate, but smaller than boulders. They can be used as stepping-stones to ascend or descend a slope. However, if a talus rock slips away it can produce more injury than scree because of its size.

h. Scree. Scree is small rocks that are from pebble size to dinner plate size. Running down scree is an effective method of descending in a hurry. One can run at full stride without worry the whole scree field is moving with you. Climbers must beware of larger rocks that may be solidly planted under the scree. Ascending scree is a tedious task. The scree does not provide a solid platform and will only slide under foot. If possible, avoid scree when ascending.

4-4. CROSS-COUNTRY MOVEMENT

Soldiers must know the terrain to determine the feasible routes for cross-country movement when no roads or trails are available.

a. A pre-operations intelligence effort should include topographic and photographic map coverage as well as detailed weather data for the area of operations. When planning mountain operations, additional information may be needed about size, location, and characteristics of landforms; drainage; types of rock and soil; and the density and distribution of vegetation. Control must be decentralized to lower levels because of varied terrain, erratic weather, and communication problems inherent to mountainous regions.

b. Movement is often restricted due to terrain and weather. The erratic weather requires that soldiers be prepared for wide variations in temperature, types, and amounts of precipitation.

(1) Movement above the timberline reduces the amount of protective cover available at lower elevations. The logistical problem is important; therefore, each man must be self-sufficient to cope with normal weather changes using materials from his rucksack.

(2) Movement during a storm is difficult due to poor visibility and bad footing on steep terrain. Although the temperature is often higher during a storm than during clear weather, the dampness of rain and snow and the penetration of wind cause soldiers to chill quickly. Although climbers should get off the high ground and seek shelter

and warmth, during severe mountain storms, capable commanders may use reduced visibility to achieve tactical surprise.

c. When the tactical situation requires continued movement during a storm, the following precautions should be observed:
   - Maintain visual contact.
   - Keep warm. Maintain energy and body heat by eating and drinking often; carry food that can be eaten quickly and while on the move.
   - Keep dry. Wear wet-weather clothing when appropriate, but do not overdress, which can cause excessive perspiration and dampen clothing. As soon as the objective is reached and shelter secured, put on dry clothing.
   - . Do not rush. Hasty movement during storms leads to breaks in contact and accidents.
   - If lost, stay warm, dry, and calm.
   - Do not use ravines as routes of approach during a storm as they often fill with water and are prone to flash floods.
   - Avoid high pinnacles and ridgelines during electrical storms.
   - Avoid areas of potential avalanche or rock-fall danger.

4-5. COVER AND CONCEALMENT

When moving in the mountains, outcroppings, boulders, heavy vegetation, and intermediate terrain can provide cover and concealment. Digging fighting positions and temporary fortifications is difficult because soil is often thin or stony. The selection of dug-in positions requires detailed planning. Some rock types, such as volcanic tuff, are easily excavated. In other areas, boulders and other loose rocks can be used for building hasty fortifications. In alpine environments, snow and ice blocks may be cut and stacked to supplement dug-in positions. As in all operations, positions and routes must be camouflaged to blend in with the surrounding terrain to prevent aerial detection.

4-6. OBSERVATION

Observation in mountains varies because of weather and ground cover. The dominating height of mountainous terrain permits excellent long-range observation. However, rapidly changing weather with frequent periods of high winds, rain, snow, sleet, hail, and fog can limit visibility. The rugged nature of the terrain often produces dead space at midranges.

a. Low cloud cover at higher elevations may neutralize the effectiveness of Observation Points (Ops) established on peaks or mountaintops. High wind speeds and sound often mask the noises of troop movement. Several OPs may need to be established laterally, in depth, and at varying altitudes to provide visual coverage of the battle area.
b. Conversely, the nature of the terrain can be used to provide concealment from observation. This concealment can be obtained in the dead space. Mountainous regions are subject to intense shadowing effects when the sun is low in relatively clear skies. The contrast from lighted to shaded areas causes visual acuity in the shaded regions to be considerably reduced. These shadowed areas can provide increased concealment when combined with other camouflage and should be considered in maneuver plans.

MOUNTAIN WEATHER

Most people subconsciously "forecast" the weather. If they look outside and see dark clouds they may decide to take rain gear. If an unexpected wind strikes, people glance to the sky for other bad signs. A conscious effort to follow weather changes will ultimately lead to a more accurate forecast. An analysis of mountain weather and how it is affected by mountain terrain shows that such weather is prone to patterns and is usually severe, but patterns are less obvious in mountainous terrain than in other areas. Conditions greatly change with altitude, latitude, and exposure to atmospheric winds and air masses. Mountain weather can be extremely erratic. It varies from stormy winds to calm, and from extreme cold to warmth within a short time or with a minor shift in locality. The severity and variance of the weather causes it to have a major impact on military operations.

4-7. CONSIDERATIONS FOR PLANNING

Mountain weather can be either a dangerous obstacle to operations or a valuable aid, depending on how well it is understood and to what extent advantage is taken of its peculiar characteristics.

a. Weather often determines the success or failure of a mission since it is highly changeable. Military operations plans must be flexible, especially in planning airmobile and airborne operations. The weather must be anticipated to allow enough time for planning so that the leaders of subordinate units can use their initiative in turning an important weather factor in their favor. The clouds that often cover the tops of mountains and the fogs that cover valleys are an excellent means of concealing movements that normally are made during darkness or in smoke. Limited visibility can be used as a combat multiplier.

b. The safety or danger of almost all high mountain regions, especially in winter, depends upon a change of a few degrees of temperature above or below the freezing point. Ease and speed of travel depend mainly on the weather. Terrain that can be crossed swiftly and safely one day may become impassable or highly dangerous the next due to snowfall, rainfall, or a rise in temperature. The reverse can happen just as quickly. The prevalence of avalanches depends on terrain, snow conditions, and weather factors.

c. Some mountains, such as those found in desert regions, are dry and barren with temperatures ranging from extreme heat in the summer to extreme cold in the winter. In tropical regions, lush jungles with heavy seasonal rains and little temperature variation often cover mountains. High rocky crags with glaciated peaks can be found in mountain ranges at most latitudes along the western portion of the Americas and Asia.

d. Severe weather may decrease morale and increase basic survival problems. These problems can be minimized when men have been trained to accept the weather by being self-sufficient. Mountain soldiers properly equipped and trained can use the weather to their advantage in combat operations.

4-8. MOUNTAIN AIR

High mountain air is dry and may be drier in the winter. Cold air has a reduced capacity to hold water vapor. Because of this increased dryness, equipment does not rust as quickly and organic material decomposes slowly. The dry air also requires soldiers to increase consumption of water. The reduced water vapor in the air causes an increase in evaporation of moisture from the skin and in loss of water through transpiration in the respiratory system. Due to the cold, most soldiers do not naturally consume the quantity of fluids they would at higher temperatures and must be encouraged to consciously increase their fluid intake.

a. Pressure is low in mountainous areas due to the altitude. The barometer usually drops 2.5 centimeters for every 300 meters gained in elevation (3 percent).

b. The air at higher altitudes is thinner as atmospheric pressure drops with the increasing altitude. The altitude has a natural filtering effect on the sun's rays. Rays are absorbed or reflected in part by the molecular content of the atmosphere. This effect is greater at lower altitudes. At higher altitudes, the thinner, drier air has a reduced molecular content and, consequently, a reduced filtering effect on the sun's rays. The intensity of both visible and ultraviolet rays is greater with increased altitude. These conditions increase the chance of sunburn, especially when combined with a snow cover that reflects the rays upward.

4-9. WEATHER CHARACTERISTICS

The earth is surrounded by an atmosphere that is divided into several layers. The world's weather systems are in the lower of these layers known as the "troposphere." This layer reaches as high as 40,000 feet. Weather is a result of an atmosphere, oceans, land masses, unequal heating and cooling from the sun, and the earth's rotation. The weather found in any one place depends on many things such as the air temperature, humidity (moisture content), air pressure (barometric pressure), how it is being moved, and if it is being lifted or not.

a. Air pressure is the "weight" of the atmosphere at any given place. The higher the pressure, the better the weather will be. With lower air pressure, the weather will more than likely be worse. In order to understand this, imagine that the air in the atmosphere acts like a liquid. Areas with a high level of this "liquid" exert more pressure on an area and are called high-pressure areas. Areas with a lower level are called low-pressure areas. The average air pressure at sea level is 29.92 inches of mercury (hg) or 1,013 millibars (mb). The higher in altitude, the lower the pressure.

(1) ***High Pressure.*** The characteristics of a high-pressure area are as follows:

- The airflow is clockwise and out.
- Otherwise known as an "anticyclone".
- Associated with clear skies.
- Generally the winds will be mild.
- Depicted as a blue "H" on weather maps.

(2) ***Low Pressure.*** The characteristics of a low-pressure area are as follows:

- The airflow is counterclockwise and in.
- Otherwise known as a "cyclone".

- Associated with bad weather.
- Depicted as a red "L" on weather maps.

b. Air from a high-pressure area is basically trying to flow out and equalize its pressure with the surrounding air. Low pressure, on the other hand, is building up vertically by pulling air in from outside itself, which causes atmospheric instability resulting in bad weather.

c. On a weather map, these differences in pressure are depicted as isobars. Isobars resemble contour lines and are measured in either millibars or inches of mercury. The areas of high pressure are called "ridges" and lows are called "troughs."

4-10. WIND

In high mountains, the ridges and passes are seldom calm; however, strong winds in protected valleys are rare. Normally, wind speed increases with altitude since the earth's frictional drag is strongest near the ground. This effect is intensified by mountainous terrain. Winds are accelerated when they converge through mountain passes and canyons. Because of these funneling effects, the wind may blast with great force on an exposed mountainside or summit. Usually, the local wind direction is controlled by topography.

a. The force exerted by wind quadruples each time the wind speed doubles; that is, wind blowing at40 knots pushes four times harder than a wind blowing at 20 knots. With increasing wind strength, gusts become more important and may be 50 percent higher than the average wind speed. When wind strength increases to a hurricane force of 64 knots or more, soldiers should lay on the ground during gusts and continue moving during lulls. If a hurricane- force wind blows where there is sand or snow, dense clouds fill the air. The rocky debris or chunks of snow crust are hurled near the surface. During the winter season, or at high altitudes, commanders must be constantly aware of the wind-chill factor and associated cold-weather injuries.

b. Winds are formed due to the uneven heating of the air by the sun and rotation of the earth. Much of the world's weather depends on a system of winds that blow in a set direction.

c. Above hot surfaces, air expands and moves to colder areas where it cools and becomes denser, and sinks to the earth's surface. The results are a circulation of air from the poles along the surface of the earth to the equator, where it rises and moves to the poles again.

d. Heating and cooling together with the rotation of the earth causes surface winds. In the Northern Hemisphere, there are three prevailing winds:
   (1) ***Polar Easterlies.*** These are winds from the polar region moving from the east. This is air that has cooled and settled at the poles.
   (2) ***Prevailing Westerlies.*** These winds originate from approximately 30 degrees north latitude from the west. This is an area where prematurely cooled air, due to the earth's rotation, has settled to the surface.
   (3) ***Northeast Tradewinds.*** These are winds that originate from approximately 30 degrees north from the northeast.

e. The jet stream is a long meandering current of high-speed winds often exceeding 250 miles per hour near the transition zone between the troposphere and the stratosphere known as the tropopause. These winds blow from a generally westerly direction dipping down and picking up air masses from the tropical regions and going north and bringing down air masses from the polar regions.

f. The patterns of wind mentioned above move air. This air comes in parcels called "air masses." These air masses can vary from the size of a small town to as large as a country. These air masses are named from where they originate:
   - Maritime over water.
   - Continental over land.
   - Polar north of 60 degrees north latitude.
   - Tropical south of 60 degrees north latitude.

   Combining these parcels of air provides the names and description of the four types of air masses:
   - Continental Polar cold, dry air mass.
   - Maritime Polar cold, wet air mass.
   - Maritime Tropical warm, wet air mass.
   - Continental Tropical warm, dry air mass.

g. Two types of winds are peculiar to mountain environments, but do not necessarily affect the weather.
   (1) ***Anabatic Wind (Valley Winds).*** These winds blow up mountain valleys to replace warm rising air and are usually light winds.

(2) ***Katabatic Wind (Mountain Wind).*** These winds blow down mountain valley slopes caused by the cooling of air and are occasionally strong winds.

4-11. HUMIDITY

Humidity is the amount of moisture in the air. All air holds water vapor even if it cannot be seen. Air can hold only so much water vapor; however, the warmer the air, the more moisture it can hold. When air can hold all that it can the air is "saturated" or has 100 percent relative humidity.

a. If air is cooled beyond its saturation point, the air will release its moisture in one form or another(clouds, fog, dew, rain, snow, and so on). The temperature at which this happens is called the "condensation point". The condensation point varies depending on the amount of water vapor contained in the air and the temperature of the air. If the air contains a great deal of water, condensation can occur at a temperature of 68 degrees Fahrenheit, but if the air is dry and does not hold much moisture, condensation may not form until the temperature drops to 32 degrees Fahrenheit or even below freezing.

b. The adiabatic lapse rate is the rate at which air cools as it rises or warms as it descends. This rate varies depending on the moisture content of the air. Saturated (moist) air will warm and cool approximately 3.2 degrees Fahrenheit per 1,000 feet of elevation gained or lost. Dry air will warm and cool approximately 5.5 degrees Fahrenheit per 1,000 feet of elevation gained or lost.

4-12. CLOUD FORMATION

Clouds are indicators of weather conditions. By reading cloud shapes and patterns, observers can forecast weather with little need for additional equipment such as a barometer, wind meter, and thermometer. Anytime air is lifted or cooled beyond its saturation point (100 percent relative humidity), clouds are formed. The four ways air gets lifted and cooled beyond its saturation point are as follows.

a. **Convective Lifting**. This effect happens due to the sun's heat radiating off the Earth's surface causing air currents (thermals) to rise straight up and lift air to a point of saturation.

b. **Frontal Lifting**. A front is formed when two air masses of different moisture content and temperature collide. Since air masses will not mix, warmer air is forced aloft over the colder air mass. From there it is cooled and then reaches its saturation point. Frontal lifting creates the majority of precipitation.

c. **Cyclonic Lifting**. An area of low pressure pulls air into its center from all over in a counterclockwise direction. Once this air reaches the center of the low pressure, it has nowhere to go but up. Air continues to lift until it reaches the saturation point.

d. **Orographic Lifting**. This happens when an air mass is pushed up and over a mass of higher ground such as a mountain. Air is cooled due to the adiabatic lapse rate until the air's saturation point is reached.

4-13. TYPES OF CLOUDS

Clouds are one of the signposts to what is happening with the weather. Clouds can be described in many ways. They can be classified by height or appearance, or even by the amount of area covered vertically or horizontally. Clouds are classified into five categories: low-, mid-, and high-level clouds; vertically-developed clouds; and less common clouds.

a. **Low-Level Clouds**. Low-level clouds (0 to 6,500 feet) are either cumulus or stratus (Figures 4-1 and 4-2). Low-level clouds are mostly composed of water droplets since their bases lie below 6,500 feet. When temperatures are cold enough, these clouds may also contain ice particles and snow.

Figure 4-1: Cumulus clouds.

Figure 4-2. Stratus clouds.

(1) The two types of precipitating low-level clouds are nimbostratus and stratocumulus (Figures 4-3 and 4-4).

Figure 4-3: Nimbostratus clouds.

Figure 4-4. Stratocumulus clouds.

(a) Nimbostratus clouds are dark, low-level clouds accompanied by light to moderately falling precipitation. The sun or moon is not visible through nimbostratus clouds, which distinguishes them from mid-level altostratus clouds. Because of the fog and falling precipitation commonly found beneath and around nimbostratus clouds, the cloud base is typically extremely diffuse and difficult to accurately determine.

(b) Stratocumulus clouds generally appear as a low, lumpy layer of clouds that is sometimes accompanied by weak precipitation. Stratocumulus vary in color from dark gray to light gray and may appear as rounded masses with breaks of clear sky in between. Because the individual elements of stratocumulus are larger than those of altocumulus, deciphering between the two cloud types is easier. With your arm extended toward the sky, altocumulus elements are about the size of a thumbnail while stratocumulus are about the size of a fist.

(2) Low-level clouds may be identified by their height above nearby surrounding relief of known elevation. Most precipitation originates from low-level clouds because rain or snow usually evaporate before reaching the ground from higher clouds. Low-level clouds usually indicate impending precipitation, especially if the cloud is more than 3,000 feet thick. (Clouds that appear dark at their bases are more than 3,000 feet thick.)

b. **Mid-Level Clouds**. Mid-level clouds (between 6,500 to 20,000 feet) have a prefix of alto. Middle clouds appear less distinct than low clouds because of their height. Alto clouds with sharp edges are warmer because they are composed mainly of water droplets. Cold clouds, composed mainly of ice crystals and usually colder than -30 degrees F, have distinct edges that grade gradually into the surrounding sky. Middle clouds usually indicate fair weather, especially if they are rising over time. Lowering middle clouds indicate potential storms, though usually hours away. There are two types of mid-level clouds, altocumulus and altostratus clouds (Figures 4-5 and 4-6).

Figure 4-5: Altocumulus.

Figure 4-6. Altostratus.

(1) Altocumulus clouds can appear as parallel bands or rounded masses. Typically a portion of an altocumulus cloud is shaded, a characteristic which makes them distinguishable from high-level cirrocumulus. Altocumulus clouds usually form in advance of a cold front. The presence of altocumulus clouds on a warm humid summer morning is commonly followed by thunderstorms later in the day. Altocumulus clouds that are scattered rather than even, in a blue sky, are called "fair weather" cumulus and suggest arrival of high pressure and clear skies.

(2) Altostratus clouds are often confused with cirrostratus. The one distinguishing feature is that a halo is not observed around the sun or moon. With altostratus, the sun or moon is only vaguely visible and appears as if it were shining through frosted glass.

c. **High-Level Clouds**. High-level clouds (more than 20,000 feet above ground level) are usually frozen clouds, indicating air temperatures at that elevation below -30 degrees Fahrenheit, with a fibrous structure and blurred outlines. The sky is often covered with a thin veil of cirrus that partly obscures the sun or, at night, produces a ring of light around the moon. The arrival of cirrus indicates moisture aloft and the approach of a traveling storm system. Precipitation is often 24 to 36 hours away. As the storm approaches, the cirrus thickens and lowers, becoming altostratus and eventually stratus. Temperatures are warm, humidity rises, and winds become southerly or southeasterly. The two types of high-level clouds are cirrus and cirrostratus (Figure 4-7 and Figure 4-8).

Figure 4-7. Cirrus.

Figure 4-8. Cirrostratus.

(1) Cirrus clouds are the most common of the high-level clouds. Typically found at altitudes greater than 20,000 feet, cirrus are composed of ice crystals that form when super-cooled water droplets freeze. Cirrus clouds generally occur in fair weather and point in the direction of air movement at their elevation. Cirrus can be observed in a variety of shapes and sizes. They can be nearly straight, shaped like a comma, or seemingly all tangled together. Extensive cirrus clouds are associated with an approaching warm front.

(2) Cirrostratus clouds are sheet-like, high-level clouds composed of ice crystals. They are relatively transparent and can cover the entire sky and be up to several thousand feet thick. The sun or moon can be seen through cirrostratus. Sometimes the only indication of cirrostratus clouds is a halo around the sun or moon. Cirrostratus clouds tend to thicken as a warm front approaches, signifying an increased production of ice crystals. As a result, the halo gradually disappears and the sun or moon becomes less visible.

d. Vertical-Development Clouds. Clouds with vertical development can grow to heights in excess of 39,000 feet, releasing incredible amounts of energy. The two types of clouds with vertical development are fair weather cumulus and cumulonimbus.

(1) Fair weather cumulus clouds have the appearance of floating cotton balls and have a lifetime of 5 to 40 minutes. Known for their flat bases and distinct outlines, fair weather cumulus exhibit only slight vertical growth, with the cloud tops designating the limit of the rising air. Given suitable conditions, however, these clouds can later develop into towering cumulonimbus clouds associated with powerful thunderstorms. Fair weather cumulus clouds are fueled by buoyant bubbles of air known as thermals that rise up from the earth's surface. As the air rises, the water vapor cools and condenses forming water droplets. Young fair weather cumulus clouds have sharply defined edges and bases while the edges of older clouds appear more ragged, an artifact of erosion. Evaporation along the cloud edges cools the surrounding air, making it heavier and producing sinking motion

outside the cloud. This downward motion inhibits further convection and growth of additional thermals from down below, which is why fair weather cumulus typically have expanses of clear sky between them. Without a continued supply of rising air, the cloud begins to erode and eventually disappears.

(2) Cumulonimbus clouds are much larger and more vertically developed than fair weather cumulus (Figure 4-9). They can exist as individual towers or form a line of towers called a squall line. Fueled by vigorous convective updrafts, the tops of cumulonimbus clouds can reach 39,000 feet or higher. Lower levels of cumulonimbus clouds consist mostly of water droplets while at higher elevations, where the temperatures are well below freezing, ice crystals dominate the composition. Under favorable conditions, harmless fair weather cumulus clouds can quickly develop into large cumulonimbus associated with powerful thunderstorms known as super-cells. Super-cells are large thunderstorms with deep rotating updrafts and can have a lifetime of several hours. Supercells produce frequent lightning, large hail, damaging winds, and tornadoes. These storms tend to develop during the afternoon and early evening when the effects of heating from the sun are the strongest.

Figure 4-9: Cumulonimbus.

e. **Other Cloud Types**. These clouds are a collection of miscellaneous types that do not fit into the previous four groups. They are orographic clouds, lenticulars, and contrails.

(1) Orographic clouds develop in response to the forced lifting of air by the earth's topography. Air passing over a mountain oscillates up and down as it moves downstream. Initially, stable air encounters a mountain, is lifted upward, and cools. If the air cools to its saturation temperature during this process, the water vapor condenses and becomes visible as a cloud. Upon reaching the mountain top, the air is heavier than the environment and will sink down the other side, warming as it descends. Once the air returns to its original height, it has the same buoyancy as the surrounding air. However, the air does not stop immediately because it still has momentum carrying it downward. With continued descent, the air becomes warmer then the surrounding air and accelerates back upwards towards its original height. Another name for this type of cloud is the lenticular cloud.

(2) Lenticular clouds are cloud caps that often form above pinnacles and peaks, and usually indicate higher winds aloft (Figure 4-10). Cloud caps with a lens shape, similar to a "flying saucer," indicate extremely high winds (over 40 knots). Lenticulars should always be watched for changes. If they grow and descend, bad weather can be expected.

Figure 4-10: Lenticular.

(3) Contrails are clouds that are made by water vapor being inserted into the upper atmosphere by the exhaust of jet engines (Figure 4-11). Contrails evaporate rapidly in fair weather. If it takes longer than two hours for contrails to evaporate, then there is impending bad weather (usually about 24 hours prior to a front).

Figure 4-11: Contrails.

f. **Cloud Interpretation**. Serious errors can occur in interpreting the extent of cloud cover, especially when cloud cover must be reported to another location. Cloud cover always appears greater on or near the horizon, especially if the sky is covered with cumulus clouds, since the observer is looking more at the sides of the clouds rather than between them. Cloud cover estimates should be restricted to sky areas more than 40 degrees above the horizon that is, to the local sky. Assess the sky by dividing the 360 degrees of sky around you into eighths. Record the coverage in eighths and the types of clouds observed.

4-14. FRONTS

Fronts occur when two air masses of different moisture and temperature contents meet. One of the indicators that a front is approaching is the progression of the clouds. The four types of fronts are warm, cold, occluded, and stationary.

a. **Warm Front**. A warm front occurs when warm air moves into and over a slower or stationary cold air mass. Because warm air is less dense, it will rise up and over the cooler air. The cloud types seen when a warm front approaches are cirrus, cirrostratus, nimbostratus (producing rain), and fog. Occasionally, cumulonimbus clouds will be seen during the summer months.

b. **Cold Front**. A cold front occurs when a cold air mass overtakes a slower or stationary warm air mass. Cold air, being denser than warm air, will force the warm air up. Clouds observed will be cirrus, cumulus, and then cumulonimbus producing a short period of showers.

c. **Occluded Front**. Cold fronts generally move faster than warm fronts. The cold fronts eventually overtake warm fronts and the warm air becomes progressively lifted from the surface. The zone of division between cold air ahead and cold air behind is called a "cold occlusion." If the air behind the front is warmer than the air ahead, it is a warm occlusion. Most land areas experience more occlusions than other types of fronts. The cloud progression observed will be cirrus, cirrostratus, altostratus, and nimbostratus. Precipitation can be from light to heavy.

d. **Stationary Front**. A stationary front is a zone with no significant air movement. When a warm or cold front stops moving, it becomes a stationary front. Once this boundary begins forward motion, it once again becomes a warm or cold front. When crossing from one side of a stationary front to another, there is typically a noticeable temperature change and shift in wind direction. The weather is usually clear to partly cloudy along the stationary front.

4-15. TEMPERATURE

Normally, a temperature drop of 3 to 5 degrees Fahrenheit for every 1,000 feet gain in altitude is encountered in motionless air. For air moving up a mountain with condensation occurring (clouds, fog, and precipitation), the temperature of the air drops 3.2 degrees Fahrenheit with every 1,000 feet of elevation gain. For air moving up a mountain with no clouds forming, the temperature of the air drops 5.5 degrees Fahrenheit for every 1,000 feet of elevation gain.

a. An expedient to this often occurs on cold, clear, calm mornings. During a troop movement or climb started in a valley, higher temperatures may often be encountered as altitude is gained. This reversal of the normal cooling with elevation is called temperature inversion. Temperature inversions are caused when mountain air is cooled by ice, snow, and heat loss through thermal radiation. This cooler, denser air settles into the valleys and low areas. The inversion continues until the sun warms the surface of the earth or a moderate wind causes a mixing of the warm and cold layers. Temperature inversions are common in the mountainous regions of the arctic, subarctic, and mid-latitudes.

b. At high altitudes, solar heating is responsible for the greatest temperature contrasts. More sunshine and solar heat are received above the clouds than below. The important effect of altitude is that the sun's rays pass through less of the atmosphere and more direct heat is received than at lower levels, where solar radiation is absorbed and reflected by dust and water vapor. Differences of 40 to 50 degrees Fahrenheit may occur between surface temperatures in the shade and surface temperatures in the sun. This is particularly true for dark metallic objects. The difference in temperature felt on the skin between the sun and shade is normally 7 degrees Fahrenheit. Special care must be taken to avoid sunburn and snow blindness. Besides permitting rapid heating, the clear air at high altitudes also favors rapid cooling at night. Consequently, the temperature rises fast after sunrise and drops quickly after sunset. Much of the chilled air drains downward, due to convection currents, so that the differences between day and night temperatures are greater in valleys than on slopes.

c. Local weather patterns force air currents up and over mountaintops. Air is cooled on the windward side of the mountain as it gains altitude, but more slowly (3.2 degrees Fahrenheit per 1,000 feet) if clouds are forming due to heat release when water vapor becomes liquid. On the leeward side of the mountain, this heat gained from the condensation on the windward side is added to the normal heating that occurs as the air descends and air pressure increases. Therefore, air and winds on the leeward slope are considerably warmer than on the windward slope, which is referred to as Chinook winds. The heating and cooling of the air affects planning considerations primarily with regard to the clothing and equipment needed for an operation.

4-16. WEATHER FORECASTING

The use of a portable aneroid barometer, thermometer, wind meter, and hygrometer help in making local weather forecasts. Reports from other localities and from any weather service, including USAF, USN, or the National Weather Bureau, are also helpful. Weather reports should be used in conjunction with the locally observed current weather situation to forecast future weather patterns.

a. Weather at various elevations may be quite different because cloud height, temperature, and barometric pressure will all be different. There may be overcast and rain in a lower area, with mountains rising above the low overcast into warmer clear weather.

b. To be effective, a forecast must reach the small-unit leaders who are expected to utilize weather conditions for assigned missions. Several different methods can be used to create a forecast. The method a forecaster chooses depends upon the forecaster's experience, the amount of data available, the level of difficulty that the forecast situation presents, and the degree of accuracy needed to make the forecast. The five ways to forecast weather are:

(1) ***Persistence Method.*** "Today equals tomorrow" is the simplest way of producing a forecast. This method assumes that the conditions at the time of the forecast

will not change; for example, if today was hot and dry, the persistence method predicts that tomorrow will be the same.

(2) ***Trends Method.*** "Nowcasting" involves determining the speed and direction of fronts, high- and low-pressure centers, and clouds and precipitation. For example, if a cold front moves 300 miles during a 24-hour period, we can predict that it will travel 300 miles in another 24-hours.

(3) ***Climatology Method.*** This method averages weather statistics accumulated over many years. This only works well when the pattern is similar to the following years.

(4) ***Analog Method.*** This method examines a day's forecast and recalls a day in the past when the weather looked similar (an analogy). This method is difficult to use because finding a perfect analogy is difficult.

(5) ***Numerical Weather Prediction.*** This method uses computers to analyze all weather conditions and is the most accurate of the five methods.

SECTION 2: MOUNTAIN HAZARDS

Hazards can be termed natural (caused by natural occurrence), manmade (caused by an individual, such as lack of preparation, carelessness, improper diet, equipment misuse), or as a combination (human trigger). There are two kinds of hazards while in the mountains subjective and objective. Combinations of objective and subjective hazards are referred to as cumulative hazards.

4-17. SUBJECTIVE HAZARDS

Subjective hazards are created by humans; for example, choice of route, companions, overexertion, dehydration, climbing above one's ability, and poor judgment.

a. **Falling.** Falling can be caused by carelessness, over-fatigue, heavy equipment, bad weather, overestimating ability, a hold breaking away, or other reasons.

b. **Bivouac Site.** Bivouac sites must be protected from rockfall, wind, lightning, avalanche run-out zones, and flooding (especially in gullies). If the possibility of falling exists, rope in, the tent and all equipment may have to be tied down.

c. **Equipment.** Ropes are not total security; they can be cut on a sharp edge or break due to poor maintenance, age, or excessive use. You should always pack emergency and bivouac equipment even if the weather situation, tour, or a short climb is seemingly low of dangers.

4-18. OBJECTIVE HAZARDS

Objective hazards are caused by the mountain and weather and cannot be influenced by man; for example, storms, rockfalls, icefalls, lightning, and so on.

a. **Altitude.** At high altitudes (especially over 6,500 feet), endurance and concentration is reduced. Cut down on smoking and alcohol. Sleep well, acclimatize slowly, stay hydrated, and be aware of signs and symptoms of high-altitude illnesses. Storms can form quickly and lightning can be severe.

b. **Visibility.** Fog, rain, darkness, and or blowing snow can lead to disorientation. Take note of your exact position and plan your route to safety before visibility decreases. Cold combined with fog can cause a thin sheet of ice to form on rocks (verglas). Whiteout conditions can be extremely dangerous. If you must move under these conditions, it is best to rope up. Have the point man move to the end of the rope. The second man will use the first man as an aiming point with the compass. Use a route sketch and march table. If the tactical situation does not require it, plan route so as not to get caught by darkness.

c. **Gullies.** Rock, snow, and debris are channeled down gullies. If ice is in the gully, climbing at night may be better because the warming of the sun will loosen stones and cause rockfalls.

d. **Rockfall.** Blocks and scree at the base of a climb can indicate recurring rockfall. Light colored spots on the wall may indicate impact chips of falling rock. Spring melt or warming by the sun of the rock/ice/snow causes rockfall.

e. **Avalanches.** Avalanches are caused by the weight of the snow overloading the slope. (Refer to paragraph 2-4 for more detailed information on avalanches.)

f. **Hanging Glaciers and Seracs.** Avoid, if at all possible, hanging glaciers and seracs. They will fall without warning regardless of the time of day or time of year. One cubic meter of glacier ice weighs 910 kilograms (about 2,000 pounds). If you must cross these danger areas, do so quickly and keep an interval between each person.

g. **Crevasses.** Crevasses are formed when a glacier flows over a slope and makes a bend, or when a glacier separates from the rock walls that enclose it. A slope of only two to three degrees is enough to form a crevasse. As this slope increases from 25 to 30 degrees, hazardous icefalls can be formed. Likewise, as a glacier makes a bend, it is likely that crevasses will form at the outside of the bend. Therefore, the safest route on a glacier would be to the inside of bends, and away from steep slopes and icefalls. Extreme care must be taken when moving off of or onto the glacier because of the moat that is most likely to be present.

4-20. WEATHER HAZARDS

Weather conditions in the mountains may vary from one location to another as little as 10 kilometers apart. Approaching storms may be hard to spot if masked by local peaks. A clear, sunny day in July could turn into a snowstorm in less than an hour. Always pack some sort of emergency gear.

a. Winds are stronger and more variable in the mountains; as wind doubles in speed, the force quadruples.

b. Precipitation occurs more on the windward side than the leeward side of ranges. This causes more frequent and denser fog on the windward slope.

c. Above approximately 8,000 feet, snow can be expected any time of year in the temperate climates.

d. Air is dryer at higher altitudes, so equipment does not rust as quickly, but dehydration is of greater concern.

e. Lightning is frequent, violent, and normally attracted to high points and prominent features in mountain storms. Signs indicative of thunderstorms are tingling of the skin, hair standing on end, humming of metal objects, crackling, and a bluish light (St. Elmo's fire) on especially prominent metal objects (summit crosses and radio towers).

(1) Avoid peaks, ridges, rock walls, isolated trees, fixed wire installations, cracks that guide water, cracks filled with earth, shallow depressions, shallow overhangs, and rock needles. Seek shelter around dry, clean rock without cracks; in scree fields; or in deep indentations

(depressions, caves). Keep at least half a body's length away from a cave wall and opening.

(2) Assume a one-point-of-contact body position. Squat on your haunches or sit on a rucksack or rope. Pull your knees to your chest and keep both feet together. If half way up the rock face, secure yourself with more than one point lightning can burn through rope. If already rappelling, touch the wall with both feet together and hurry to the next anchor.

f. During and after rain, expect slippery rock and terrain in general and adjust movement accordingly. Expect flash floods in gullies or chimneys. A climber can be washed away or even drowned if caught in a gully during a rainstorm. Be especially alert for falling objects that the rain has loosened.

g. Dangers from impending high winds include frostbite (from increased wind-chill factor), windburn, being blown about (especially while rappelling), and debris being blown about. Wear protective clothing and plan the route to be finished before bad weather arrives.

h. For each 100-meter rise in altitude, the temperature drops approximately one degree Fahrenheit. This can cause hypothermia and frostbite even in summer, especially when combined with wind, rain, and snow. Always wear or pack appropriate clothing.

i. If it is snowing, gullies may contain avalanches or snow sloughs, which may bury the trail. Snowshoes or skis may be needed in autumn or even late spring. Unexpected snowstorms may occur in the summer with accumulations of 12 to 18 inches; however, the snow quickly melts.

j. Higher altitudes provide less filtering effects, which leads to greater ultraviolet (UV) radiation intensity. Cool winds at higher altitudes may mislead one into underestimating the sun's intensity, which can lead to sunburns and other heat injuries. Use sunscreen and wear hat and sunglasses, even if overcast. Drink plenty of fluids.

4-21. AVALANCHE HAZARDS

Avalanches occur when the weight of accumulated snow on a slope exceeds the cohesive forces that hold the snow in place. (Table 4-1 shows an avalanche hazard evaluation checklist.)

AVALANCHE HAZARD EVALUATION CHECKLIST

Critical Data PARAMETERS:	KEY INFORMATION	Hazard Rating G	Y	R
TERRAIN: *Is the terrain capable of producing an avalanche?*				
	-Slope angle (steep enough to slide? prime time?)	☐	☐	☐
	-Slope aspect (leeward, shadowed, or extremely sunny?)	☐	☐	☐
	-Slope configuration (anchoring? shape?)	☐	☐	☐
	Overall Terrain Rating:	☐	☐	☐
SNOWPACK: *Could the snow fail?*				
	-Slab Configuration (slab? depth and distribution?)	☐	☐	☐
	-Bonding Ability (weak layer? tender spots?)	☐	☐	☐
	-Sensitivity (how much force to fail? shear tests? clues?)	☐	☐	☐
	Overall Snowpack Rating:	☐	☐	☐
Weather: *Is the weather contributing to instability?*				
	-Precipitation (type, amount, intensity? added weight?)	☐	☐	☐
	-Wind (snow transport? amount and rate of deposition?)	☐	☐	☐
	-Temperature (storm trends? effects on snowpack?)	☐	☐	☐
	Overall Weather Rating:	☐	☐	☐
Human: *What are your alternatives and their possible consequences?*				
	-Attitude (toward life? risk? goals? assumptions?)	☐	☐	☐
	-Technical Skill Level (traveling? evaluating aval. hazard?)	☐	☐	☐
	-Strength/Equipment (strength? prepared for the worst?)	☐	☐	☐
	Overall Human Rating:	☐	☐	☐

Decision/Action:
Overall Hazard Rating/GO or NO Go? **GO** ☐ **or NOGO** ☐

***HAZARD LEVEL SYMBOLS:**
R = Red light (stop/dangerous)
G = Green light (go/OK)
Y = Yellow light (caution/potentially dangerous).

Table 4-1: Avalanche hazard evaluation checklist.

a. **Slope Stability**. Slope stability is the key factor in determining the avalanche danger.
   (1) ***Slope Angle.*** Slopes as gentle as 15 degrees have avalanched. Most avalanches occur on slopes between 30 and 45 degrees. Slopes above 60 degrees often do not build up significant quantities of snow because they are too steep.

(2) ***Slope Profile.*** Dangerous slab avalanches are more likely to occur on convex slopes, but may occur on concave slopes.

(3) ***Slope Aspect.*** Snow on north facing slopes is more likely to slide in midwinter. South facing slopes are most dangerous in the spring and on sunny, warm days. Slopes on the windward side are generally more stable than leeward slopes.

(4) ***Ground Cover.*** Rough terrain is more stable than smooth terrain. On grassy slopes or scree, the snowpack has little to anchor to.

b. **Triggers.** Various factors trigger avalanches.

(1) ***Temperature.*** When the temperature is extremely low, settlement and adhesion occur slowly. Avalanches that occur during extreme cold weather usually occur during or immediately following a storm. At a temperature just below freezing, the snowpack stabilizes quickly. At temperatures above freezing, especially if temperatures rise quickly, the potential for avalanche is high. Storms with a rise in temperature can deposit dry snow early, which bonds poorly with the heavier snow deposited later. Most avalanches occur during the warmer midday.

(2) ***Precipitation.*** About 90 percent of avalanches occur during or within twenty-four hours after a snowstorm. The rate at which snow falls is important. High rates of snowfall (2.5 centimeters per hour or greater), especially when accompanied by wind, are usually responsible for major periods of avalanche activity. Rain falling on snow will increase its weight and weakens the snowpack.

(3) ***Wind.*** Sustained winds of 15 miles per hour and over transport snow and form wind slabs on the lee side of slopes.

(4) ***Weight.*** Most victims trigger the avalanches that kill them.

(5) ***Vibration.*** Passing helicopters, heavy equipment, explosions, and earth tremors have triggered avalanches.

c. **Snow Pits.** Snow pits can be used to determine slope stability.

(1) Dig the snow pit on the suspect slope or a slope with the same sun and wind conditions. Snow deposits may vary greatly within a few meters due to wind and sun variations. (On at least one occasion, a snow pit dug across

the fall line triggered the suspect slope). Dig a2-meter by 2-meter pit across the fall line, through all the snow, to the ground. Once the pit is complete, smooth the face with a shovel.

(2) Conduct a shovel shear test.

(a) A shovel shear test puts pressure on a representative sample of the snowpack. The core of this test is to isolate a column of the snowpack from three sides. The column should be of similar size to the blade of the shovel. Dig out the sides of the column without pressing against the column with the shovel (this affects the strength). To isolate the rear of the column, use a rope or string to saw from side to side to the base of the column.

(b) If the column remained standing while cutting the rear, place the shovel face down on the top of the column. Tap with varying degrees of strength on the shovel to see what force it takes to create movement on the bed of the column. The surface that eventually slides will be the layer to look at closer. This test provides a better understanding of the snowpack strength. For greater results you will need to do this test in many areas and formulate a scale for the varying methods of tapping the shovel.

(3) Conduct a Rutschblock test. To conduct the test, isolate a column slightly longer than the length of your snowshoes or skis (same method as for the shovel shear test). One person moves on their skis or snowshoes above the block without disturbing the block. Once above, the person carefully places one showshoe or ski onto the block with no body weight for the first stage of the test. The next stage is adding weight to the first leg. Next, place the other foot on the block. If the block is still holding up, squat once, then twice, and so on. The remaining stage is to jump up and land on the block.

d. **Types of Snow Avalanches.** There are two types of snow avalanches: loose snow (point) and slab.

(1) Loose snow avalanches start at one point on the snow cover and grow in the shape of an inverted "V." Although they happen most frequently during the winter snow season, they can occur at any time of the year in the

mountains. They often fall as many small sluffs during or shortly after a storm. This process removes snow from steep upper slopes and either stabilize slower slopes or loads them with additional snow.

(2) Wet loose snow avalanches occur in spring and summer in all mountain ranges. Large avalanches of this type, lubricated and weighed down by melt water or rain can travel long distances and have tremendous destructive power. Coastal ranges that have high temperatures and frequent rain are the most common areas for this type of avalanche.

(3) Slab avalanches occur when cohesive snow begins to slide on a weak layer. The fracture line where the moving snow breaks away from the snowpack makes this type of avalanche easy to identify. Slab release is rapid. Although any avalanche can kill you, slab avalanches are generally considered more dangerous than loose snow avalanches.

(a) Most slab avalanches occur during or shortly after a storm when slopes are loaded with new snow at a critical rate. The old rule of never travel in avalanche terrain for a few days after a storm still holds true.

(b) As slabs become harder, their behavior becomes more unpredictable; they may allow several people to ski across before releasing. Many experts believe they are susceptible to rapid temperature changes. Packed snow expands and contracts with temperature changes. For normal density, settled snow, a drop in temperature of 10 degrees Celsius (18 degrees Fahrenheit) would cause a snow slope 300 meters wide to contract 2 centimeters. Early ski mountaineers in the Alps noticed that avalanches sometimes occurred when shadows struck a previously sun-warmed slope.

d. **Protective Measures.** Avoiding known or suspected avalanche areas is the easiest method of protection. Other measures include:

(1) ***Personal Safety.*** Remove your hands from ski pole wrist straps. Detach ski runaway cords. Prepare to discard equipment. Put your hood on. Close up your clothing to prepare for hypothermia. Deploy avalanche cord. Make avalanche probes and shovels accessible. Keep your pack

on at all times do not discard. Your pack can act as a flotation device, as well as protect your spine.

(2) ***Group Safety.*** Send one person across the suspect slope at a time with the rest of the group watching. All members of the group should move in the same track from safe zone to safe zone.

e. **Route Selection.** Selecting the correct route will help avoid avalanche prone areas, which is always the best choice. Always allow a wide margin of safety when making your decision.

(1) The safest routes are on ridge tops, slightly on the windward side; the next safest route is out in the valley, far from the bottom of slopes.

(2) Avoid cornices from above or below. Should you encounter a dangerous slope, either climb to the top of the slope or descend to the bottom well out of the way of the runout zone. If you must traverse, pick a line where you can traverse downhill as quickly as possible. When you must ascend a dangerous slope, climb to the side of the avalanche path, and not directly up the center.

(3) Take advantage of dense timber, ridges, or rocky outcrops as islands of safety. Use them for lunch and rest stops. Spend as little time as possible on open slopes.

(4) Since most avalanches occur within twenty-four hours of a storm and or at midday, avoid moving during these periods. Moving at night is tactically sound and may be safer.

f. **Stability Analysis.** Look for nature's billboards on slopes similar to the one you are on.

(1) ***Evidence of Avalanching.*** Look for recent avalanches and for signs of wind-loading and wind-slabs.

(2) ***Fracture Lines.*** Avoid any slopes showing cracks.

(3) ***Sounds.*** Beware of hollow sounds a "whumping" noise. They may suggest a radical settling of the snowpack.

g. **Survival**. People trigger avalanches that bury people. If these people recognized the hazard and chose a different route, they would avoid the avalanche. The following steps should be followed if caught in an avalanche.

(1) Discard equipment. Equipment can injure or burden you; discarded equipment will indicate your position to rescuers.

(2) Swim or roll to stay on top of the snow. FIGHT FOR YOUR LIFE. Work toward the edge of the avalanche. If you feel your feet touch the ground, give a hard push and try to "pop out" onto the surface.

(3) If your head goes under the snow, shut your mouth, hold your breath, and position your hands and arms to form an air pocket in front of your face. Many avalanche victims suffocate by having their mouths and noses plugged with snow.

(4) When you sense the slowing of the avalanche, you must try your hardest to reach the surface. Several victims have been found quickly because a hand or foot was sticking above the surface.

(5) When the snow comes to rest it sets up like cement and even if you are only partially buried, it may be impossible to dig yourself out. Don't shout unless you hear rescuers immediately above you; in snow, no one can hear you scream. Don't struggle to free yourself you will only waste energy and oxygen.

(6) Try to relax. If you feel yourself about to pass out, do not fight it. The respiration of an unconscious person is shallower, their pulse rate declines, and the body temperature is lowered, all of which reduce the amount of oxygen needed.

4-22. ACUTE MOUNTAIN SICKNESS

In addition to dangers caused by dehydration, sunburn, hypothermia, and other cold-weather problems, a high altitude can have physiological effects in itself. Acute mountain sickness is a temporary illness that may affect both the beginner and experienced climber.

Soldiers are subject to this sickness in altitudes as low as 5,000 feet. Incidence and severity increases with altitude, and when quickly transported to high altitudes. Disability and ineffectiveness can occur in 50 to 80 percent of the troops who are rapidly brought to altitudes above 10,000 feet. At lower altitudes, or where ascent to altitudes is gradual, most personnel can complete assignments with moderate effectiveness and little discomfort.

a. Personnel arriving at moderate elevations (5,000 to 8,000 feet) usually feel well for the first few hours; a feeling of exhilaration or well-being is not unusual. There may be an initial awareness of breathlessness upon exertion and a need for frequent pauses to rest. Irregular breathing can occur, mainly during sleep; these changes may cause apprehension. Severe symptoms may begin 4 to 12 hours after arrival at higher altitudes with symptoms of nausea, sluggishness, fatigue, headache, dizziness, insomnia, depression, uncaring attitude, rapid and labored breathing, weakness, and loss of appetite.

b. A headache is the most noticeable symptom and may be severe. Even when a headache is not present, some loss of appetite and a decrease in tolerance for food occurs. Nausea, even without food intake, occurs and leads to less food intake. Vomiting may occur and contribute to dehydration. Despite fatigue, personnel are unable to sleep. The symptoms usually develop and increase to a peak by the second day. They gradually subside over the next several days so that the total course of AMS may extend from five to seven days. In some instances, the headache may become incapacitating and the soldier should be evacuated to a lower elevation.

c. Treatment for AMS includes the following:
   - Oral pain medications such as ibuprofen or aspirin.
   - Rest.
   - Frequent consumption of liquids and light foods in small amounts.
   - Movement to lower altitudes (at least 1,000 feet) to alleviate symptoms, which provides for a more gradual acclimatization.
   - Realization of physical limitations and slow progression.
   - Practice of deep-breathing exercises.
   - Use of acetazolamide in the first 24 hours for mild to moderate cases.

d. AMS is nonfatal, although if left untreated or further ascent is attempted, development of high-altitude pulmonary edema (HAPE) and or high-altitude cerebral edema (HACE) can be seen. A severe persistence of symptoms may identify soldiers who acclimatize poorly and, thus, are more prone to other types of mountain sickness.

4-23. CHRONIC MOUNTAIN SICKNESS

Although not commonly seen in mountaineers, chronic mountain sickness (CMS) (or Monge's disease) can been seen in people who live at sufficiently high altitudes (usually at or above 10,000 feet) over a period of several years. CMS is a right-sided heart failure characterized by chronic pulmonary edema that is caused by years of strain on the right ventricle.

4-24. UNDERSTANDING HIGH-ALTITUDE ILLNESSES

As altitude increases, the overall atmospheric pressure decreases. Decreased pressure is the underlying source of altitude illnesses. Whether at sea level or 20,000 feet the surrounding atmosphere has the same percentage of oxygen. As pressure decreases the body has a much more difficult time passing oxygen from the lungs to the red blood cells and thus to the tissues of the body. This lower pressure means lower oxygen levels in the blood and increased carbon dioxide levels. Increased carbon dioxide levels in the blood cause a systemic vasodilatation, or expansion of blood vessels. This increased vascular size stretches the vessel walls causing leakage of the fluid portions of the blood into the interstitial spaces, which leads to cerebral edema or HACE. Unless treated, HACE will continue to progress due to the decreased atmospheric pressure of oxygen. Further ascent will hasten the progression of HACE and could possibly cause death.

While the body has an overall systemic vasodilatation, the lungs initially experience pulmonary vasoconstriction. This constricting of the vessels in the lungs causes increased workload on the right ventricle, the chamber of the heart that receives de-oxygenated blood from the right atrium and pushes it to the lungs to be re-oxygenated. As the right ventricle works harder to force blood to the lungs, its overall output is decreased thus decreasing the overall pulmonary perfusion. Decreased pulmonary perfusion causes decreased cellular respiration the transfer of oxygen from the alveoli to the red blood cells. The body is now experiencing increased carbon dioxide levels due to the decreased oxygen levels, which now causes pulmonary vasodilatation. Just as in HACE, this expanding of the vascular structure causes leakage into interstitial space resulting in pulmonary edema or HAPE. As the edema or fluid in the lungs increases, the capability to pass oxygen to the red blood cells decreases thus creating a vicious cycle, which can quickly become fatal if left untreated.

4-25. HIGH-ALTITUDE PULMONARY EDEMA

HAPE is a swelling and filling of the lungs with fluid, caused by rapid ascent. It occurs at high altitudes and limits the oxygen supply to the body.

a. HAPE occurs under conditions of low oxygen pressure, is encountered at high elevations (over 8,000 feet), and can occur in healthy soldiers. HAPE may be considered a form of, or manifestation of, AMS since it occurs during the period of susceptibility to this disorder.

b. HAPE can cause death. Incidence and severity increase with altitude. Except for acclimatization to altitude, no known factors indicate resistance or immunity. Few cases have been reported after 10 days at high altitudes. When remaining at the same altitude, the incidence of HAPE is less frequent than that of AMS. No common indicator dictates how a soldier will react from one exposure to another. Contributing factors are:
   - A history of HAPE.
   - A rapid or abrupt transition to high altitudes.
   - Strenuous physical exertion.
   - Exposure to cold.
   - Anxiety.

c. Symptoms of AMS can mask early pulmonary difficulties. Symptoms of HAPE include:
   - Progressive dry coughing with frothy white or pink sputum (this is usually a later sign) and then coughing up of blood.
   - Cyanosis a blue color to the face, hands, and feet.
   - An increased ill feeling, labored breathing, dizziness, fainting, repeated clearing of the throat, and development of a cough.
   - Respiratory difficulty, which may be sudden, accompanied by choking and rapid deterioration.
   - Progressive shortness of breath, rapid heartbeat (pulse 120 to 160), and coughing (out of contrast to others who arrived at the same time to that altitude).
   - Crackling, cellophane-like noises (rales) in the lungs caused by fluid buildup (a stethoscope is usually needed to hear them).
   - Unconsciousness, if left untreated. Bubbles form in the nose and mouth, and death results.

d. HAPE is prevented by good nutrition, hydration, and gradual ascent to altitude (no more than 1,000 to 2,000 feet per day to an area of sleep). A rest day, with no gain in altitude or heavy physical exertion, is planned for every 3,000 feet of altitude gained. If a soldier develops symptoms despite precautions, immediate descent is mandatory where he receives prompt treatment, rest, warmth, and oxygen. He is quickly evacuated to lower altitudes as a litter patient. A descent of 300meters may help; manual descent is not delayed to await air evacuation. If untreated, HAPE may become irreversible and cause death. Cases that are recognized early and treated promptly may expect to recover with no aftereffects. Soldiers who have had previous attacks of HAPE are prone to second attacks.

e. Treatment of HAPE includes:
- Immediate descent (2,000 to 3,000 feet minimum) if possible; if not, then treatment in a monoplace hyperbaric chamber.
- Rest (litter evacuation).
- Supplemental oxygen if available.
- Morphine for the systemic vasodilatation and reduction of preload. This should be carefully considered due to the respiratory depressive properties of the drug.
- Furosemide (Lasix), which is a diuretic, given orally can also be effective.
- The use of mannitol should not be considered due to the fact that it crystallizes at low temperatures. Since almost all high-altitude environments are cold, using mannitol could be fatal.
- Nifidipine (Procardia), which inhibits calcium ion flux across cardiac and smooth muscle cells, decreasing contractility and oxygen demand. It may also dilate coronary arteries and arterioles.
- Diphenhydramine (Benadryl), which can help alleviate the histamine response that increases mucosal secretions.

4-26. HIGH-ALTITUDE CEREBRAL EDEMA

HACE is the accumulation of fluid in the brain, which results in swelling and a depression of brain function that may result in death. It is caused by a rapid ascent to altitude without progressive acclimatization. Prevention of HACE is the same as for HAPE. HAPE and HACE may occur inexperienced, well-acclimated mountaineers without warning or obvious predisposing conditions. They can be fatal; when the first symptoms occur, immediate descent is mandatory.

a. Contributing factors include rapid ascent to heights over 8,000 feet and aggravation by overexertion.

b. Symptoms of HACE include mild personality changes, paralysis, stupor, convulsions, coma, inability to concentrate, headaches, vomiting, decrease in urination, and lack of coordination. The main symptom of HACE is a severe headache. A headache combined with any other physical or psychological disturbances should be assumed to be manifestations of HACE. Headaches may be accompanied by a loss of coordination, confusion, hallucinations, and unconsciousness. These maybe combined with symptoms of HAPE. The victim is often mistakenly left alone since others may think he is only irritable or temperamental; no one should ever be ignored. The symptoms may rapidly progress to death. Prompt descent to a lower altitude is vital.

c. Preventive measures include good eating habits, maintaining hydration, and using a gradual ascent to altitude. Rest, warmth, and oxygen at lower elevations enhance recovery. Left untreated, HACE can cause death.

d. Treatment for HACE includes:
   - Dexamethasone injection immediately followed by oral dexamethasone.
   - Supplemental oxygen.
   - Rapid descent and medical attention.
   - Use of a hyberbaric chamber if descent is delayed.

4-27. HYDRATION IN HAPE AND HACE

HAPE and HACE cause increased proteins in the plasma, or the fluid portion of the blood, which in turn increases blood viscosity. Increased viscosity increases vascular pressure. Vascular leakage caused by stretching of the vessel walls is made worse because of this increased vascular pressure. From this, edema, both cerebral and pulmonary, occurs. Hydration simply decreases viscosity.

SECTION 3: MOUNTAINEERING EQUIPMENT

EQUIPMENT DESCRIPTION AND MAINTENANCE

With mountainous terrain encompassing a large portion of the world's land mass, the proper use of mountaineering equipment will enhance a unit's combat capability and provide a combat multiplier. The equipment described in this chapter is produced by many different manufacturers; however, each item is produced and tested to extremely high standards to ensure safety when being used correctly. The weak link in the safety chain is the user. Great care in performing preventative maintenance checks and services and proper training in the use of the equipment is paramount to ensuring safe operations.

The manufacturers of each and every piece of equipment provide recommendations on how to use and care for its product. It is imperative to follow these instructions explicitly.

4-28. FOOTWEAR

In temperate climates a combination of footwear is most appropriate to accomplish all tasks.

a. The hot weather boot provides an excellent all-round platform for movement and climbing techniques and should be the boot of choice when the weather permits. The intermediate cold weather boot provides an acceptable platform for operations when the weather is less than ideal. These two types of boots issued together will provide the unit with the footwear necessary to accomplish the majority of basic mountain missions.

b. Mountain operations are encumbered by extreme cold, and the extreme cold weather boot (with vapor barrier) provides an adequate platform for many basic mountain missions. However, plastic mountaineering boots should be incorporated into training as soon as possible. These boots provide a more versatile platform for any condition that would be encountered in the mountains, while keeping the foot dryer and warmer.

c. Level 2 and level 3 mountaineers will need mission-specific footwear that is not currently available in the military supply system. The two types of footwear they will need are climbing shoes and plastic mountaineering boots.

(1) Climbing shoes are made specifically for climbing vertical or near vertical rock faces.
These shoes are made with a soft leather upper, a lace-up configuration, and a smooth "sticky rubber" sole (Figure 4-12). The smooth "sticky rubber" sole is the key to the climbing shoe, providing greater friction on the surface of the rock, allowing the climber access to more difficult terrain.

(2) The plastic mountaineering boot is a double boot system (Figure 4-12). The inner boot provides support, as well as insulation against the cold. The inner boot may or may not come with a breathable membrane. The outer boot is a molded plastic (usually with a lace-up configuration) with a lug sole. The welt of the boot is molded in such a way that crampons, ski bindings, and snowshoes are easily attached and detached.

> ✍ **NOTE**
> Maintenance of all types of footwear must closely follow the manufacturers' recommendations.

Figure 4-12: Climbing shoes and plastic mountaineering boots.

4-29. CLOTHING

Clothing is perhaps the most underestimated and misunderstood equipment in the military inventory. The clothing system refers to every piece of clothing placed against the skin, the insulation layers, and the outer most garments, which protect the soldier from the elements. When clothing is worn properly, the soldier is better able to accomplish his tasks. When worn improperly, he is, at best, uncomfortable and, at worst, develops hypothermia or frostbite.

a. **Socks.** Socks are one of the most under-appreciated part of the entire clothing system. Socks are extremely valuable in many respects, if worn correctly. As a system, socks provide cushioning for the foot, remove excess moisture, and provide insulation from cold temperatures. Improper wear and excess moisture are the biggest causes of hot spots and blisters.

Regardless of climatic conditions, socks should always be worn in layers.

(1) The first layer should be a hydrophobic material that moves moisture from the foot surface to the outer sock.

(2) The outer sock should also be made of hydrophobic materials, but should be complimented with materials that provide cushioning and abrasion resistance.

(3) A third layer can be added depending upon the climatic conditions.

(a) In severe wet conditions, a waterproof type sock can be added to reduce the amount of water that would saturate the foot. This layer would be worn over the first two layers if conditions were extremely wet.

(b) In extremely cold conditions a vapor barrier sock can be worn either over both of the original pairs of socks or between the hydrophobic layer and the insulating layer. If the user is wearing VB boots, the vapor barrier sock is not recommended.

b. **Underwear.** Underwear should also be made of materials that move moisture from the body. Many civilian companies manufacture this type of underwear. The primary material in this product is polyester, which moves moisture from the body to the outer layers keeping the user drier and more comfortable in all climatic conditions. In colder environments, several pairs of long underwear of different thickness should be made available. A lightweight set coupled with a heavyweight set will provide a multitude of layering combinations.

c. **Insulating Layers.** Insulating layers are those layers that are worn over the underwear and under the outer layers of clothing. Insulating layers provide additional warmth when the weather turns bad. For the most part, today's insulating layers will provide for easy moisture movement as well as trap air to increase the insulating factor. The insulating layers that are presently available are referred to as pile or fleece. The ECWCS (Figure 4-13) also incorporates the field jacket and field pants liner as additional insulating layers. However, these two components do not move moisture as effectively as the pile or fleece.

Figure 4-13. Extreme cold weather clothing system.

d. **Outer Layers.** The ECWCS provides a jacket and pants made of a durable waterproof fabric. Both are constructed with a nylon shell with a laminated breathable membrane attached. This membrane allows the garment to release moisture to the environment while the nylon shell provides a degree of water resistance during rain and snow. The nylon also acts as a barrier to wind, which helps the garment retain the warm air trapped by the insulating layers. Leaders at all levels must understand the importance of wearing the ECWCS correctly.

> ✍ **NOTE:**
> Cotton layers must not be included in any layer during operations in a cold environment.

e. **Gaiters.** Gaiters are used to protect the lower leg from snow and ice, as well as mud, twigs, and stones. The use of waterproof fabrics or other breathable materials laminated to the nylon makes the gaiter an integral component of the cold weather clothing system. Gaiters are not presently fielded in the standard ECWCS and, in most cases, will need to be locally purchased. Gaiters are available in three styles (Figure 4-14).

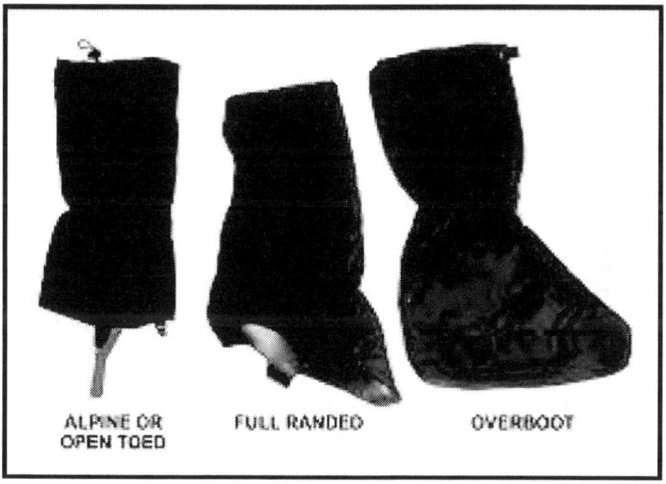

Figure 4-14: Three types of gaiters.

(1) The most common style of gaiter is the open-toed variety, which is a nylon shell that mayor may not have a breathable material laminated to it. The open front allows

the boot to slip easily into it and is closed with a combination of zipper, hook-pile tape, and snaps. It will have an adjustable neoprene strap that goes under the boot to keep it snug to the boot. The length should reach to just below the knee and will be kept snug with a drawstring and cord lock.

(2) The second type of gaiter is referred to as a full or randed gaiter. This gaiter completely covers the boot down to the welt. It can be laminated with a breathable material and can also be insulated if necessary. This gaiter is used with plastic mountaineering boots and should be glued in place and not removed.

(3) The third type of gaiter is specific to high-altitude mountaineering or extremely cold temperatures and is referred to as an overboot. It is worn completely over the boot and must be worn with crampons because it has no traction sole.

f. **Hand Wear.** During operations in mountainous terrain the use of hand wear is extremely important. Even during the best climatic conditions, temperatures in the mountains will dip below the freezing point. While mittens are always warmer than gloves, the finger dexterity needed to do most tasks makes gloves the primary cold weather hand wear (Figure 4-15).

Figure 4-15: Hand wear.

(1) The principals that apply to clothing also apply to gloves and mittens. They should provide moisture transfer from the skin to the outer layers the insulating layer must insulate the hand from the cold and move moisture to the outer layer. The outer layer must be weather resistant and breathable. Both gloves and mittens should be required for all soldiers during mountain operations, as well as replacement liners for both. This will provide enough flexibility to accomplish all tasks and keep the users' hands warm and dry.

(2) Just as the clothing system is worn in layers, gloves and mittens work best using the same principle. Retention cords that loop over the wrist work extremely well when the wearer needs to remove the outer layer to accomplish a task that requires fine finger dexterity. Leaving the glove or mitten dangling from the wrist ensures the wearer knows where it is at all times.

g. **Headwear.** A large majority of heat loss (25 percent) occurs through the head and neck area. The most effective way to counter heat loss is to wear a hat. The best hat available to the individual soldier through the military supply system is the black watch cap. Natural fibers, predominately wool, are acceptable but can be bulky and difficult to fit under a helmet. As with clothes and hand wear, man-made fibers are preferred. For colder climates a neck gaiter can be added. The neck gaiter is a tube of man-made material that fits around the neck and can reach up over the ears and nose (Figure 4-16). For extreme cold, a balaclava can be added. This covers the head, neck, and face leaving only a slot for the eyes (Figure 4-16). Worn together the combination is warm and provides for moisture movement, keeping the wearer drier and warmer.

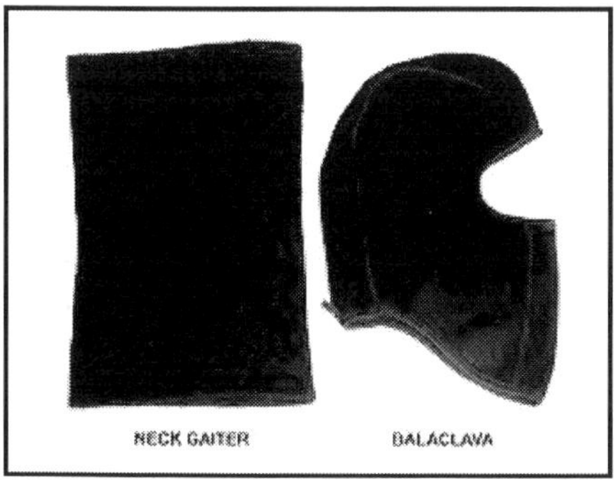

Figure 4-16: Neck gaiter and balaclava.

h. **Helmets.** The Kevlar ballistic helmet can be used for most basic mountaineering tasks. It must befitted with parachute retention straps and the foam impact pad (Figure 4-17). The level 2 and 3 mountaineer will need a lighter weight helmet for specific climbing scenarios. Several civilian manufacturers produce an effective helmet. Whichever helmet is selected, it should be designed specifically for mountaineering and adjustable so the user can add a hat under it when needed.

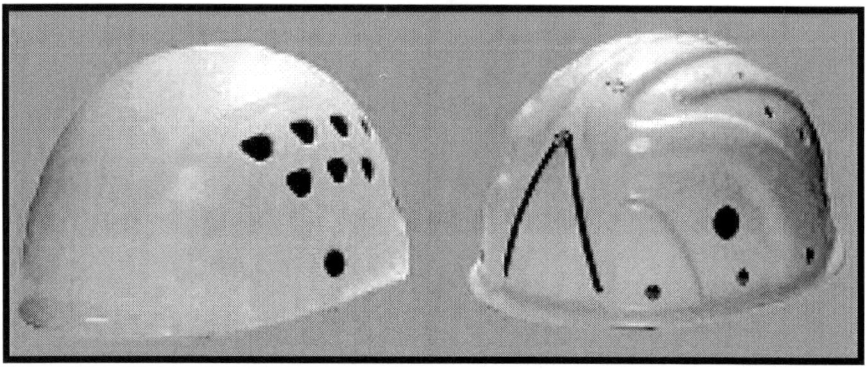

Figure 4-17: Helmets.

i. **Eyewear**. The military supply system does not currently provide adequate eyewear for mountaineering. Eyewear is divided into two categories: glacier glasses and goggles (Figure 4-18) Glacier glasses are sunglasses that cover the entire eye socket. Many operations in the mountains occur above the tree line or on ice and snow surfaces where the harmful UV rays of the sun can bombard the eyes from every angle increasing the likelihood of snow blindness. Goggles for mountain operations should be antifogging. Double or triple lenses work best. UV rays penetrate clouds so the goggles should be UV protected. Both glacier glasses and goggles are required equipment in the mountains. The lack of either one can lead to severe eye injury or blindness.

Figure 4-18: Glacier glasses and goggles.

j. **Maintenance of Clothing.** Clothing and equipment manufacturers provide specific instructions for proper care. Following these instructions is necessary to ensure the equipment works as intended.

4-30. CLIMBING SOFTWARE

Climbing software refers to rope, cord, webbing, and harnesses. All mountaineering specific equipment, to include hardware (see paragraph 4-31), should only be used if it has the UIAA certificate of safety. UIAA is the organization that oversees the testing of mountaineering equipment. It is based in Paris, France, and comprises several commissions. The safety commission has established standards for mountaineering and climbing equipment that have become well recognized throughout the world. Their work continues as new equipment develops and is brought into common use. Community Europe (CE) recognizes UIAA testing standards and, as the broader-based testing facility for the combined European economy, meets or exceeds the UIAA standards for all climbing and mountaineering equipment produced in Europe. European norm (EN) and CE have been combined to make combined European norm (CEN). While the United States has no specific standards, American manufacturers have their equipment tested by UIAA to ensure safe operating tolerances.

a. **Ropes and Cord.** Ropes and cords are the most important pieces of mountaineering equipment and proper selection deserves careful thought. These items are your lifeline in the mountains, so selecting the right type and size is of the utmost importance. All ropes and cord used in mountaineering and climbing today are constructed with the same basic configuration. The construction technique is referred to as Kernmantle, which is, essentially, a core of nylon fibers protected by a woven sheath, similar to parachute or 550 cord (Figure 4-19).

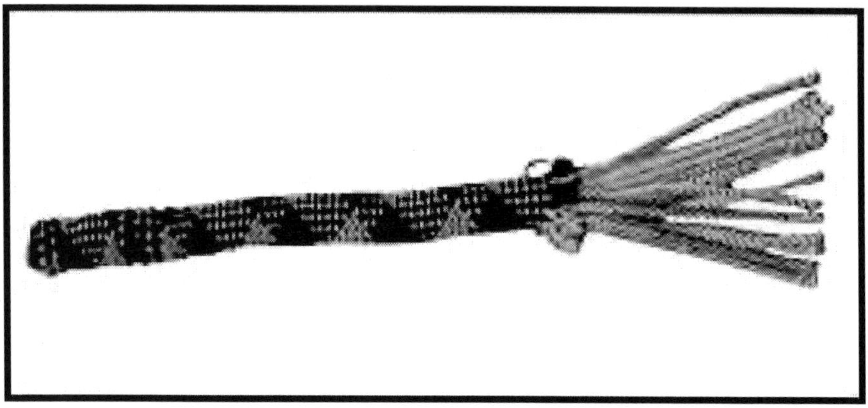

Figure 4-19: Kernmantle construction.

(1) Ropes come in two types: static and dynamic. This refers to their ability to stretch under tension. A static rope has very little stretch, perhaps as little as one to two percent, and is best used in rope installations. A dynamic rope is most useful for climbing and general mountaineering. Its ability to stretch up to 1/3 of its overall length makes it the right choice any time the user might take a fall. Dynamic and static ropes come in various diameters and lengths. For most military applications, a standard 10.5- or 11-millimeter by 50-meterdynamic rope and 11-millimeter by 45-meter static rope will be sufficient.

(2) When choosing dynamic rope, factors affecting rope selection include intended use, impact force, abrasion resistance, and elongation. Regardless of the rope chosen, it should be UIAA certified.

(3) Cord or small diameter rope is indispensable to the mountaineer. Its many uses make it a valuable piece of

equipment. All cord is static and constructed in the same manner as larger rope. If used for Prusik knots, the cord's diameter should be 5 to 7 millimeters when used on an 11-mm rope.

b. **Webbing and Slings.** Loops of tubular webbing or cord, called slings or runners, are the simplest pieces of equipment and some of the most useful. The uses for these simple pieces are endless, and they are a critical link between the climber, the rope, carabiners, and anchors. Runners are predominately made from either 9/16-inch or 1-inch tubular webbing and are either tied or sewn by a manufacturer (Figure 3-9). Runners can also be made from a high-performance fiber known as spectra, which is stronger, more durable, and less susceptible to ultraviolet deterioration. Runners should be retired regularly following the same considerations used to retire a rope. For most military applications, a combination of different lengths of runners is adequate.

(1) Tied runners have certain advantages over sewn runners they are inexpensive to make, can be untied and threaded around natural anchors, and can be untied and retied to other pieces of webbing to create extra-long runners.

(2) Sewn runners have their own advantages they tend to be stronger, are usually lighter, and have less bulk than the tied version. They also eliminate a major concern with the homemade knotted runner the possibility of the knot untying. Sewn runners come in four standard lengths: 2 inches, 4 inches, 12 inches, and 24 inches. They also come in three standard widths: 9/16 inch, 11/16 inch, and 1 inch.

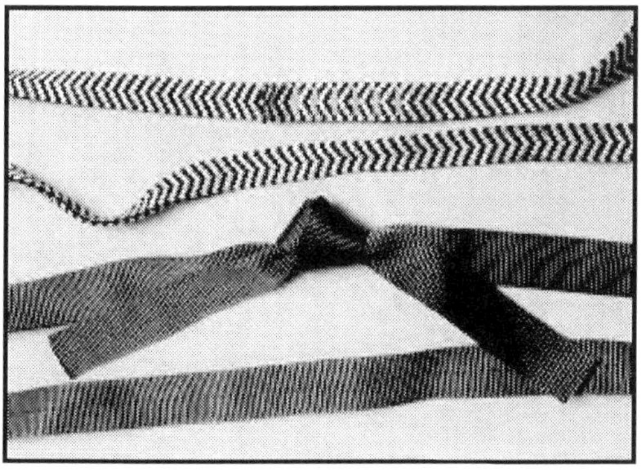

Figure 4-20. Tied or sewn runners.

c. **Harnesses.** Years ago climbers secured themselves to the rope by wrapping the rope around their bodies and tying a bowline-on-a-coil. While this technique is still a viable way of attaching to a rope, the practice is no longer encouraged because of the increased possibility of injury from a fall. The bowline-on-a-coil is best left for low-angle climbing or an emergency situation where harness material is unavailable. Climbers today can select from a wide range of manufactured harnesses. Fitted properly, the harness should ride high on the hips and have snug leg loops to better distribute the force of a fall to the entire pelvis. This type of harness, referred to as a seat harness, provides a comfortable seat for rappelling (Figure 4-21).

(1) Any harness selected should have one very important feature a double-passed buckle. This is a safety standard that requires the waist belt to be passed over and back through the main buckle a second time. At least 2 inches of the strap should remain after double-passing the buckle.

(2) Another desirable feature on a harness is adjustable leg loops, which allows a snug fit regardless of the number of layers of clothing worn. Adjustable leg loops allow the soldier to make a latrine call without removing the harness or untying the rope.

(3) Equipment loops are desirable for carrying pieces of climbing equipment. For safety purposes always follow the manufacturer's directions for tying-in.

(4) A field-expedient version of the seat harness can be constructed by using 22 feet of either1-inch or 2-inch (preferred) tubular webbing (Figure 4-21). Two double-overhand knots form the leg loops, leaving 4 to 5 feet of webbing coming from one of the leg loops. The leg loops should just fit over the clothing. Wrap the remaining webbing around the waist ensuring the first wrap is routed through the 6- to 10-inch long strap between the double-overhand knots. Finish the waist wrap with a water knot tied as tightly as possible. With the remaining webbing, tie a square knot without safeties over the water knot ensuring a minimum of 4inches remains from each strand of webbing.

(5) The full body harness incorporates a chest harness with a seat harness (Figure 3-10). This type of harness has a higher tie-in point and greatly reduces the chance of flipping backward during a fall. This is the only type of harness that is approved by the UIAA. While these harnesses are safer, they do present several disadvantages they are more expensive, are more restrictive, and increase the difficulty of adding or removing clothing. Most mountaineers prefer to incorporate a separate chest harness with their seat harness when warranted.

(6) A separate chest harness can be purchased from a manufacturer, or a field-expedient version can be made from either two runners or a long piece of webbing. Either chest harness is then attached to the seat harness with a carabiner and a length of webbing or cord.

Figure 4-21: Seat harness, field-expedient harness, and full body harness.

4-31. CLIMBING HARDWARE

Climbing hardware refers to all the parts and pieces that allow the trained mountain soldier to accomplish many tasks in the mountains. The importance of this gear to the mountaineer is no less than that of the rifle to the infantryman.

a. **Carabiners.** One of the most versatile pieces of equipment available to the mountaineer is the carabiner. This simple piece of gear is the critical connection between the climber, his rope, and the protection attaching him to the mountain. Carabiners must be strong enough to hold hard falls, yet light enough for the climber to easily carry a quantity of them. Today's high tech metal alloys allow carabiners to meet both of these requirements. Steel is still widely used, but is not preferred for general mountaineering, given other options. Basic carabiner construction affords the user several different shapes. The oval, the D-shaped, and the pear-shaped carabiner are just some of the types currently available. Most models can be made with or without a locking mechanism for the gate opening (Figure 4-22). If the carabiner does have a locking mechanism, it is usually referred to as a locking carabiner. When using a carabiner, great care should be taken to avoid loading the carabiner on its minor axis and to avoid three-way loading (Figure 4-23).

> ✍ **NOTE:**
> Great care should be used to ensure all carabiner gates are closed and locked during use.

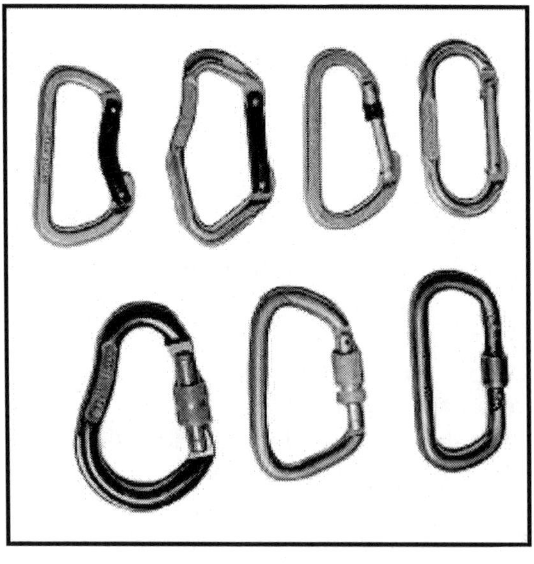

Figure 4-22. Nonlocking and locking carabiners.

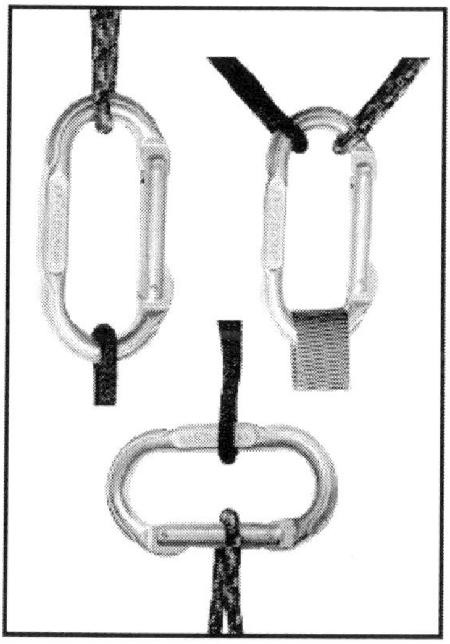

Figure 4-23. Major and minor axes and three-way loading.

(1) The major difference between the oval and the D-shaped carabiner is strength. Because of the design of the D-shaped carabiner, the load is angled onto the spine of the carabiner thus keeping it off the gate. The down side is that racking any gear or protection on the D-shaped carabiner is difficult because the angle of the carabiner forces all the gear together making it impossible to separate quickly.

(2) The pear-shaped carabiner, specifically the locking version, is excellent for clipping a descender or belay device to the harness. They work well with the munter hitch belaying knot.

(3) Regardless of the type chosen, all carabiners should be UIAA tested. This testing is extensive and tests the carabiner in three ways along its major axis, along its minor axis, and with the gate open.

b. **Pitons.** A piton is a metal pin that is hammered into a crack in the rock. They are described by their thickness, design, and length (Figure 4-24). Pitons provide a secure anchor for a rope attached by a carabiner. The many different kinds of pitons include: vertical, horizontal, wafer, and angle. They are made of malleable steel, hardened steel, or other alloys. The strength of the piton is determined by its placement rather than its rated tensile strength. The two most common types of pitons are: blades, which hold when wedged into tight-fitting cracks, and angles, which hold blade compression when wedged into a crack.

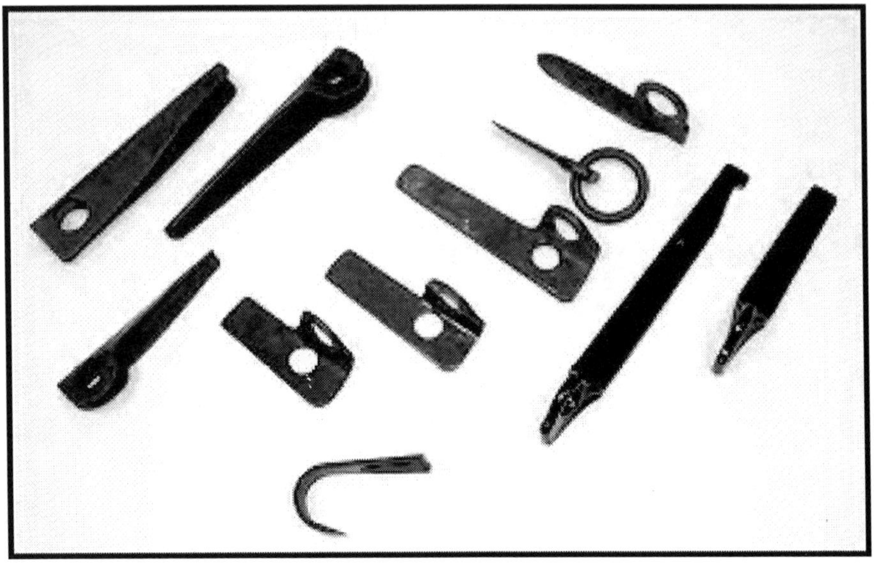

Figure 4-24. Various pitons.

(1) ***Vertical Pitons.*** On vertical pitons, the blade and eye are aligned. These pitons are used in flush, vertical cracks.
(2) ***Horizontal Pitons.*** On horizontal pitons, the eye of the piton is at right angles to the blade. These pitons are used in flush, horizontal cracks and in offset or open-book type vertical or horizontal cracks. They are recommended for use in vertical cracks instead of vertical pitons because the torque on the eye tends to wedge the piton into place. This provides more holding power than the vertical piton under the same circumstances.
(3) ***Wafer Pitons.*** These pitons are used in shallow, flush cracks. They have little holding power and their weakest points are in the rings provided for the carabiner.
(4) ***Knife Blade Pitons.*** These are used in direct-aid climbing. They are small and fit into thin, shallow cracks. They have a tapered blade that is optimum for both strength and holding power.
(5) ***Realized Ultimate Reality Pitons.*** Realized ultimate reality pitons (RURPs) are hatchet-shaped pitons about

1-inch square. They are designed to bite into thin, shallow cracks.

(6) ***Angle Pitons.*** These are used in wide cracks that are flush or offset. Maximum strength is attained only when the legs of the piton are in contact with the opposite sides of the crack.

(7) ***Bong Pitons.*** These are angle pitons that are more than 3.8 centimeters wide. Bongs are commonly made of steel or aluminum alloy and usually contain holes to reduce weight and accommodate carabiners. They have a high holding power and require less hammering than other pitons.

(8) ***Skyhook (Cliffhangers).*** These are small hooks that cling to tiny rock protrusions, ledges, or flakes. Skyhooks require constant tension and are used in a downward pull direction. The curved end will not straighten under body weight. The base is designed to prevent rotation and aid stability.

c. **Piton Hammers.** A piton hammer has a flat metal head; a handle made of wood, metal, or fiberglass; and a blunt pick on the opposite side of the hammer (Figure 4-25). A safety lanyard of nylon cord, webbing, or leather is used to attach it to the climber The lanyard should be long enough to allow for full range of motion. Most hammers are approximately 25.5 centimeters long and weigh 12 to 25 ounces. The primary use for a piton hammer is to drive pitons, to be used as anchors,

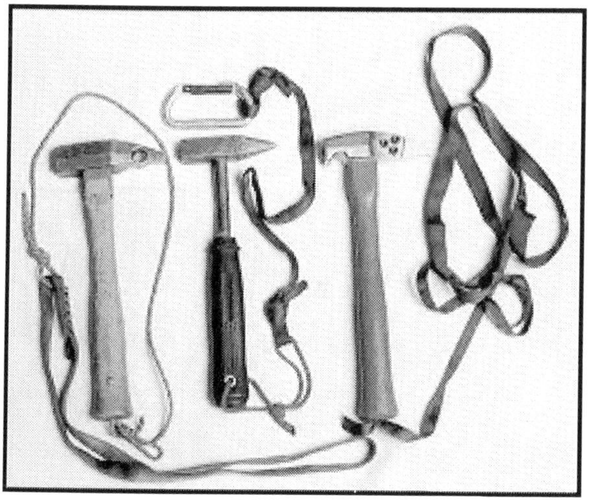

Figure 4-25: Piton hammer.

into the rock. The piton hammer can also be used to assist in removing pitons, and in cleaning cracks and rock surfaces to prepare for inserting the piton. The type selected should suit individual preference and the intended use.

d. **Chocks.** "Chocks" is a generic term used to describe the various types of artificial protection other than bolts or pitons. Chocks are essentially a tapered metal wedge constructed in various sizes to fit different sized openings in the rock (Figure 4-26). The design of a chock will determine whether it fits into one of two categories wedges or cams. A wedge holds by wedging into a constricting crack in the rock. A cam holds by slightly rotating in a crack, creating a camming action that lodges the chock in the crack or pocket. Some chocks are manufactured to perform either in the wedging mode or the camming mode. One of the chocks that falls into the category of both a wedge and cam is the hexagonal-shaped or "hex" chock. This type of chock is versatile and comes with either a cable loop or is tied with cord or webbing. All chocks come in different sizes to fit varying widths of cracks. Most chocks come with a wired loop that is stronger than cord and allows for easier placement. Bigger chocks can be threaded with cord or webbing if the user ties the chock himself. Care should be taken to place tubing in the chock before threading the cord. The cord used with chocks is designed to be stiffer and stronger than regular cord and is typically made of Kevlar. The advantage of using a chock rather than a piton is that a climber can carry many different sizes and use them repeatedly.

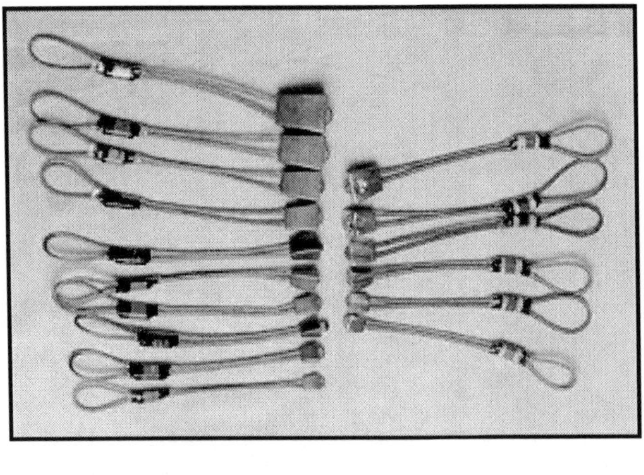

Figure 4-26: Chocks.

e. **Three-Point Camming Device.** The three-point camming device's unique design allows it to be used both as a camming piece and a wedging piece (Figure 4-27). Because of this design it is extremely versatile and, when used in the camming mode, will fit a wide range of cracks. The three-point camming device comes in several different sizes with the smaller sizes working in pockets that no other piece of gear would fit in.

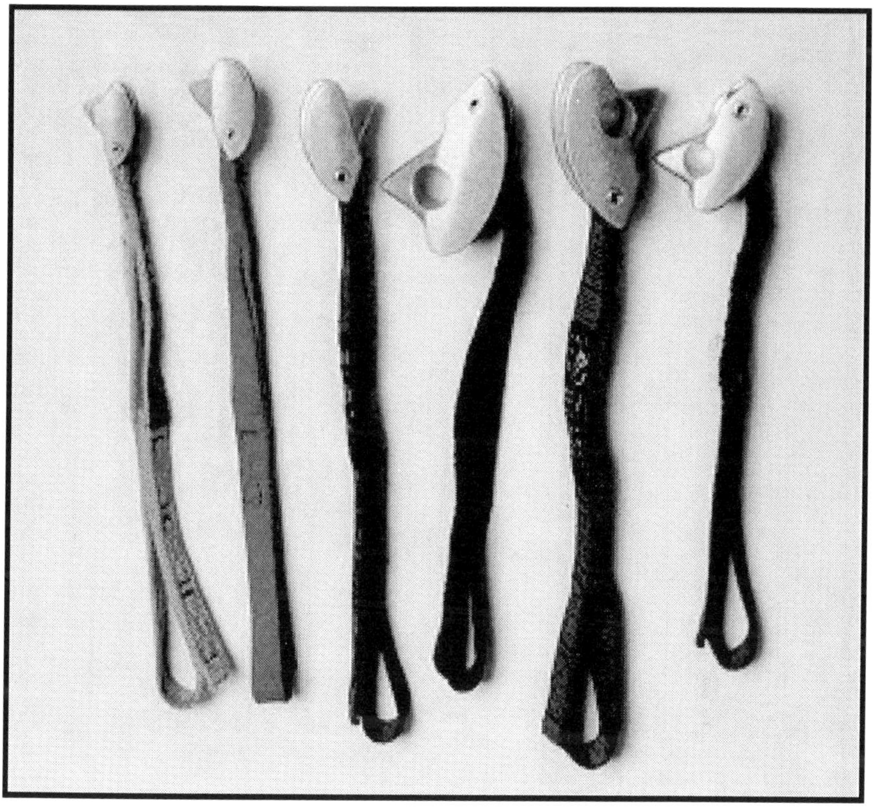

Figure 4-27: Three-point camming device.

f. **Spring-Loaded Camming Devices.** Spring-loaded camming devices (SLCDs) (Figure 4-28) provide convenient, reliable placement in cracks where standard chocks are not practical (parallel or flaring cracks or cracks under roofs). SLCDs have three or four cams rotating around a single or double axis with a rigid or semi-rigid point of attachment. These are placed quickly and easily, saving time and effort. SLCDs are available in many sizes to accommodate different size cracks. Each fits a wide range of crack widths due to the rotating cam heads. The shafts may be rigid metal or semi-rigid cable loops. The flexible cable reduces the risk of stem breakage over an edge in horizontal placements.

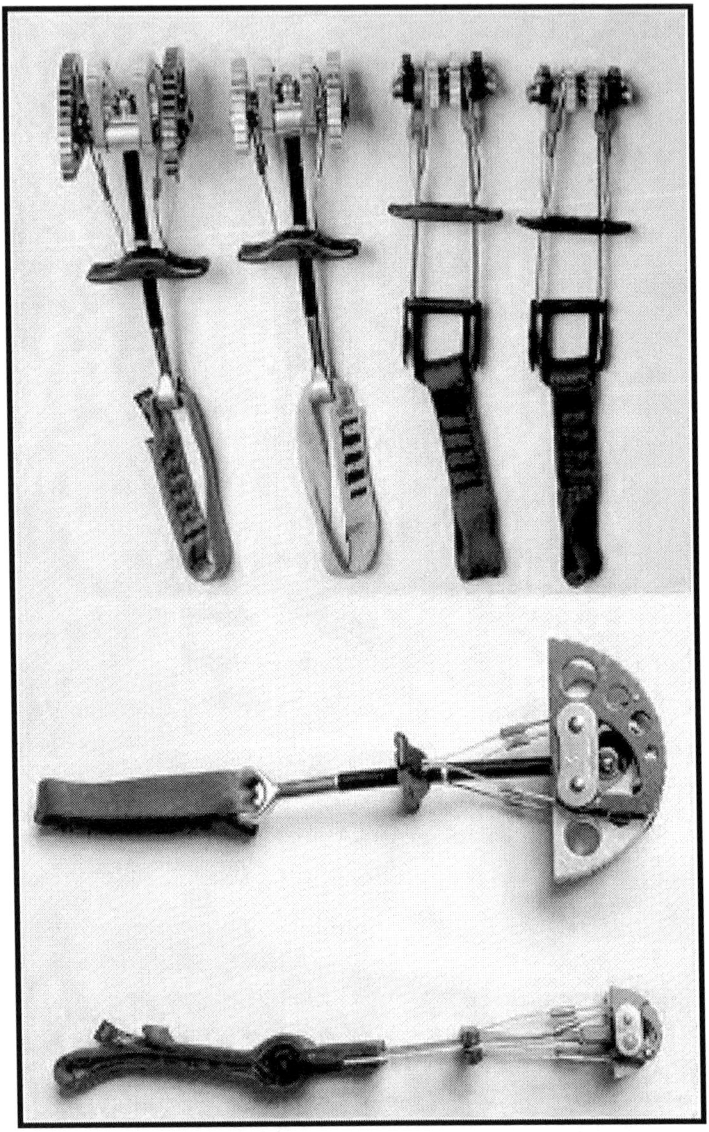

Figure 4-28: Spring-loaded camming devices.

g. **Chock Picks.** Chock picks are primarily used to extract chocks from rock when they become severely wedged (Figure 4-29). They are also handy to clean cracks with. Made from thin metal, they can be purchased or homemade. When using a chock pick to extract a chock be sure no force is applied directly to the cable juncture. One end of the chock pick should have a hook to use on jammed SLCDs.

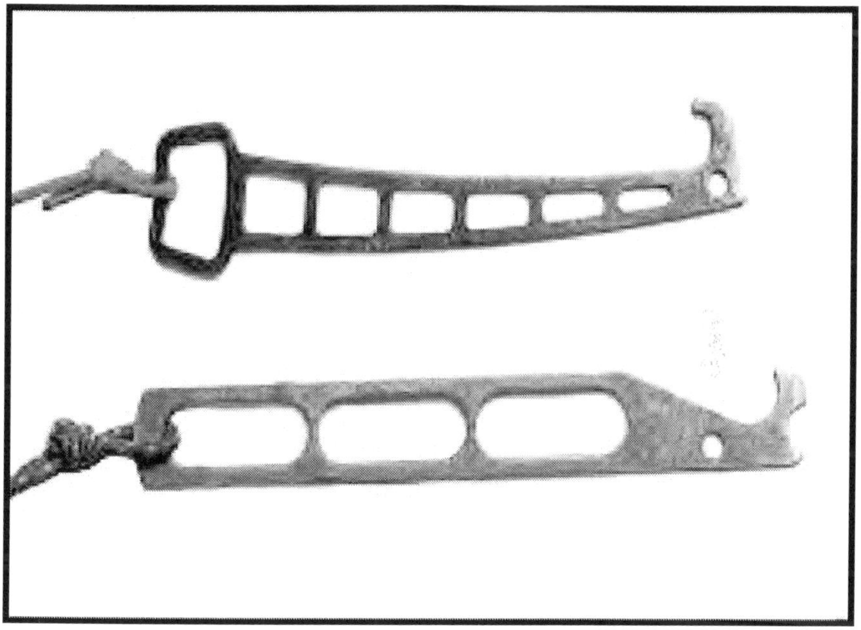

Figure 4-29: Chock picks.

h. **Bolts.** Bolts are screw-like shafts made from metal that are drilled into rock to provide protection (Figure 4-30). The two types are contraction bolts and expansion bolts. Contraction bolts are squeezed together when driven into a rock. Expansion bolts press around a surrounding sleeve to form a snug fit into a rock. Bolts require drilling a hole into a rock, which is time-consuming, exhausting, and extremely noisy. Once emplaced, bolts are the most secure protection for a multidirectional pull. Bolts should be used only when chocks and pitons cannot be emplaced. A bolt is hammered only when it is the nail or self-driving type.

(1) A hanger (for carabiner attachment) and nut are placed on the bolt. The bolt is then inserted and driven into the hole. Because of this requirement, a hand drill must be carried in addition to a piton hammer. Hand drills (also called star drills) are available in different sizes, brands, and weights. A hand drill should have a lanyard to prevent loss.

(2) Self-driving bolts are quicker and easier to emplace. These require a hammer, bolt driver, and drilling anchor, which is driven into the rock. A bolt and carrier are then secured to the emplaced drilling anchor. All metal surfaces should be smooth and free of rust, corrosion, dirt, and moisture. Burrs, chips, and rough spots should be filed smooth and wire-brushed or rubbed clean with steel wool. Items that are cracked or warped indicate excessive wear and should be discarded.

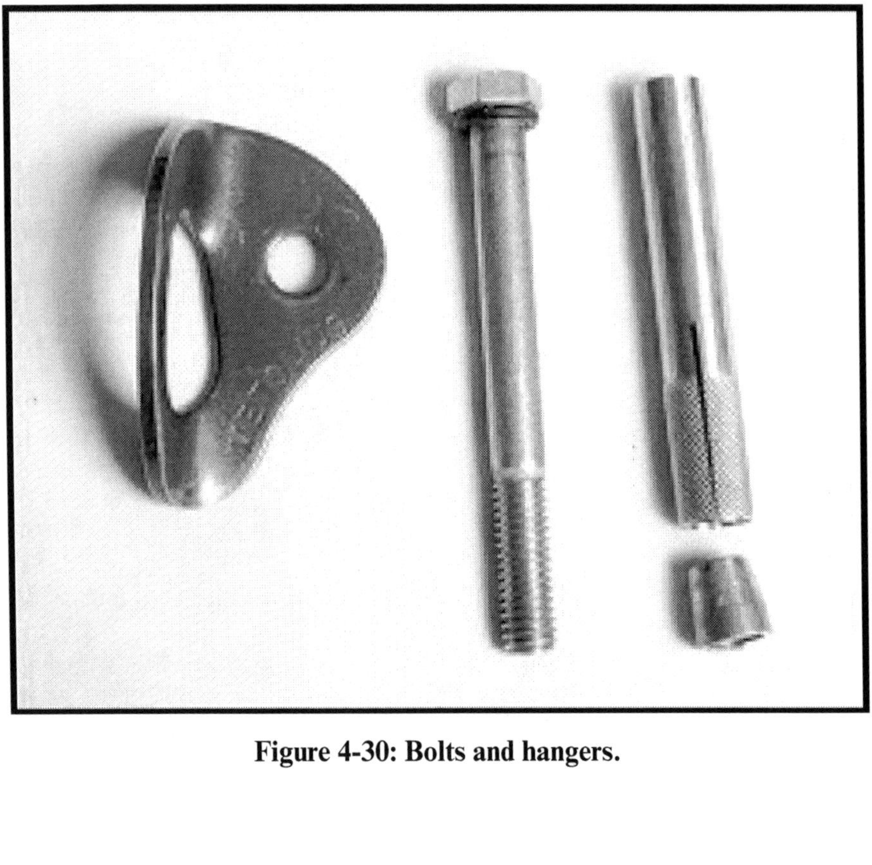

Figure 4-30: Bolts and hangers.

i. **Belay Devices.** Belay devices range from the least equipment intensive (the body belay) to high-tech metal alloy pieces of equipment. Regardless of the belay device chosen, the basic principal remains the same friction around or through the belay device controls the ropes' movement. Belay devices are divided into three categories: the slot, the tuber, and the mechanical camming device (Figure 4-31).

(1) The slot is a piece of equipment that attaches to a locking carabiner in the harness; a bight of rope slides through the slot and into the carabiner for the belay. The most common slot type belay device is the Sticht plate.

(2) The tuber is used exactly like the slot but its shape is more like a cone or tube.

(3) The mechanical camming device is a manufactured piece of equipment that attaches to the harness with a locking carabiner. The rope is routed through this device so that when force is applied the rope is crammed into a highly frictioned position.

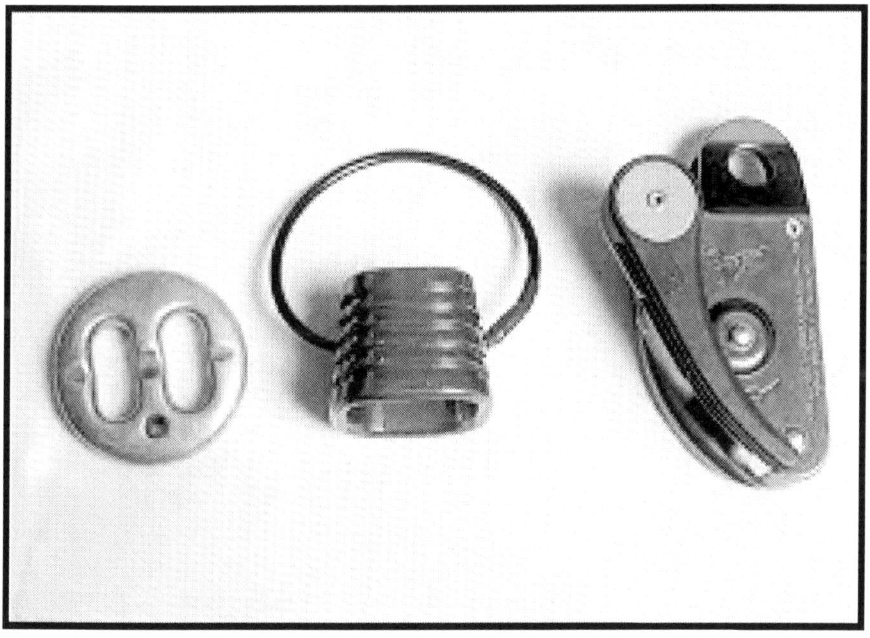

Figure 4-31: Slot, tuber, mechanical camming device.

j. **Descenders.** One piece of equipment used for generations as a descender is the carabiner. A figure-eight is another useful piece of equipment and can be used in conjunction with the carabiner for descending (Figure 4-32).

> ✍ **NOTE:**
> All belay devices can also be used as descending devices.

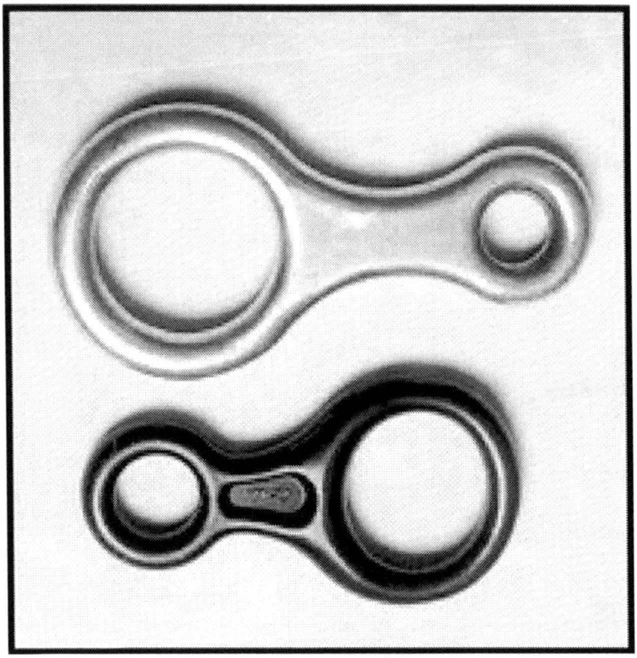

Figure 4-33: Figure-eights.

k. **Ascenders.** Ascenders may be used in other applications such as a personal safety or hauling line cam. All modern ascenders work on the principle of using a cam-like device to allow movement in one direction. Ascenders are primarily made of metal alloys and come in a variety of sizes (Figure 4-34). For difficult vertical terrain, two ascenders work best. For lower angle movement, one ascender is sufficient. Most manufacturers make ascenders as a right and left-handed pair.

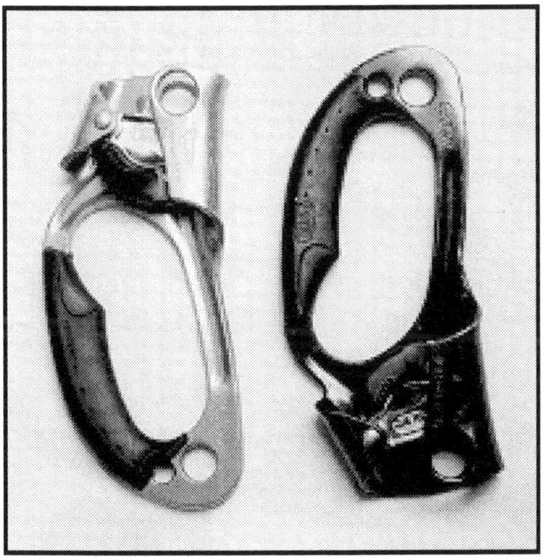

Figure 4-34: Ascenders.

l. **Pulleys.** Pulleys are used to change direction in rope systems and to create mechanical advantage in hauling systems. A pulley should be small, lightweight, and strong. They should accommodate the largest diameter of rope being used. Pulleys are made with several bearings, different-sized sheaves (wheel), and metal alloy sideplates (Figure 4-35). Plastic pulleys should always be avoided. The sideplate should rotate on the pulley axle to allow the pulley to be attached at any point along the rope. For best results, the sheave diameter must be at least four times larger than the rope's diameter to maintain high rope strength.

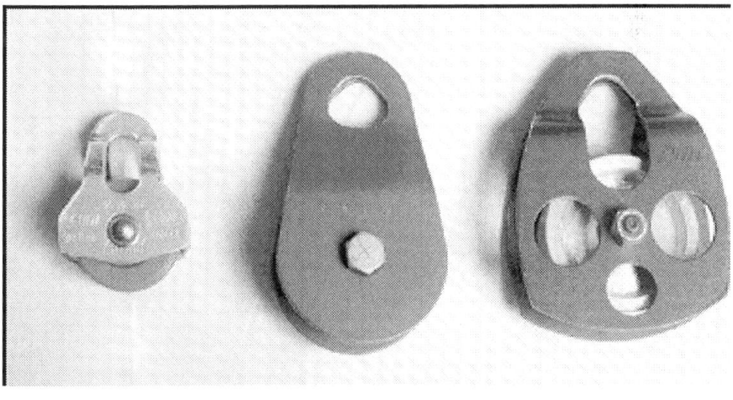

Figure 4-35: Pulley.

4-32. SNOW AND ICE CLIMBING HARDWARE

Snow and ice climbing hardware is the equipment that is particular to operations in some mountainous terrain. Specific training on this type of equipment is essential for safe use. Terrain that would otherwise be inaccessible-snowfields, glaciers, and frozen waterfalls-can now be considered avenues of approach using the snow and ice climbing gear listed in this paragraph.

a. **Ice Ax.** The ice ax is one of the most important tools for the mountaineer operating on snow or ice. The climber must become proficient in its use and handling. The versatility of the ax lends itself to balance, step cutting, probing, self-arrest, belays, anchors, direct-aid climbing, and ascending and descending snow and ice covered routes.

(1) Several specific parts comprise an ice ax: the shaft, head (pick and adze), and spike (Figure 4-36).

(a) The shaft (handle) of the ax comes in varying lengths (the primary length of the standard mountaineering ax is 70 centimeters). It can be made of fiberglass, hollow aluminum, or wood; the first two are stronger, therefore safer for mountaineering.

(b) The head of the ax, which combines the pick and the adze, can have different configurations. The pick should be curved slightly and have teeth at least one-fourth of its length. The adze, used for chopping, is perpendicular to the shaft. It can be flat or curved along its length and straight or rounded from side to side. The head can be of one-piece construction or have replaceable picks and adzes. The head should have a hole directly above the shaft to allow for a leash to be attached.

(c) The spike at the bottom of the ax is made of the same material as the head and comes in a variety of shapes.

(2) As climbing becomes more technical, a shorter ax is much more appropriate, and adding a second tool is a must when the terrain becomes vertical. The shorter ax has all the attributes of the longer ax, but it is anywhere from 40 to 55 centimeters long and can have a straight or bent shaft depending on the preference of the user.

b. **Ice Hammer.** The ice hammer is as short or shorter than the technical ax (Figure 3-24). It is used for pounding protection into the ice or pitons into the rock. The only difference between the ice ax and the ice hammer is the ice hammer has a hammerhead instead of an adze. Most of the shorter ice tools have a hole in the shaft to which a leash is secured, which provides a more secure purchase in the ice.

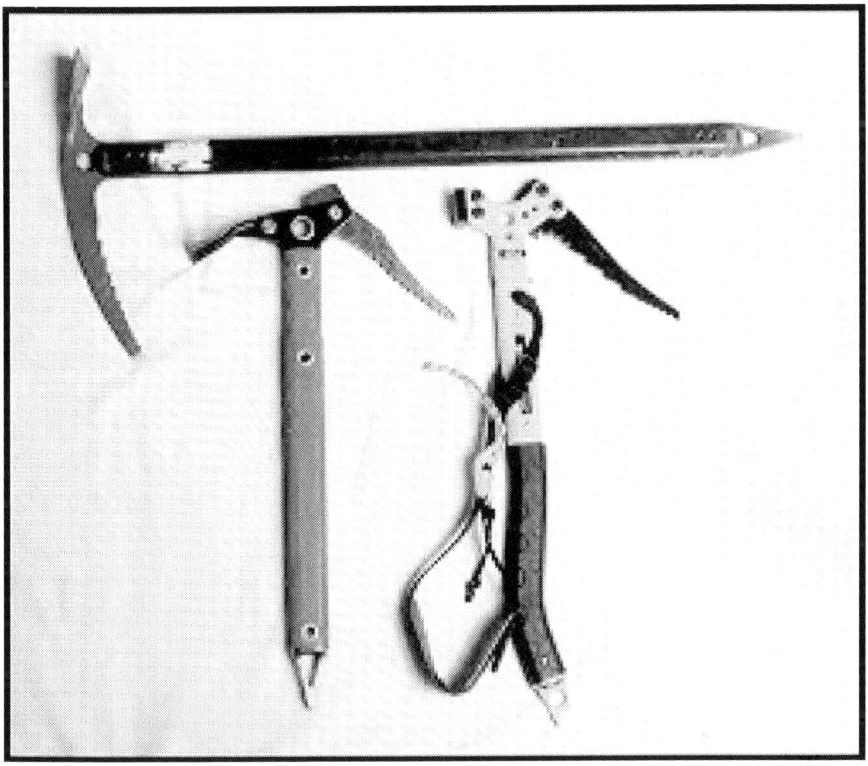

Figure 4-36: Ice ax and ice hammers.

c. **Crampons.** Crampons are used when the footing becomes treacherous. They have multiple spikes on the bottom and spikes protruding from the front (Figure 4-37). Two types of crampons are available: flexible and rigid. Regardless of the type of crampon chosen, fit is the most important factor associated with crampon wear. The crampon should fit snugly on the boot with a minimum of1 inch of front point protruding. Straps should fit snugly around the foot and any long, loose ends should be trimmed. Both flexible and rigid crampons come in pairs, and any tools needed for adjustment will be provided by the manufacturer.

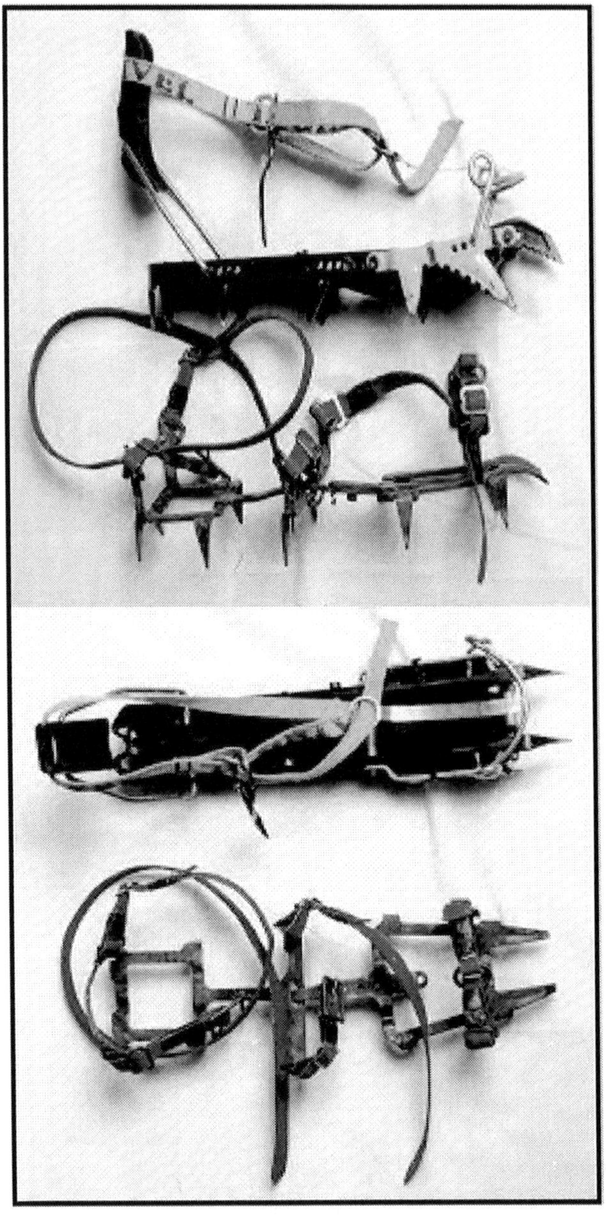

Figure 4-37: Crampons.

(1) The hinged or flexible crampon is best used when no technical ice climbing will be done. It is designed to be used with soft, flexible boots, but can be attached to plastic mountaineering boots. The flexible crampon gets its name from the flexible hinge on the crampon itself. All flexible crampons are adjustable for length while some allow for width adjustment. Most flexible crampons will attach to the boot by means of a strap system. The flexible crampon can be worn with a variety of boot types.

(2) The rigid crampon, as its name implies, is rigid and does not flex. This type of crampon is designed for technical ice climbing, but can be used on less vertical terrain. The rigid crampon can only be worn with plastic mountaineering boots. Rigid crampons will have a toe and heel bail attachment with a strap that wraps around the ankle.

d. **Ice Screws.** Ice screws provide artificial protection for climbers and equipment for operations in icy terrain. They are screwed into ice formations. Ice screws are made of chrome-molybdenum steel and vary in lengths from 11 centimeters to 40 centimeters (Figure 4-38). The eye is permanently affixed to the top of the ice screw. The tip consists of milled or

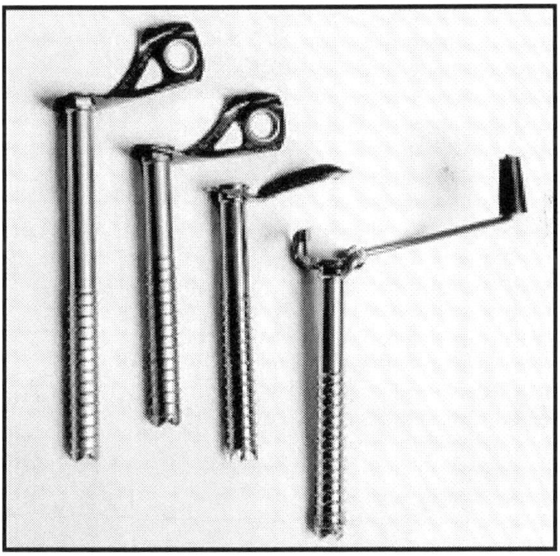

Figure 4-38: Ice screws.

hand-ground teeth, which create sharp points to grab the ice when being emplaced. The ice screw has right-hand threads to penetrate the ice when turned clockwise.

(1) When selecting ice screws, choose a screw with a large thread count and large hollow opening. The close threads will allow for ease in turning and better strength. The large hollow opening will allow snow and ice to slide through when turning.
  - Type I is 17 centimeters in length with a hollow inner tube.
  - Type II is 22 centimeters in length with a hollow inner tube.
  - Other variations are hollow alloy screws that have a tapered shank with external threads, which are driven into ice and removed by rotation.

(2) Ice screws should be inspected for cracks, bends, and other deformities that may impair strength or function. If any cracks or bends are noticed, the screw should be turned in. A file may be used to sharpen the ice screw points. Steel wool should be rubbed on rusted surfaces and a thin coat of oil applied when storing steel ice screws.

> ✍ **NOTE:**
> Ice screws should always be kept clean and dry. The threads and teeth should be protected and kept sharp for ease of application.

e. **Ice Pitons.** Ice pitons are used to establish anchor points for climbers and equipment when conducting operations on ice. They are made of steel or steel alloys (chrome-molybdenum), and are available in various lengths and diameters (Figure 4-39). They are tubular with a hollow core and are hammered into ice with an ice hammer. The eye is permanently fixed to the top of the ice piton. The tip may be beveled to help grab the ice to facilitate insertion. Ice pitons are extremely strong when placed properly in hard ice. They can, however, pull out easily on warm days and require a considerable amount of effort to extract in cold temperatures.

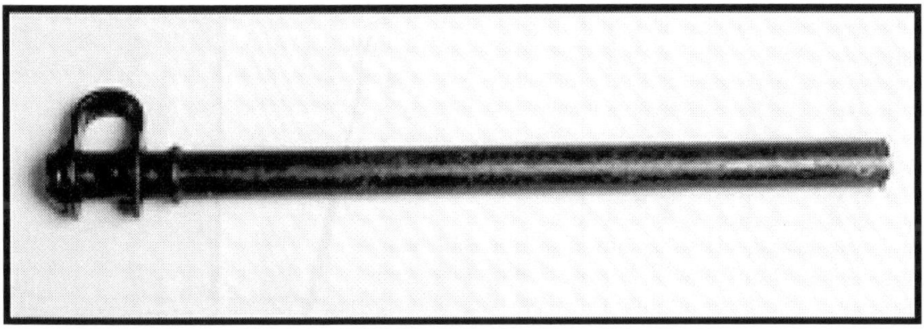

Figure 4-39: Ice piton.

f. **Wired Snow Anchors.** The wired snow anchor (or fluke) provides security for climbers and equipment in operations involving steep ascents by burying the snow anchor into deep snow (Figure 4-40). The fluted anchor portion of the snow anchor is made of aluminum. The wired portion is made of either galvanized steel or stainless steel. Fluke anchors are available in various sizes their holding ability generally increases with size. They are available with bent faces, flanged sides, and fixed cables. Common types are:
   - Type I is 22 by 14 centimeters. Minimum breaking strength of the swaged wire loop is 600 kilograms.
   - Type II is 25 by 20 centimeters. Minimum breaking strength of the swaged wire loop is 1,000 kilograms.

   The wired snow anchor should be inspected for cracks, broken wire strands, and slippage of the wire through the swage. If any cracks, broken wire strands, or slippage is noticed, the snow anchor should be turned in.

g. **Snow Picket.** The snow picket is used in constructing anchors in snow and ice (Figure 4-40). The snow picket is made of a strong aluminum alloy 3 millimeters thick by 4 centimeters wide, and 45 to 90 centimeters long. They can be angled or T-section stakes. The picket should be inspected for bends, chips, cracks, mushrooming ends, and other deformities. The ends should be filed smooth. If bent or cracked, the picket should be turned in for replacement.

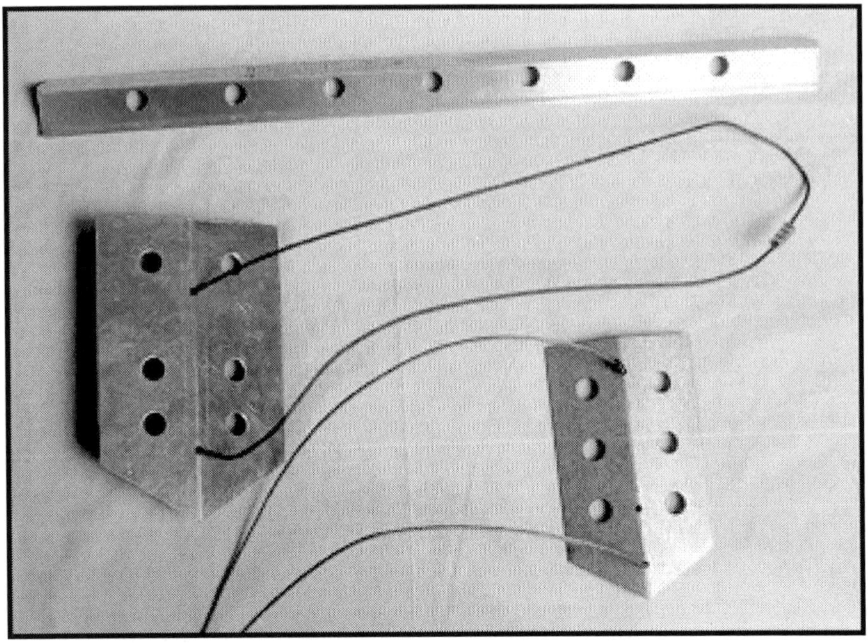

Figure 4-40: Snow anchors, flukes, and pickets.

4-33. SUSTAINABILITY EQUIPMENT

This paragraph describes all additional equipment not directly involved with climbing. This equipment is used for safety (avalanche equipment, wands), bivouacs, movement, and carrying gear. While not all of it will need to be carried on all missions, having the equipment available and knowing how to use it correctly will enhance the unit's capability in mountainous terrain.

a. **Snow Saw.** The snow saw is used to cut into ice and snow. It can be used in step cutting, in shelter construction, for removing frozen obstacles, and for cutting snow stability test pits. The special tooth design of the snow saw easily cuts into frozen snow and ice. The blade is a rigid aluminum alloy of high strength about 3 millimeters thick and 38 centimeters long with a pointed end to facilitate entry on the forward stroke. The handle is either wooden or plastic and is riveted to the blade for a length of about 50 centimeters. The blade should be inspected for rust, cracks, warping, burrs, and missing or dull teeth. A file can repair most defects, and steel wool can be rubbed on rusted areas. The handle should be inspected for cracks, bends, and stability. On folding models, the hinge and nuts should be secure. If the saw is beyond repair, it should not be used.

b. **Snow Shovel.** The snow shovel is used to cut and remove ice and snow. It can be used for avalanche rescue, shelter construction, step cutting, and removing obstacles. The snow shovel is made of a special, lightweight aluminum alloy. The handle should be telescopic, folding, or removable to be compact when not in use. The shovel should have a flat or rounded bottom and be of strong construction. The shovel should be inspected for cracks, bends, rust, and burrs. A file and steel wool can remove rust and put an edge on the blade of the shovel. The handle should be inspected for cracks, bends, and stability. If the shovel is beyond repair, it should be turned in.

c. **Wands.** Wands are used to identify routes, crevasses, snowbridges, caches, and turns on snow and glaciers. Spacing of wands depends on the number of turns, number of hazards identified, weather conditions (and visibility), and number of teams in the climbing party. Carry too many wands is better than not having enough if they become lost. Wands are 1 to 1.25 meters long and made of lightweight bamboo or plastic shafts pointed on one end with a plastic or nylon flag (bright enough in color to see at a distance) attached to the other end. The shafts should be inspected for cracks, bends, and deformities. The flag should be inspected for tears, frays, security to the shaft, fading, and discoloration. If any defects are discovered, the wands should be replaced.

d. **Avalanche rescue equipment.** Avalanche rescue equipment (Figure 4-41) includes the following:

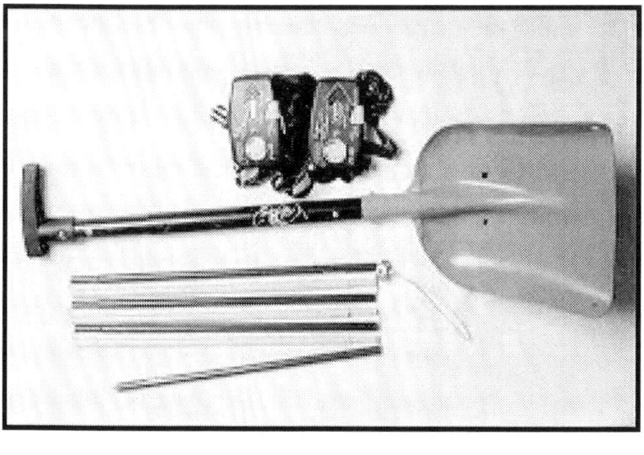

Figure 4-41: Avalanche rescue equipment.

(1) ***Avalanche Probe.*** Although ski poles may be used as an emergency probe when searching for a victim in an avalanche, commercially manufactured probes are better for a thorough search. They are 9-millimeter thick shafts made of an aluminum alloy, which can be joined to probe up to 360 centimeters. The shafts must be strong enough to probe through avalanche debris. Some manufacturers of ski poles design poles that are telescopic and mate with other poles to create an avalanche probe.

(2) ***Avalanche Transceivers.*** These are small, compact radios used to identify avalanche burial sites. They transmit electromagnetic signals that are picked up by another transceiver on the receive mode.

e. **Packs.** Many types and brands of packs are used for mountaineering. The two most common types are internal and external framed packs.

(1) Internal framed packs have a rigid frame within the pack that help it maintain its shape and hug the back. This assists the climber in keeping their balance as they climb or ski. The weight in an internal framed pack is carried low on the body assisting with balance. The body-hugging nature of this type pack also makes it uncomfortable in warm weather.

(2) External framed packs suspend the load away from the back with a ladder-like frame. The frame helps transfer the weight to the hips and shoulders easier, but can be cumbersome when balance is needed for climbing and skiing.

(3) Packs come in many sizes and should be sized appropriately for the individual according to manufacturer's specifications. Packs often come with many unneeded features. A good rule of thumb is: The simpler the pack, the better it will be.

f. **Stoves.** When selecting a stove one must define its purpose will the stove be used for heating, cooking or both? Stoves or heaters for large elements can be large and cumbersome. Stoves for smaller elements might just be used for cooking and making water, and are simple and lightweight. Stoves are a necessity in mountaineering for cooking and making water from snow and ice. When choosing a stove, factors that should be considered are weight, altitude and temperature where it will be used, fuel availability, and its reliability.

(1) There are many choices in stove design and in fuel types. White gas, kerosene, and butane are the common fuels used. All stoves require a means of pressurization to force the fuel to the burner. Stoves that burn white gas or kerosene have a hand pump to generate the pressurization and butane stoves have pressurized cartridges. All stoves need to vaporize the liquid fuel before it is burned. This can be accomplished by burning a small amount of fuel in the burner cup assembly, which will vaporize the fuel in the fuel line.

(2) Stoves should be tested and maintained prior to a mountaineering mission. They should be easy to clean and repair during an operation. The reliability of the stove has a huge impact on the success of the mission and the morale of personnel.

g. **Tents.** When selecting a tent, the mission must be defined to determine the number of people the tent will accommodate. The climate the tents will be used in is also of concern. A tent used for warmer temperatures will greatly differ from tents used in a colder, harsher environment. Manufacturers of tents offer many designs of different sizes, weights, and materials.

(1) Mountaineering tents are made out of a breathable or weatherproof material. A single-wall tent allows for moisture inside the tent to escape through the tent's material. A double-wall tent has a second layer of material (referred to as a fly) that covers the tent. The fly protects against rain and snow and the space between the fly and tent helps moisture to escape from inside. Before using a new tent, the seams should be treated with seam sealer to prevent moisture from entering through the stitching.

(2) The frame of a tent is usually made of an aluminum or carbon fiber pole. The poles are connected with an elastic cord that allows them to extend, connect, and become long and rigid. When the tent poles are secured into the tent body, they create the shape of the tent.

(3) Tents are rated by a "relative strength factor," the speed of wind a tent can withstand before the frame deforms. Temperature and expected weather for the mission should be determined before choosing the tent.

h. **Skis.** Mountaineering skis are wide and short. They have a binding that pivots at the toe and allows for the heel to be free for uphill travel or locked for downhill. Synthetic skins with fibers on the bottom can be attached to the bottom of the ski and allow the ski to travel forward and prevent slipping backward. The skins aid in traveling uphill and slow down the rate of descents. Wax can be applied to the ski to aid in ascents instead of skins. Skis can decrease the time needed to reach an objective depending on the ability of the user. Skis can make crossing crevasses easier because of the load distribution, and they can become a makeshift stretcher for casualties. Ski techniques can be complicated and require thorough training for adequate proficiency.

i. **Snowshoes.** Snowshoes are the traditional aid to snow travel that attach to most footwear and have been updated into small, lightweight designs that are more efficient than older models. Snowshoes offer a large displacement area on top of soft snow preventing tiresome post-holing. Some snowshoes come equipped with a crampon like binding that helps in ascending steep snow and ice. Snowshoes are slower than skis, but are better suited for mixed terrain, especially if personnel are not experienced with the art of skiing. When carrying heavy packs, snowshoes can be easier to use than skis.

j. **Ski poles.** Ski poles were traditionally designed to assist in balance during skiing. They have become an important tool in mountaineering for aid in balance while hiking, snowshoeing, and carrying heavy packs. They can take some of the weight off of the lower body when carrying a heavy pack. Some ski poles are collapsible for ease of packing when not needed (Figure 4-42). The basket at the bottom prevents the pole from plunging deep into the snow and, on some models, can be detached so the pole becomes an avalanche or crevasse probe. Some ski poles come with a self-arrest grip, but should not be the only means of protection on technical terrain.

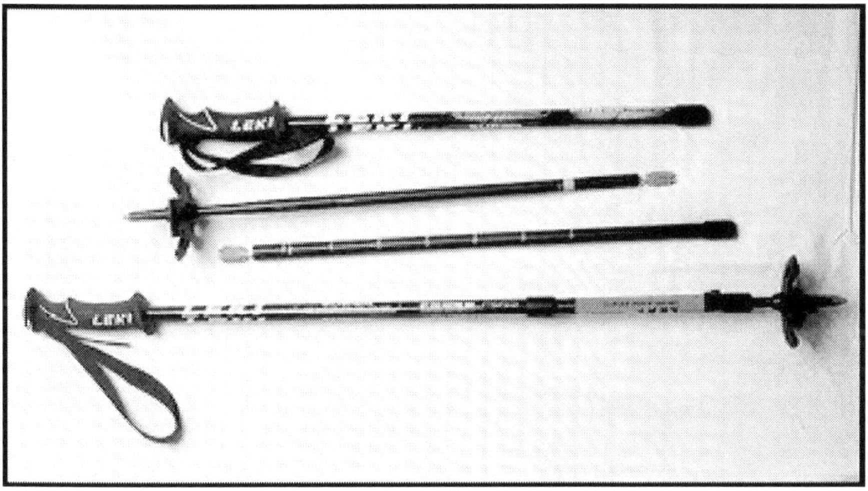

Figure 4-42: Collapsible ski poles.

k. **Sleds.** Sleds vary greatly in size, from the squad-size Ahkio, a component of the 10-man arctic tent system, to the one-person skow. Regardless of the size, sleds are an invaluable asset during mountainous operations when snow and ice is the primary surface on which to travel. Whichever sled is chosen, it must be attachable to the person or people that will be pulling it. Most sleds are constructed using fiberglass bottoms with or without exterior runners. Runners will aid the sled's ability to maintain a true track in the snow. The sled should also come with a cover of some sort whether nylon or canvas, a cover is essential for keeping the components in the sled dry. Great care should be taken when packing the sled, especially when hauling fuel. Heavier items should be carried towards the rear of the sled and lighter items towards the front.

l. **Headlamps.** A headlamp is a small item that is not appreciated until it is needed. It is common to need a light source and the use of both hands during limited light conditions in mountaineering operations. A flashlight can provide light, but can be cumbersome when both hands are needed. Most headlamps attach to helmets by means of elastic bands.

(1) When choosing a headlamp, ensure it is waterproof and the battery apparatus is small. All components should be reliable in extreme weather conditions. When the light is being packed, care should be taken that the switch doesn't accidentally activate and use precious battery life.

(2) The battery source should complement the resupply available. Most lights will accept alkaline, nickel-cadmium, or lithium batteries. Alkaline battery life diminishes quickly in cold temperatures, nickel-cadmium batteries last longer in cold but require a recharging unit, and lithium batteries have twice the voltage so modifications are required.

EQUIPMENT PACKING

Equipment brought on a mission is carried in the pack, worn on the body, or hauled in a sled (in winter).Obviously, the rucksack and sled (or Ahkio) can hold much more than a climber can carry. They would be used for major bivouac gear, food, water, first aid kits, climbing equipment, foul weather shells, stoves, fuel, ropes, and extra ammunition and demolition materials, if needed.

4-. CHOICE OF EQUIPMENT

Mission requirements and unit SOP will influence the choice of gear carried but the following lists provide a sample of what should be considered during mission planning.

a. **Personal Gear.** Personal gear includes emergency survival kit containing signaling material, fire starting material, food procurement material, and water procurement material. Pocket items should include a knife, whistle, pressure bandage, notebook with pen or pencil, sunglasses, sunblock and lip protection, map, compass and or altimeter.
b. **Standard Gear.** Standard gear that can be individually worn or carried includes cushion sole socks; combat boots or mountain boots, if available; BDU and cap; LCE with canteens, magazine pouches, and first aid kit; individual weapon; a large rucksack containing waterproof coat and trousers, polypropylene top, sweater, or fleece top; helmet; poncho; and sleeping bag.

> ⚠ **CAUTION**
>
> Cotton clothing, due to its poor insulating and moisture-wicking characteristics, is virtually useless in most mountain climates, the exception being hot, desert, or jungle mountain environments. Cotton clothing should be replaced with synthetic fabric clothing.

c. **Mountaineering Equipment and Specialized Gear.** This gear includes:
   - Sling rope or climbing harness.
   - Utility cord(s).
   - Nonlocking carabiners.
   - Locking carabiner(s).
   - Rappelling gloves.
   - Rappel/belay device.
   - Ice ax.
   - Crampons.
   - Climbing rope, one per climbing team.
   - Climbing rack, one per climbing team.

d. **Day Pack.** When the soldier plans to be away from the bivouac site for the day on a patrol or mountaineering mission, he carries a light day pack. This pack should contain the following items:
   - Extra insulating layer: polypropylene, pile top, or sweater.
   - Protective layer: waterproof jacket and pants, rain suit, or poncho.
   - First aid kit.
   - Flashlight or headlamp.
   - Canteen.
   - Cold weather hat or scarf.
   - Rations for the time period away from the base camp.
   - Survival kit.
   - Sling rope or climbing harness.
   - Carabiners.
   - Gloves.
   - Climbing rope, one per climbing team.
   - Climbing rack, one per climbing team.

e. **Squad or Team Safety Pack.** When a squad-sized element leaves the bivouac site, squad safety gear should be carried in addition to individual day packs. This can either be loaded into one rucksack or cross-loaded among the squad members. In the event of an injury, casualty evacuation, or unplanned bivouac, these items may make the difference between success and failure of the mission.
   - Sleeping bag.
   - Sleeping mat.
   - Squad stove.
   - Fuel bottle.

f. **The Ten Essentials.** Regardless of what equipment is carried, the individual military mountaineer should always carry the "ten essentials" when moving through the mountains.

(1) ***Map.***

(2) ***Compass, Altimeter, and or GPS.***

(3) ***Sunglasses and Sunscreen.***

(a) In alpine or snow-covered sub-alpine terrain, sunglasses are a vital piece of equipment for preventing snow blindness. They should filter 95 to 100 percent of ultraviolet light. Side shields, which minimize the light entering from the side, should permit ventilation to help prevent lens fogging. At least one extra pair of sunglasses should be carried by each independent climbing team.

(b) Sunscreens should have an SPF factor of 15 or higher. For lip protection, a total UV blocking lip balm that resists sweating, washing, and licking is best. This lip protection should be carried in the chest pocket or around the neck to allow frequent reapplication.

(4) ***Extra Food.*** One day's worth extra of food should be carried in case of delay caused by bad weather, injury, or navigational error.

(5) ***Extra Clothing.*** The clothing used during the active part of a climb, and considered to be the basic climbing outfit, includes socks, boots, underwear, pants, blouse, sweater or fleece jacket, hat, gloves or mittens, and foul weather gear (waterproof, breathable outerwear or waterproof rain suit).

(a) Extra clothing includes additional layers needed to make it through the long, inactive hours of an unplanned bivouac. Keep in mind the season when selecting this gear.

- Extra underwear to switch out with sweat-soaked underwear.
- Extra hats or balaclavas.
- Extra pair of heavy socks.
- Extra pair of insulated mittens or gloves.
- In winter or severe mountain conditions, extra insulation for the upper body and the legs.

(b) To back up foul weather gear, bring a poncho or extra-large plastic trash bag. A reflective emergency space blanket can be used for hypothermia first aid and emergency shelter. Insulated foam pads prevent heat loss while sitting or lying on snow. Finally, a bivouac sack can help by protecting insulating layers from the weather, cutting the wind, and trapping essential body heat inside the sack.

(6) ***Headlamp and or Flashlight.*** Headlamps provide the climber a hands-free capability, which is important while climbing, working around the camp, and employing weapons systems. Miniature flashlights can be used, but commercially available headlamps are best. Red lens covers can be fabricated for tactical conditions. Spare batteries and spare bulbs should also be carried.

(7) ***First-aid Kit.*** Decentralized operations, the mountain environment steep, slick terrain and loose rock combined with heavy packs, sharp tools, and fatigue requires each climber to carry his own first-aid kit. Common mountaineering injuries that can be expected are punctures and abrasions with severe bleeding, a broken bone, serious sprain, and blisters. Therefore, the kit should contain at least enough material to stabilize these conditions. Pressure dressings, gauze pads, elastic compression wrap, small adhesive bandages, butterfly bandages, moleskin, adhesive tape, scissors, cleanser, latex gloves and splint material (if above tree line) should all be part of the kit.

(8) ***Fire Starter.*** Fire starting material is key to igniting wet wood for emergency campfires. Candles, heat tabs, and canned heat all work. These can also be used for quick warming of water or soup in a canteen cup. In alpine zones above tree line with no available firewood, a stove works as an emergency heat source.

(9) ***Matches and Lighter.*** Lighters are handy for starting fires, but they should be backed up by matches stored in a waterproof container with a strip of sandpaper.

(10) ***Knife.*** A multipurpose pocket tool should be secured with cord to the belt, harness, or pack.

g. Other Essential Gear. Other essential gear may be carried depending on mission and environmental considerations.

(1) ***Water and Water Containers.*** These include wide-mouth water bottles for water collection; camel-back type water

holders for hands-free hydration; and a small length of plastic tubing for water procurement at snow-melt seeps and rainwater puddles on bare rock.

(2) ***Ice Ax.*** The ice ax is essential for travel on snowfields and glaciers as well as snow-covered terrain in spring and early summer. It helps for movement on steep scree and on brush and heather covered slopes, as well as for stream crossings.

(3) ***Repair Kit.*** A repair kit should include:
- Stove tools and spare parts.
- Duct tape.
- Patches.
- Safety pins.
- Heavy-duty thread.
- Awl and or needles.
- Cord and or wire.
- Small pliers (if not carrying a multipurpose tool).
- Other repair items as needed.

(4) ***Insect Repellent.***

(5) ***Signaling Devices.***

(6) ***Snow Shovel.***

4-35. TIPS ON PACKING

When loading the internal frame pack the following points should be considered.

a. In most cases, speed and endurance are enhanced if the load is carried more by the hips (using the waist belt) and less by the shoulders and back. This is preferred for movement over trails or less difficult terrain. By packing the lighter, more compressible items (sleeping bag, clothing) in the bottom of the rucksack and the heavier gear (stove, food, water, rope, climbing hardware, extra ammunition) on top, nearer the shoulder blades, the load is held high and close to the back, thus placing the most weight on the hips.

b. In rougher terrain it pays to modify the pack plan. Heavy articles of gear are placed lower in the pack and close to the back, placing more weight on the shoulders and back. This lowers the climber's center of gravity and helps him to better keep his balance.

c. Equipment that may be needed during movement should be arranged for quick access using either external pockets or placing immediately underneath the top flap of the pack. As much as possible, this placement should be standardized across the team so that necessary items can be quickly reached without unnecessary unpacking of the pack in emergencies.

d. The pack and its contents should be soundly waterproofed. Clothing and sleeping bag are separately sealed and then placed in the larger wet weather bag that lines the rucksack. Zip-lock plastic bags can be used for small items, which are then organized into color-coded stuff sacks. A few extra-large plastic garbage bags should be carried for a variety of uses spare waterproofing, emergency bivouac shelter, and water procurement, among others.
e. The ice ax, if not carried in hand, should be stowed on the outside of the pack with the spike up and the adze facing forward or to the outside, and be securely fastened. Mountaineering packs have ice ax loops and buckle fastening systems for this. If not, the ice ax is placed behind one of the side pockets, as stated above, and then tied in place.
f. Crampons should be secured to the outside rear of the pack with the points covered.

SECTION 4: ROPE MANAGEMENT AND KNOTS

The rope is a vital piece of equipment to the mountaineer. When climbing, rappelling, or building various installations, the mountaineer must know how to properly use and maintain this piece of equipment. If the rope is not managed or maintained properly, serious injury may occur. This chapter discusses common rope terminology, management techniques, care and maintenance procedures, and knots.

PREPARATION, CARE AND MAINTENANCE, INSPECTION, TERMINOLOGY

The service life of a rope depends on the frequency of use, applications (rappelling, climbing, rope installations), speed of descent, surface abrasion, terrain, climate, and quality of maintenance. Any rope may fail under extreme conditions (shock load, sharp edges, misuse).

4-36. PREPARATION

The mountaineer must select the proper rope for the task to be accomplished according to type, diameter, length, and tensile strength. It is important to prepare all ropes before departing on a mission. Avoid rope preparation in the field.

a. **Packaging.** New rope comes from the manufacturer in different configurations boxed on a spool in various lengths, or coiled and bound in some manner. Precut ropes are usually packaged in a protective cover such as plastic or burlap. Do not remove the protective cover until the rope is ready for use.

b. **Securing the Ends of the Rope:** If still on a spool, the rope must be cut to the desired length. All ropes will fray at the ends unless they are bound or seared. Both static and dynamic rope ends are secured in the same manner. The ends must be heated to the melting point so as to attach the inner core strands to the outer sheath. By fusing the two together, the sheath cannot slide backward or forward. Ensure that this is only done to the ends of the rope. If the rope is exposed to extreme temperatures, the sheath could be weakened, along with the inner core, reducing overall tensile strength. The ends may also be dipped in enamel or lacquer for further protection.

4-37. CARE AND MAINTENANCE

The rope is a climber's lifeline. It must be cared for and used properly. These general guidelines should be used when handling ropes.

a. Do not step on or drag ropes on the ground unnecessarily. Small particles of dirt will be ground between the inner strands and will slowly cut them.

b. While in use, do not allow the rope to come into contact with sharp edges. Nylon rope is easily cut, particularly when under tension. If the rope must be used over a sharp edge, pad the edge for protection.

c. Always keep the rope as dry as possible. Should the rope become wet, hang it in large loops off the ground and allow it to dry. Never dry a rope with high heat or in direct sunlight.

d. Never leave a rope knotted or tightly stretched for longer than necessary. Over time it will reduce the strength and life of the rope.

e. Never allow one rope to continuously rub over or against another. Allowing rope-on-rope contact with nylon rope is extremely dangerous because the heat produced by the friction will cause the nylon to melt.

f. Inspect the rope before each use for frayed or cut spots, mildew or rot, or defects in construction (new rope).
g. The ends of the rope should be whipped or melted to prevent unraveling.
h. Do not splice ropes for use in mountaineering.
i. Do not mark ropes with paints or allow them to come in contact with oils or petroleum products. Some of these will weaken or deteriorate nylon.
j. Never use a mountaineering rope for any purpose except mountaineering.
k. Each rope should have a corresponding rope log, which is also a safety record. It should annotate use, terrain, weather, application, number of falls, dates, and so on, and should be annotated each time the rope is used (Figure 4-44).

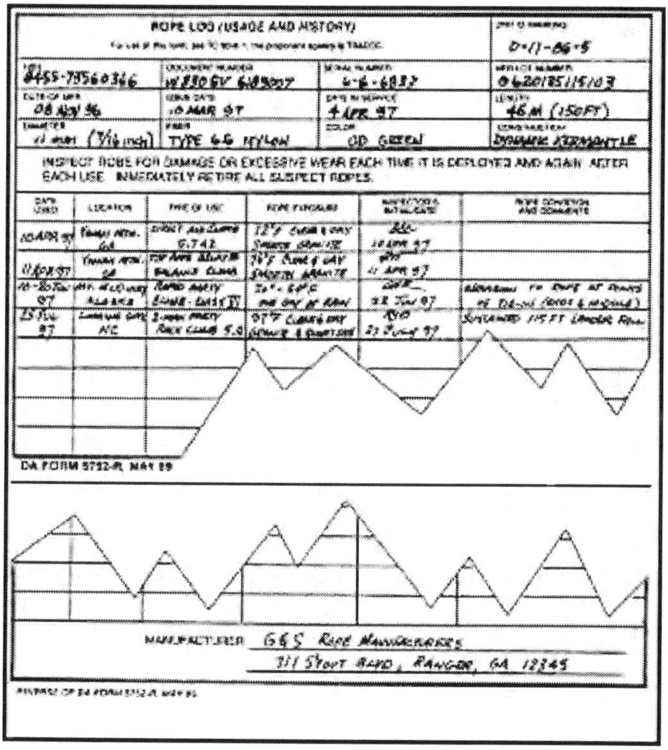

Figure 4-44: Example of completed DA Form 5752-R.

l. Never subject the rope to high heat or flame. This will significantly weaken it.
m. All ropes should be washed periodically to remove dirt and grit, and rinsed thoroughly. Commercial rope washers are made from short pieces of modified pipe that connect to any faucet. Pinholes within the pipe force water to circulate around and scrub the rope as you slowly feed it through the washer. Another method is to machine wash, on a gentle cycle, in cold water with a nylon safe soap, never bleach or harsh cleansers. Ensure that only front loading washing machine are used to wash ropes.
n. Ultraviolet radiation (sunlight) tends to deteriorate nylon over long periods of time. This becomes important if rope installations are left in place over a number of months.
o. When not in use, ropes should be loosely coiled and hung on wooden pegs rather than nails or other metal objects. Storage areas should be relatively cool with low humidity levels to prevent mildew or rotting. Rope may also be loosely stacked and placed in a rope bag and stored on a shelf. Avoid storage in direct sunlight, as the ultraviolet radiation will deteriorate the nylon over long periods

4-38. INSPECTION

Ropes should be inspected before and after each use, especially when working around loose rock or sharp edges.

a. Although the core of the kernmantle rope cannot be seen, it is possible to damage the core without damaging the sheath. Check a kernmantle rope by carefully inspecting the sheath before and after use while the rope is being coiled. When coiling, be aware of how the rope feels as it runs through the hands. Immediately note and tie off any lumps or depressions felt.
b. Damage to the core of a kernmantle rope usually consists of filaments or yarn breakage that results in a slight retraction. If enough strands rupture, a localized reduction in the diameter of the rope results in a depression that can be felt or even seen.
c. Check any other suspected areas further by putting them under tension (the weight of one person standing on a Prusik tensioning system is about maximum). This procedure will

emphasize the lump or depression by separating the broken strands and enlarging the dip. If a noticeable difference in diameter is obvious, retire the rope immediately.

d. Many dynamic kernmantle ropes are quite soft. They may retain an indention occasionally after an impact or under normal use without any trauma to the core. When damage is suspected, patiently inspect the sheath for abnormalities. Damage to the sheath does not always mean damage to the core. Inspect carefully.

4-39. TERMINOLOGY

When using ropes, understanding basic terminology is important. The terms explained in this section are the most commonly used in military mountaineering. (Figure 4-45 illustrates some of these terms.)

a. **Bight.** A bight of rope is a simple bend of rope in which the rope does not cross itself.
b. **Loop.** A loop is a bend of a rope in which the rope does cross itself.
c. **Half Hitch.** A half hitch is a loop that runs around an object in such a manner as to lock or secure itself.
d. **Turn.** A turn wraps around an object, providing 360-degree contact.
e. **Round Turn.** A round turn wraps around an object one and one-half times. A round turn is used to distribute the load over a small diameter anchor (3 inches or less). It may also be used around larger diameter anchors to reduce the tension on the knot, or provide added friction.
f. **Running End.** A running end is the loose or working end of the rope.
g. **Standing Part.** The standing part is the static, stationary, or nonworking end of the rope.
h. **Lay.** The lay is the direction of twist used in construction of the rope.
i. **Pigtail.** The pigtail (tail) is the portion of the running end of the rope between the safety knot and the end of the rope.
j. **Dress.** Dress is the proper arrangement of all the knot parts, removing unnecessary kinks, twists, and slack so that all rope parts of the knot make contact.

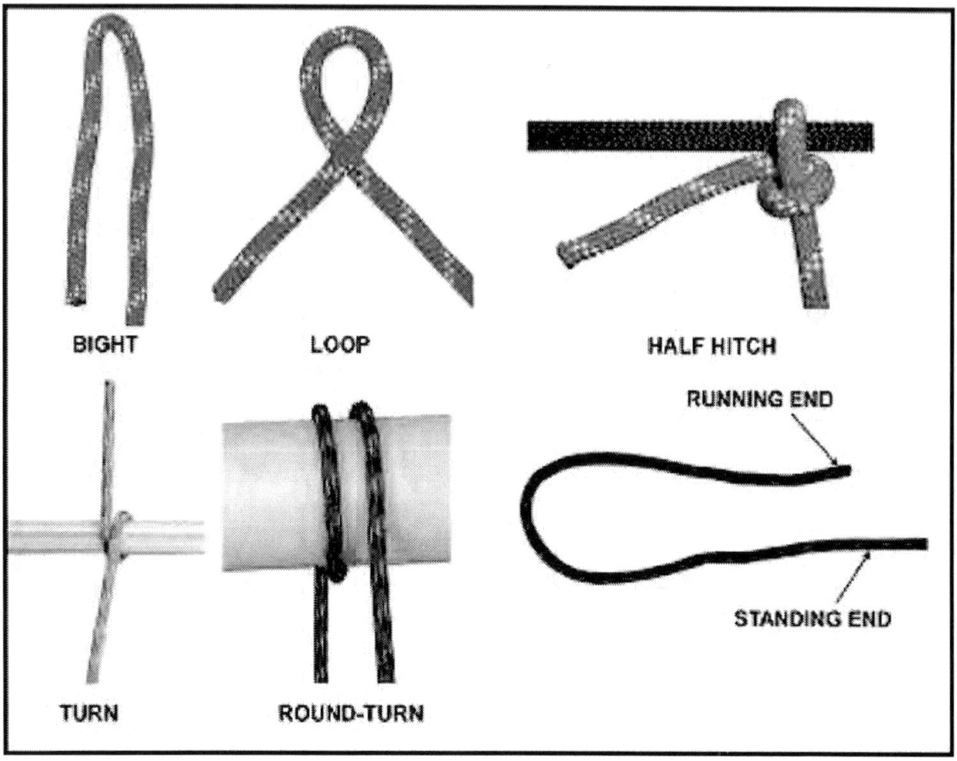

Figure 4-45: Examples of roping terminology.

COILING, CARRYING, THROWING

The ease and speed of rope deployment and recovery greatly depends upon technique and practice.

4-40. COILING AND CARRYING THE ROPE

Use the butterfly or mountain coil to coil and carry the rope. Each is easy to accomplish and results in a minimum amount of kinks, twists, and knots later during deployment.

a. **Mountain Coil**. To start a mountain coil, grasp the rope approximately 1 meter from the end with one hand. Run the other hand along the rope until both arms are outstretched. Grasping the rope firmly, bring the hands together forming a loop, which is laid in the hand closest to the end of the rope. This is repeated, forming uniform loops that run in a clockwise direction, until the rope is completely coiled. The rope may be given a 1/4 twist as each loop is formed to overcome any tendency for the rope to twist or form figure-eights.

(1) In finishing the mountain coil, form a bight approximately 30 centimeters long with the starting end of the rope and lay it along the top of the coil. Uncoil the last loop and, using this length of the rope, begin making wraps around the coil and the bight, wrapping toward the closed end of the bight and making the first wrap bind across itself so as to lock it into place. Make six to eight wraps to adequately secure the coil, and then route the end of the rope through the closed end of the bight. Pull the running end of the bight tight, securing the coil.

(2) The mountain coil may be carried either in the pack (by forming a figure eight), doubling it and placing it under the flap, or by placing it over the shoulder and under the opposite arm, slung across the chest. (Figure 4-46 shows how to coil a mountain coil.)

Figure 4-46: Mountain coil.

b. **Butterfly Coil.** The butterfly coil is the quickest and easiest technique for coiling (Figure 4-47).

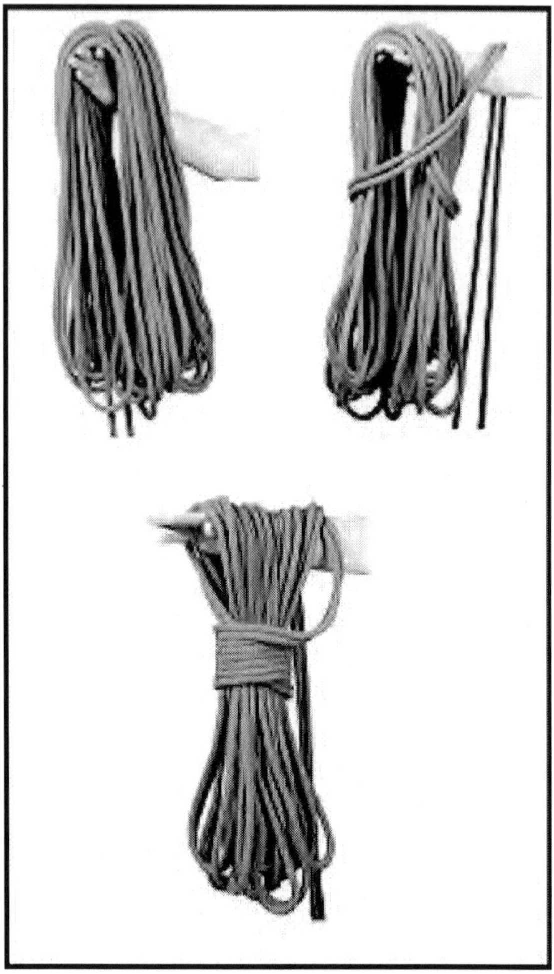

Figure 4-47: Butterfly coil.

(1) Coiling. To start the double butterfly, grasp both ends of the rope and begin back feeding. Find the center of the rope forming a bight. With the bight in the left hand, grasp both ropes and slide the right hand out until there is approximately one arm's length of rope. Place the doubled rope over the head, draping it around the neck and on top of the shoulders. Ensure that it hangs no lower than the waist. With the rest of the doubled rope in front of you, make doubled bights placing them over the head in the same manner as the first bight. Coil alternating

from side to side (left to right, right to left) while maintaining equal-length bights. Continue coiling until approximately two arm-lengths of rope remain. Remove the coils from the neck and shoulders carefully, and hold the center in one hand. Wrap the two ends around the coils a minimum of three doubled wraps, ensuring that the first wrap locks back on itself.

(2) Tie-off and Carrying. Take a doubled bight from the loose ends of rope and pass it through the apex of the coils. Pull the loose ends through the doubled bight and dress it down. Place an overhand knot in the loose ends, dressing it down to the apex of the bight securing coils. Ensure that the loose ends do not exceed the length of the coils. In this configuration the coiled rope is secure enough for hand carrying or carrying in a rucksack, or for storage. (Figure 4-48 shows a butterfly coil tie-off.)

Figure 4-48: Butterfly coil tie-off.

c. **Coiling Smaller Diameter Rope.** Ropes of smaller diameters may be coiled using the butterfly or mountain coil depending on the length of the rope. Pieces 25 feet and shorter (also known as cordage, sling rope, utility cord) may be coiled so that they can be hung from the harness. Bring the two ends of the rope together, ensuring no kinks are in the rope. Place the ends of the rope in the left hand with the two ends facing the body. Coil the doubled rope in a clockwise direction forming 6- to 8-inch coils (coils may be larger depending on the length of rope) until an approximate 12-inch bight is left. Wrap that bight around the coil, ensuring that the first wrap locks on itself. Make three or more wraps. Feed the bight up through the bights formed at the top of the coil. Dress it down tightly. Now the piece of rope may be hung from a carabiner on the harness.

d. **Uncoiling, Back-feeding, and Stacking.** When the rope is needed for use, it must be uncoiled and stacked on the ground properly to avoid kinks and snarls.

(1) Untie the tie-off and lay the coil on the ground. Back-feed the rope to minimize kinks and snarls. (This is also useful when the rope is to be moved a short distance and coiling is not desired.) Take one end of the rope in the left hand and run the right hand along the rope until both arms are outstretched. Next, lay the end of the rope in the left hand on the ground. With the left hand, re-grasp the rope next to the right hand and continue laying the rope on the ground.

(2) The rope should be laid or stacked in a neat pile on the ground to prevent it from becoming tangled and knotted when throwing the rope, feeding it to a lead climber, and so on. This technique can also be started using the right hand.

4-41. THROWING THE ROPE

Before throwing the rope, it must be properly managed to prevent it from tangling during deployment. The rope should first be anchored to prevent complete loss of the rope over the edge when it is thrown. Several techniques can be used when throwing a rope. Personal preference and situational and environmental conditions should be taken into consideration when determining which technique is best.

a. Back feed and neatly stack the rope into coils beginning with the anchored end of the rope working toward the running end. Once stacked, make six to eight smaller coils in the left hand. Pickup the rest of the larger coils in the right hand. The arm should be generally straight when throwing. The rope may be thrown underhanded or overhanded depending on obstacles around the edge of the site. Make a few preliminary swings to ensure a smooth throw. Throw the large coils in the right hand first. Throw up and out. A slight twist of the wrist, so that the palm of the hand faces up as the rope is thrown, allows the coils to separate easily without tangling. A smooth follow through is essential. When a slight tug on the left hand is felt, toss the six to eight smaller coils out. This will prevent the ends of the rope from becoming entangled with the rest of the coils as they deploy. As soon as the rope leaves the hand, the thrower should sound off with a warning of "ROPE" to alert anyone below the site.
b. Another technique may also be used when throwing rope. Anchor, back feed, and stack the rope properly as described above. Take the end of the rope and make six to eight helmet-size coils in the right hand (more may be needed depending on the length of the rope). Assume a "quarterback" simulated stance. Aiming just above the horizon, vigorously throw the rope overhanded, up and out toward the horizon. The rope must be stacked properly to ensure smooth deployment.
c. When windy weather conditions prevail, adjustments must be made. In a strong cross wind, the rope should be thrown angled into the wind so that it will land on the desired target. The stronger the wind, the harder the rope must be thrown to compensate.

KNOTS

All knots used by a mountaineer are divided into four classes: Class I joining knots, Class II anchor knots, Class III middle rope knots, and Class IV special knots. The variety of knots, bends, bights, and hitches is almost endless. These classes of knots are intended only as a general guide since some of the knots discussed may be appropriate in more than one class. The skill of knot tying can perish if not used and practiced. With experience and practice, knot tying becomes instinctive and helps the mountaineer in many situations.

4-49. SQUARE KNOT

The square knot is used to tie the ends of two ropes of equal diameter (Figure 4-49). It is a joining knot.

a. **Tying the Knot.**

STEP 1. Holding one working end in each hand, place the working end in the right hand over the one in the left hand.

STEP 2. Pull it under and back over the top of the rope in the left hand.

STEP 3. Place the working end in the left hand over the one in the right hand and repeat STEP 2.

STEP 4. Dress the knot down and secure it with an overhand knot on each side of the square knot.

Figure 4-49 Square knot.

b. Checkpoints.

(1) There are two interlocking bights.

(2) The running end and standing part are on the same side of the bight formed by the other rope.

(3) The running ends are parallel to and on the same side of the standing ends with 4-inch minimum pig tails after the overhand safeties are tied.

4-50. FISHERMAN'S KNOT

The fisherman's knot is used to tie two ropes of the same or approximately the same diameter (Figure 4-50). It is a joining knot.

a. Tying the Knot.

STEP 1. Tie an overhand knot in one end of the rope.

STEP 2. Pass the working end of the other rope through the first overhand knot. Tie an overhand knot around the standing part of the first rope with the working end of the second rope.

STEP 3. Tightly dress down each overhand knot and tightly draw the knots together.

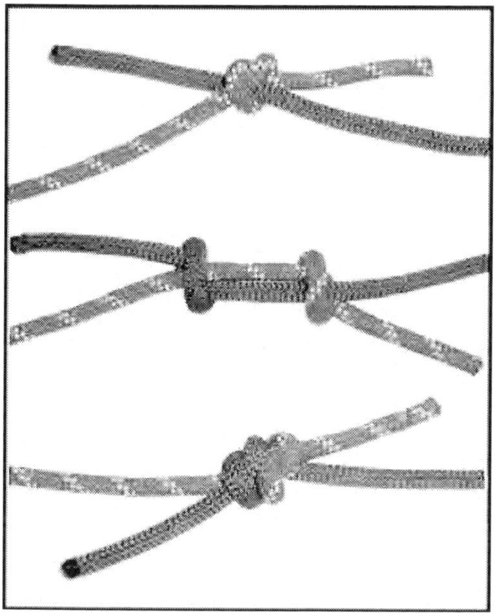

Figure 4-50: Fisherman's knot.

b. Checkpoints.

(1) The two separate overhand knots are tied tightly around the long, standing part of the opposing rope.

(2) The two overhand knots are drawn snug.

(3) Ends of rope exit knot opposite each other with 4-inch pigtails.

4-51. DOUBLE FISHERMAN'S KNOT

The double fisherman's knot (also called double English or grapevine) is used to tie two ropes of the same or approximately the same diameter (Figure 4-51). It is a joining knot.

a. **Tying the Knot.**
STEP 1. With the working end of one rope, tie two wraps around the standing part of another rope.
STEP 2. Insert the working end (STEP 1) back through the two wraps and draw it tight.
STEP 3. With the working end of the other rope, which contains the standing part (STEPS 1 and 2), tie two wraps around the standing part of the other rope (the working end in STEP 1). Insert the working end back through the two wraps and draw tight.
STEP 4. Pull on the opposing ends to bring the two knots together.

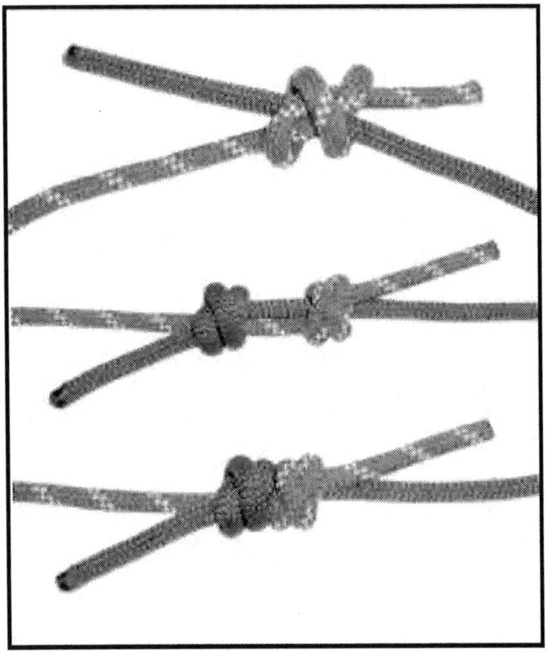

Figure 4-51: Double fisherman's knot.

b. **Checkpoints.**
(1) Two double overhand knots securing each other as the standing parts of the rope are pulled apart.
(2) Four rope parts on one side of the knot form two "x" patterns, four rope parts on the other side of the knot are parallel.
(3) Ends of rope exit knot opposite each other with 4-inch pigtails.

4-52. FIGURE-EIGHT BEND

The figure-eight bend is used to join the ends of two ropes of equal or unequal diameter within 5-mm difference (Figure 4-52).

a. **Tying the Knot.**

STEP 1. Grasp the top of a 2-foot bight.

STEP 2. With the other hand, grasp the running end (short end) and make a 360-degree turnaround the standing end.

STEP 3. Place the running end through the loop just formed creating an in-line figure eight.

STEP 4. Route the running end of the other ripe back through the figure eight starting from the original rope's running end. Trace the original knot to the standing end.

STEP 5. Remove all unnecessary twists and crossovers. Dress the knot down.

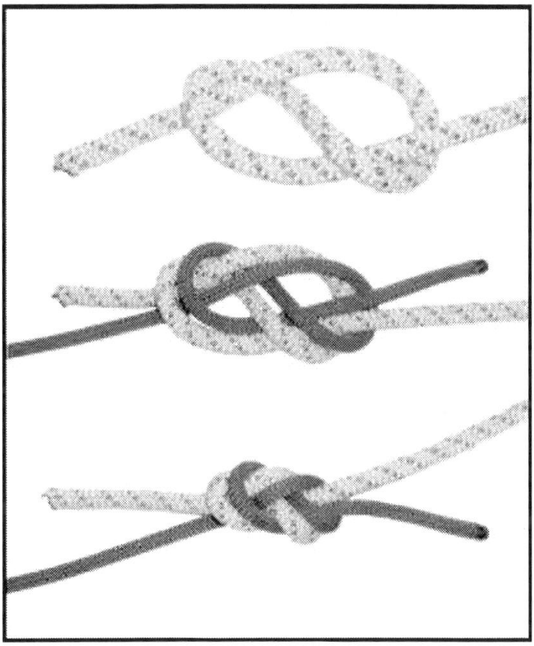

Figure 4-52: Figure-eight bend.

b. **Checkpoints.**

(1) There is a figure eight with two ropes running side by side.

(2) The running ends are on opposite sides of the knot.

(3) There is a minimum 4-inch pigtail.

4-53. WATER KNOT

The water knot is used to attach two webbing ends (Figure 4-53). It is also called a ring bend, overhand retrace, or tape knot. It is used in runners and harnesses and is a joining knot.

a. **Tying the Knot.**

STEP 1. Tie an overhand knot in one of the ends.

STEP 2. Feed the other end back through the knot, following the path of the first rope in reverse.

STEP 3. Draw tight and pull all of the slack out of the knot. The remaining tails must extend at least 4 inches beyond the knot in both directions.

Figure 4-53: Water knot.

b. **Checkpoints.**

(1) There are two overhand knots, one retracing the other.

(2) There is no slack in the knot, and the working ends come out of the knot in opposite directions.

(3) There is a minimum 4-inch pigtail.

4-54 BOWLINE

The bowline is used to tie the end of a rope around an anchor. It may also be used to tie a single fixed loop in the end of a rope (Figure 4-54). It is an anchor knot.

a. **Tying the Knot.**

STEP 1. Bring the working end of the rope around the anchor, from right to left (as the climber faces the anchor).

STEP 2. Form an overhand loop in the standing part of the rope (on the climber's right) toward the anchor.

STEP 3. Reach through the loop and pull up a bight.

STEP 4. Place the working end of the rope (on the climber's left) through the bight, and bring it back onto itself. Now dress the knot down.

STEP 5. Form an overhand knot with the tail from the bight.

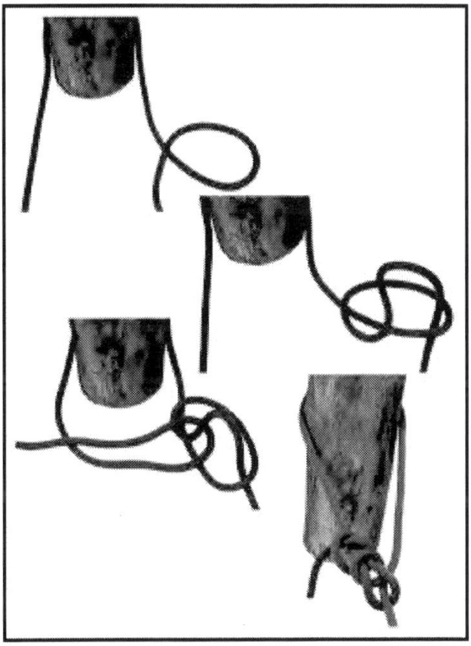

Figure 4-54: Bowline knot.

b. **Checkpoints.**

(1) The bight is locked into place by a loop.

(2) The short portion of the bight is on the inside and on the loop around the anchor (or inside the fixed loop).

(3) There is a minimum 4-inch pigtail after tying the overhand safety.

4-55. ROUND TURN AND TWO HALF HITCHES

This knot is used to tie the end of a rope to an anchor, and it must have constant tension (Figure 4-55). It is an anchor knot.

a. **Tying the Knot.**

STEP 1. Route the rope around the anchor from right to left and wrap down (must have two wraps in the rear of the anchor, and one in the front). Run the loop around the object to provide 360-degree contact, distributing the load over the anchor.
STEP 2. Bring the working end of the rope left to right and over the standing part, forming a half hitch (first half hitch).
STEP 3. Repeat STEP 2 (last half hitch has a 4 inch pigtail).
STEP 4. Dress the knot down.

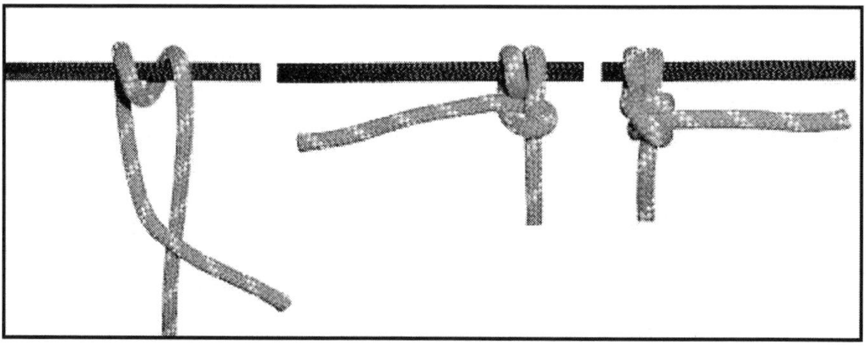

Figure 4-55: Round turn and two half hitches.

b. **Checkpoints.**

(1) A complete round turn should exist around the anchor with no crosses.
(2) Two half hitches should be held in place by a diagonal locking bar with no less than a 4-inch pigtail remaining.

4-56. FIGURE-EIGHT RETRACE (REROUTED FIGURE-EIGHT)

The figure-eight retrace knot produces the same result as a figure-eight loop. However, by tying the knot in a retrace, it can be used to fasten the rope to trees or to places where the loop cannot be used (Figure 4-56). It is also called a rerouted figure-eight and is an anchor knot.

a. **Tying the Knot.**

STEP 1. Use a length of rope long enough to go around the anchor, leaving enough rope to work with.

STEP 2. Tie a figure-eight knot in the standing part of the rope, leaving enough rope to go around the anchor. To tie a figure-eight knot form a loop in the rope, wrap the working end around the standing part, and route the working end through the loop. The finished knot is dressed loosely.
STEP 3. Take the working end around the anchor point.
STEP 4. With the working end, insert the rope back through the loop of the knot in reverse.
STEP 5. Keep the original figure eight as the outside rope and retrace the knot around the wrap and back to the long-standing part.
STEP 6. Remove all unnecessary twists and crossovers; dress the knot down.

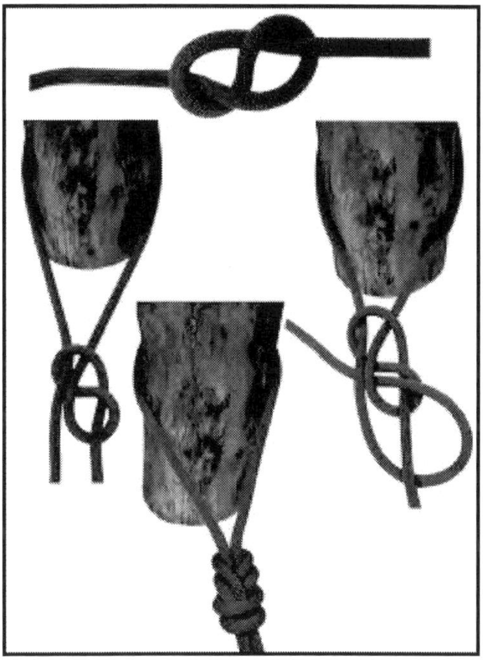

Figure 4-56. Figure-eight retrace.

b. **Checkpoints**
   (1) A figure eight with a doubled rope running side by side, forming a fixed loop around a fixed object or harness.
   (2) There is a minimum 4-inch pigtail.

4-57. CLOVE HITCH

The clove hitch is an anchor knot that can be used in the middle of the rope as well as at the end (Figure 4-57). The knot must have constant tension on it once tied to prevent slipping. It can be used as either an anchor or middle of the rope knot, depending on how it is tied.

a. **Tying the Knot.**

(1) ***Middle of the Rope.***

STEP 1. Hold rope in both hands, palms down with hands together. Slide the left hand to the left from 20 to 25 centimeters.

STEP 2. Form a loop away from and back toward the right.

STEP 3. Slide the right hand from 20 to 25 centimeters to the right. Form a loop inward and back to the left hand.

STEP 4. Place the left loop on top of the right loop. Place both loops over the anchor and pull both ends of the rope in opposite directions. The knot is tied.

> ✍ **NOTE**
>
> For instructional purposes, assume that the anchor is horizontal.

(2) **End of the Rope.**

STEP 1. Place 76 centimeters of rope over the top of the anchor. Hold the standing end in the left hand. With the right hand, reach under the horizontal anchor, grasp the working end, and bring it inward.

STEP 2. Place the working end of the rope over the standing end (to form a loop). Hold the loop in the left hand. Place the working end over the anchor from 20 to 25 centimeters to the left of the loop.

STEP 3. With the right hand, reach down to the left hand side of the loop under the anchor. Grasp the working end of the rope. Bring the working end up and outward.

STEP 4. Dress down the knot.

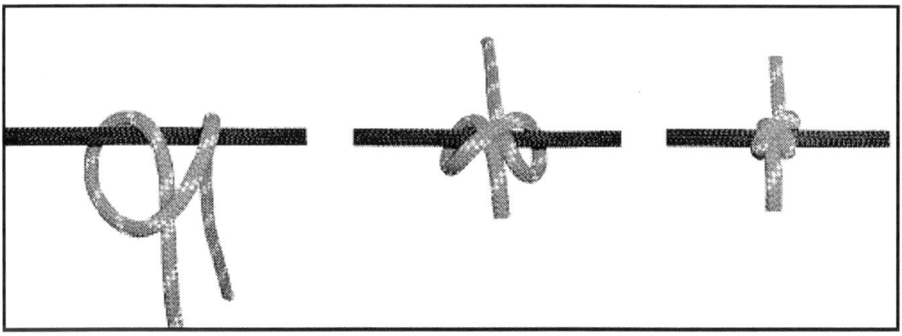

Figure 4-57: Clove hitch.

b. **Checkpoints.**
   (1) The knot has two round turns around the anchor with a diagonal locking bar.
   (2) The locking bar is facing 90 degrees from the direction of pull.
   (3) The ends exit 180 degrees from each other.
   (4) The knot has more than a 4-inch pigtail remaining.

4-58. WIREMAN'S KNOT

The wireman's knot forms a single, fixed loop in the middle of the rope (Figure 4-58). It is a middle rope knot.

a. **Tying the Knot.**

STEP 1. When tying this knot, face the anchor that the tie-off system will be tied to. Take up the slack from the anchor, and wrap two turns around the left hand (palm up) from left to right.

STEP 2. A loop of 30 centimeters is taken up in the second round turn to create the fixed loop of the knot.

STEP 3. Name the wraps from the palm to the fingertips: heel, palm, and fingertip.

STEP 4. Secure the palm wrap with the right thumb and forefinger, and place it over the heel wrap.

STEP 5. Secure the heel wrap and place it over the fingertip wrap.

STEP 6. Secure the fingertip wrap and place it over the palm wrap.

STEP 7. Secure the palm wrap and pull up to form a fixed loop.

STEP 8. Dress the knot down by pulling on the fixed loop and the two working ends.
STEP 9. Pull the working ends apart to finish the knot.

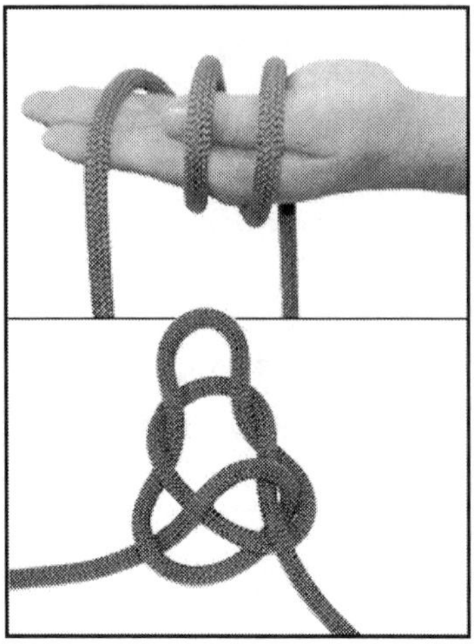

Figure 4-58: Wireman's knot.

b. **Checkpoints.**
   (1) The completed knot should have four separate bights locking down on themselves with the fixed loop exiting from the top of the knot and laying toward the near side anchor point.
   (2) Both ends should exit opposite each other without any bends.

4-59. DIRECTIONAL FIGURE-EIGHT

The directional figure-eight knot forms a single, fixed loop in the middle of the rope that lays back along the standing part of the rope (Figure 4-59). It is a middle rope knot.

a. **Tying the Knot.**
   STEP 1. Face the far side anchor so that when the knot is tied, it lays inward.
   STEP 2. Lay the rope from the far side anchor over the left palm. Make one wrap around the palm.
   STEP 3. With the wrap thus formed, tie a figure-eight knot around the standing part that leads to the far side anchor.
   STEP 4. When dressing the knot down, the tail and the bight must be together.

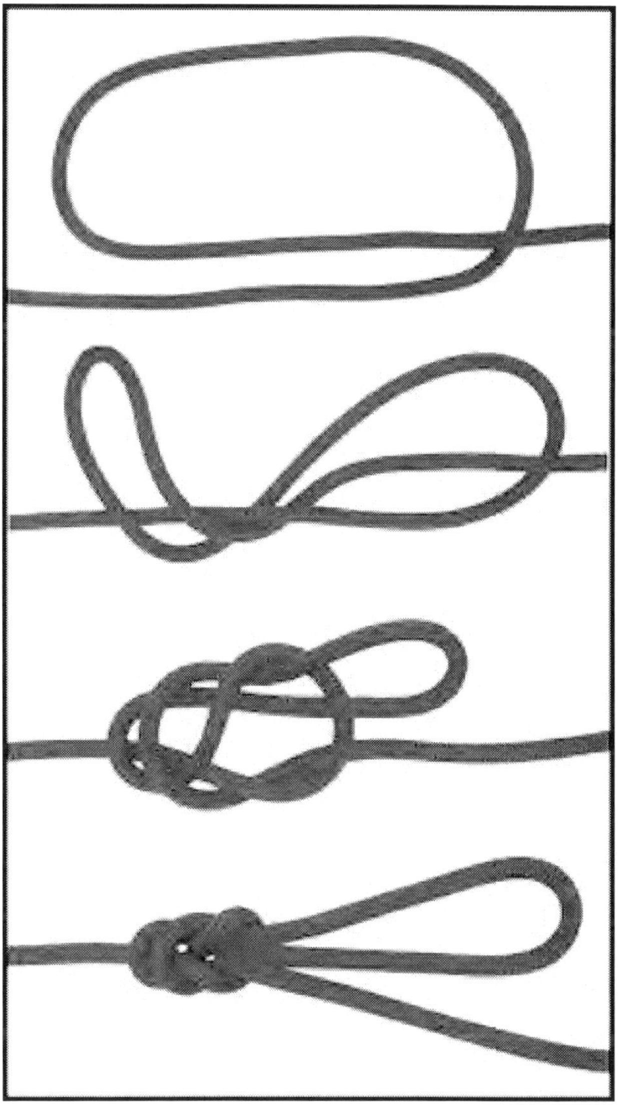

Figure 4-59. Directional figure-eight.

b. **Checkpoints.**

(1) The loop should be large enough to accept a carabiner but no larger than a helmet-size loop.

(2) The tail and bight must be together.

(3) The figure eight is tied tightly.

(4) The bight in the knot faces back toward the near side.

4-60. BOWLINE-ON-A-BIGHT (TWO-LOOP BOWLINE)

The bowline-on-a-bight is used to form two fixed loops in the middle of a rope (Figure 4-60). It is a middle rope knot.

a. **Tying the Knot.**

STEP 1. Form a bight in the rope about twice as long as the finished loops will be.

STEP 2. Tie an overhand knot on a bight.

STEP 3. Hold the overhand knot in the left hand so that the bight is running down and outward.

STEP 4. Grasp the bight with the right hand; fold it back over the overhand knot so that the overhand knot goes through the bight.

STEP 5. From the end (apex) of the bight, follow the bight back to where it forms the cross in the overhand knot. Grasp the two ropes that run down and outward and pull up, forming two loops.

STEP 6. Pull the two ropes out of the overhand knot and dress the knot down.

STEP 7. A final dress is required: grasp the ends of the two fixed loops and pull, spreading them apart to ensure the loops do not slip.

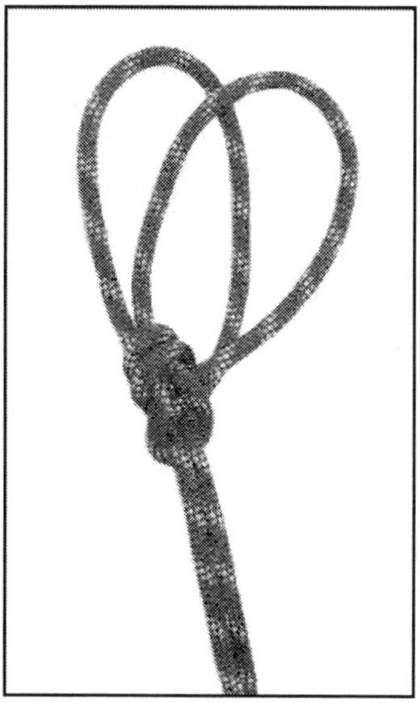

Figure 4-60: Bowline-on-a-bight.

b. **Checkpoints.**
   (1) There are two fixed loops that will not slip.
   (2) There are no twists in the knot.
   (3) A double loop is held in place by a bight.

4-61. TWO-LOOP FIGURE-EIGHT

The two-loop figure-eight is used to form two fixed loops in the middle of a rope (Figure 4-61). It is a middle rope knot.

a. **Tying the Knot.**
   STEP 1. Using a doubled rope, form an 18-inch bight in the left hand with the running end facing to the left.
   STEP 2. Grasp the bight with the right hand and make a 360-degree turn around the standing end in a counterclockwise direction.
   STEP 3. With the working end, form another bight and place that bight through the loop just formed in the left hand.
   STEP 4. Hold the bight with the left hand, and place the original bight (moving toward the left hand) over the knot.
   STEP 5. Dress the knot down.

Figure 4-61: Two-loop figure-eight.

b. **Checkpoints.**
   (1) There is a double figure-eight knot with two loops that share a common locking bar.
   (2) The two loops must be adjustable by means of a common locking bar.
   (3) The common locking bar is on the bottom of the double figure-eight knot.

4-62. FIGURE-EIGHT LOOP (FIGURE-EIGHT-ON-A-BIGHT)

The figure-eight loop, also called the figure-eight-on-a-bight, is used to form a fixed loop in a rope (Figure 4-62). It is a middle of the rope knot.

a. **Tying the Knot.**

STEP 1. Form a bight in the rope about as large as the diameter of the desired loop.

STEP 2. With the bight as the working end, form a loop in rope (standing part).

STEP 3. Wrap the working end around the standing part 360 degrees and feed the working end through the loop. Dress the knot tightly.

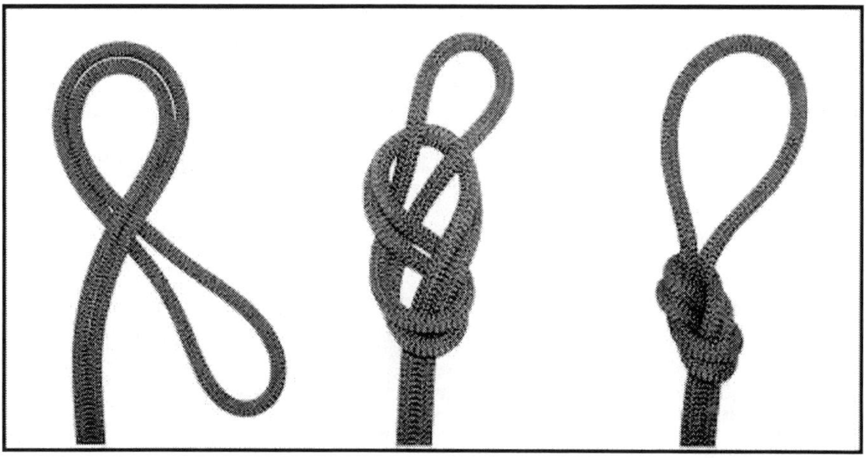

Figure 4-62: Figure-eight loop.

b. **Checkpoints.**

(1) The loop is the desired size.

(2) The ropes in the loop are parallel and do not cross over each other.

(3) The knot is tightly dressed.

4-63. PRUSIK KNOT

The Prusik knot is used to put a moveable rope on a fixed rope such as a Prusik ascent or a tightening system. This knot can be tied as a middle or end of the rope Prusik. It is a specialty knot.

a. **Tying the Knot.**

(1) ***Middle-of-the-Rope Prusik.*** The 4-63.):

STEP 1. Double the short rope, forming a bight, with the working ends even. Lay it over the long rope so that the closed end of the bight is 12 inches below the long rope and the remaining part of the rope (working ends) is the closest to the climber; spread the working end apart.

STEP 2. Reach down through the 12-inch bight. Pull up both of the working ends and lay them over the long rope. Repeat this process making sure that the working ends pass in the middle of the first two wraps. Now there are four wraps and a locking bar working across them on the long rope.

STEP 3. Dress the wraps and locking bar down to ensure they are tight and not twisted. Tying an overhand knot with both ropes will prevent the knot from slipping during periods of variable tension.

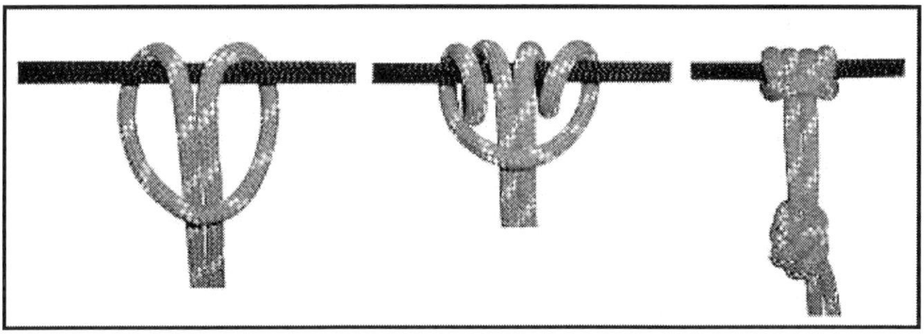

Figure 4-63: Middle-of-the-rope Prusik.

(2) ***End-of-the-Rope Prusik*** (Figure 4-64).

STEP 1. Using an arm's length of rope, and place it over the long rope.

STEP 2. Form a complete round turn in the rope.

STEP 3. Cross over the standing part of the short rope with the working end of the short rope.

STEP 4. Lay the working end under the long rope.

STEP 5. Form a complete round turn in the rope, working back toward the middle of the knot.

STEP 6. There are four wraps and a locking bar running across them on the long rope. Dress the wraps and locking bar down. Ensure they are tight, parallel, and not twisted.

STEP 7. Finish the knot with a bowline to ensure that the Prusik knot will not slip out during periods of varying tension.

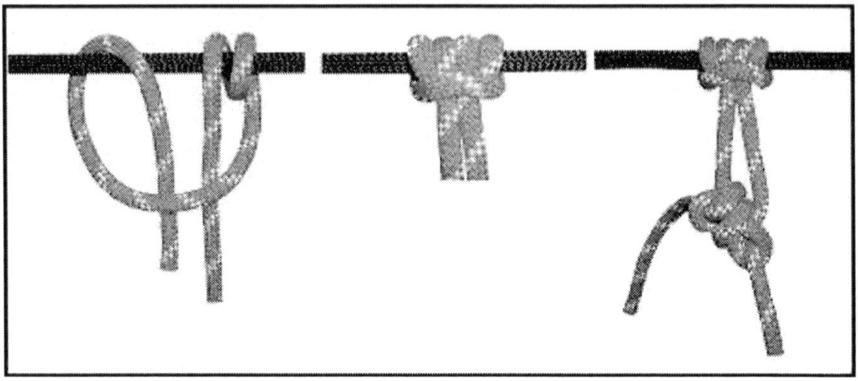

Figure 4-64: End-of-the-rope Prusik knot.

b. **Checkpoints.**
   (1) Four wraps with a locking bar.
   (2) The locking bar faces the climber.
   (3) The knot is tight and dressed down with no ropes twisted or crossed.
   (4) Other than a finger Prusik, the knot should contain an overhand or bowline to prevent slipping.

4-64. BACHMAN KNOT

The Bachman knot provides a means of using a makeshift mechanized ascender (Figure 4-65). It is a specialty knot.

a. **Tying the Knot.**

STEP 1. Find the middle of a utility rope and insert it into a carabiner.

STEP 2. Place the carabiner and utility rope next to a long climbing rope.

STEP 3. With the two ropes parallel from the carabiner, make two or more wraps around the climbing rope and through the inside portion of the carabiner.

> ✍ **NOTE**
> The rope can be tied into an etrier (stirrup) and used as a Prusik-friction principle ascender.

b. **Checkpoints.**
   (1) The bight of the climbing rope is at the top of the carabiner.
   (2) The two ropes run parallel without twisting or crossing.
   (3) Two or more wraps are made around the long climbing rope and through the inside portion of the carabiner.

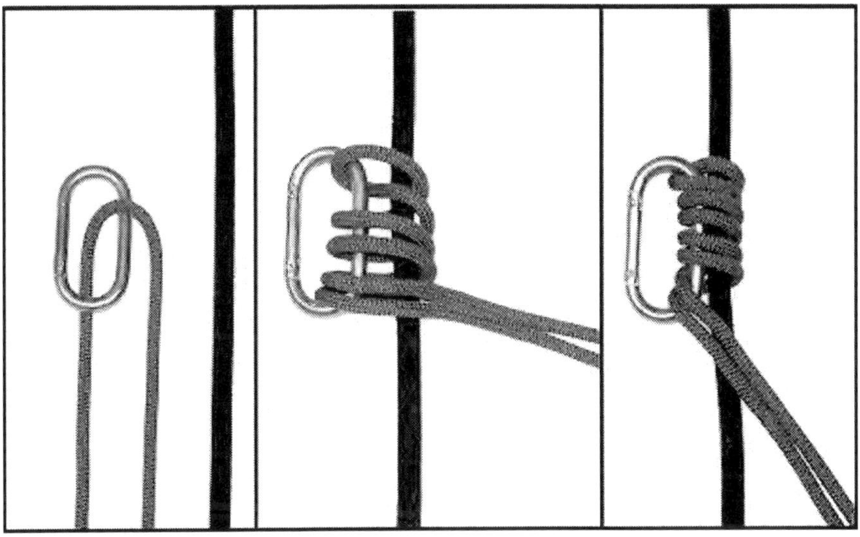

Figure 4-65. Bachman knot.

4-65. BOWLINE-ON-A-COIL

The bowline-on-a-coil is an expedient tie-in used by climbers when a climbing harness is not available (Figure 4-66). It is a specialty knot.

a. **Tying the Knot.**
   STEP 1. With the running end, place 3 feet of rope over your right shoulder. The running end is to the back of the body.
   STEP 2. Starting at the bottom of your rib cage, wrap the standing part of the rope around your body and down in a clockwise direction four to eight times.

STEP 3. With the standing portion of the rope in your left hand, make a clockwise loop toward the body. The standing portion is on the bottom.
STEP 4. Ensuring the loop does not come uncrossed, bring it up and under the coils between the rope and your body.
STEP 5. Using the standing part, bring a bight up through the loop. Grasp the running end of the rope with the right hand. Pass it through the bight from right to left and back on itself.
STEP 6. Holding the bight loosely, dress the knot down by pulling on the standing end.
STEP 7. Safety the bowline with an overhand around the top, single coil. Then, tie an overhand around all coils, leaving a minimum 4-inch pigtail.

b. **Checkpoints.**
   (1) A minimum of four wraps, not crossed, with a bight held in place by a loop.
   (2) The loop must be underneath all wraps.
   (3) A minimum 4-inch pigtail after the second overhand safety is tied.
   (4) Must be centered on the mid-line of the body.

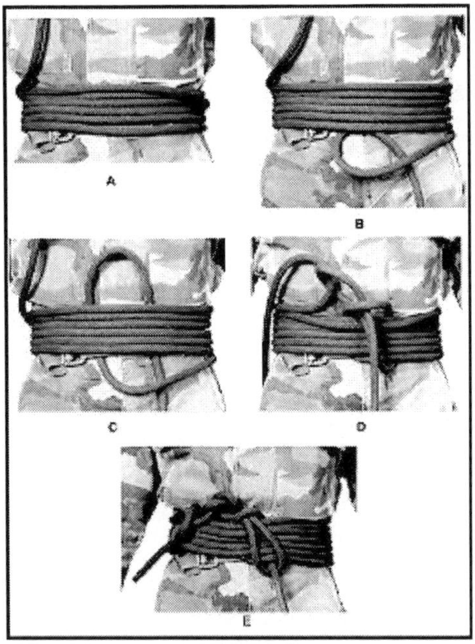

Figure 4-66: Bowline-on-a-coil.

4-66. THREE-LOOP BOWLINE

The three-loop bowline is used to form three fixed loops in the middle of a rope (Figure 4-67). It is used in a self-equalizing anchor system. It is a specialty knot.

a. **Tying the Knot.**

STEP 1. Form an approximate 24-inch bight.

STEP 2. With the right thumb facing toward the body, form a doubled loop in the standing part by turning the wrist clockwise. Lay the loops to the right.

STEP 3. With the right hand, reach down through the loops and pull up a doubled bight from the standing part of the rope.

STEP 4. Place the running end (bight) of the rope (on the left) through the doubled bight from left to right and bring it back on itself. Hold the running end loosely and dress the knot down by pulling on the standing parts.

STEP 5. Safety it off with a doubled overhand knot.

Figure 4-67: Three-loop bowline.

b. **Checkpoints.**
   (1) There are two bights held in place by two loops.
   (2) The bights form locking bars around the standing parts.
   (3) The running end (bight) must be on the inside of the fixed loops.
   (4) There is a minimum 4-inch pigtail after the double overhand safety knot is tied.

4-67. FIGURE-EIGHT SLIP KNOT

The figure eight slip knot forms an adjustable bight in a rope (Figure 4-68). It is a specialty knot.

a. **Tying the Knot.**
   STEP 1. Form a 12-inch bight in the end of the rope.
   STEP 2. Hold the center of the bight in the right hand. Hold the two parallel ropes from the bight in the left hand about 12 inches up the rope.
   STEP 3. With the center of the bight in the right hand, twist two complete turns clockwise.
   STEP 4. Reach through the bight and grasp the long, standing end of the rope. Pull another bight (from the long standing end) back through the original bight.
   STEP 5. Pull down on the short working end of the rope and dress the knot down.
   STEP 6. If the knot is to be used in a transport tightening system, take the working end of the rope and form a half hitch around the loop of the figure eight knot.

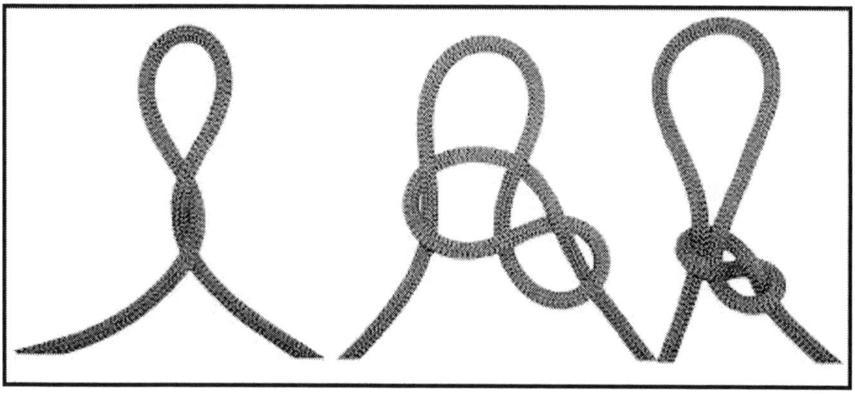

Figure 4-68: Figure-eight slip knot.

b. **Checkpoints.**
   (1) The knot is in the shape of a figure eight.
   (2) Both ropes of the bight pass through the same loop of the figure eight.
   (3) The sliding portion of the rope is the long working end of the rope.

4-68. TRANSPORT KNOT (OVERHAND SLIP KNOT/ MULE KNOT)

The transport knot is used to secure the transport tightening system (Figure 4-69). It is simply an overhand slip knot.

a. **Tying the Knot.**
STEP 1. Pass the running end of the rope around the anchor point passing it back under the standing portion (leading to the far side anchor) forming a loop.
STEP 2. Form a bight with the running end of the rope. Pass over the standing portion and down through the loop and dress it down toward the anchor point.
STEP 3. Secure the knot by tying a half hitch around the standing portion with the bight.

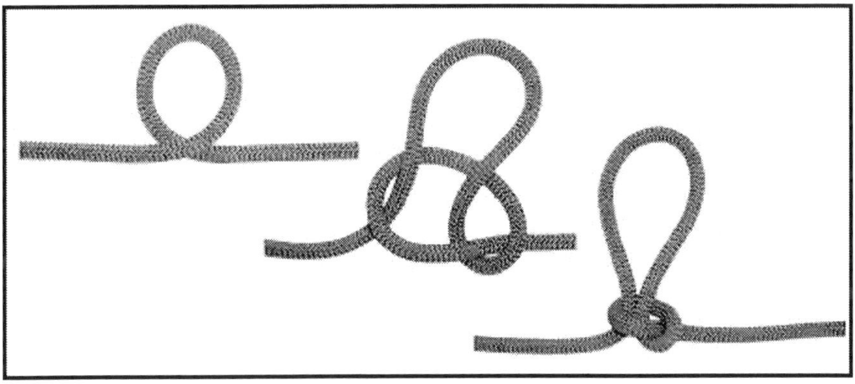

Figure 4-69. Transport knot.

b. **Check Points.**
   (1) There is a single overhand slip knot.
   (2) The knot is secured using a half hitch on a bight.
   (3) The bight is a minimum of 12 inches long.

4-69. KLEIMHIEST KNOT

The Kleimhiest knot provides a moveable, easily adjustable, high-tension knot capable of holding extremely heavy loads while being pulled tight (Figure 4-70). It is a special-purpose knot.

a. **Tying the Knot.**

STEP 1. Using a utility rope or webbing offset the ends by 12 inches. With the ends offset, find the center of the rope and form a bight. Lay the bight over a horizontal rope.

STEP 2. Wrap the tails of the utility rope around the horizontal rope back toward the direction of pull. Wrap at least four complete turns.

STEP 3. With the remaining tails of the utility rope, pass them through the bight (see STEP 1).

STEP 4. Join the two ends of the tail with a joining knot.

STEP 5. Dress the knot down tightly so that all wraps are touching.

> ✍ **NOTE**
>
> Spectra should not be used for the Kleimhiest knot. It has a low melting point and tends to slip.

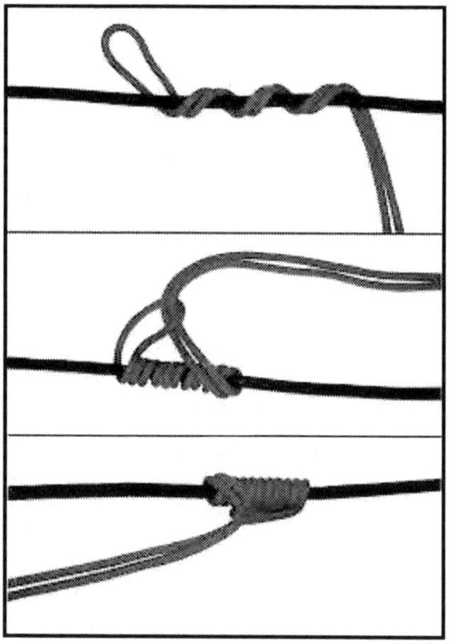

Figure 4-70: Kleimhiest knot.

b. **Checkpoints.**
   (1) The bight is opposite the direction of pull.
   (2) All wraps are tight and touching.
   (3) The ends of the utility rope are properly secured with a joining knot.

4-70. FROST KNOT

The frost knot is used when working with webbing (Figure 4-71). It is used to create the top loop of an etrier. It is a special-purpose knot.

a. **Tying the Knot.**
   STEP 1. Lap one end (a bight) of webbing over the other about 10 to 12 inches.
   STEP 2. Tie an overhand knot with the newly formed triple-strand webbing; dress tightly.

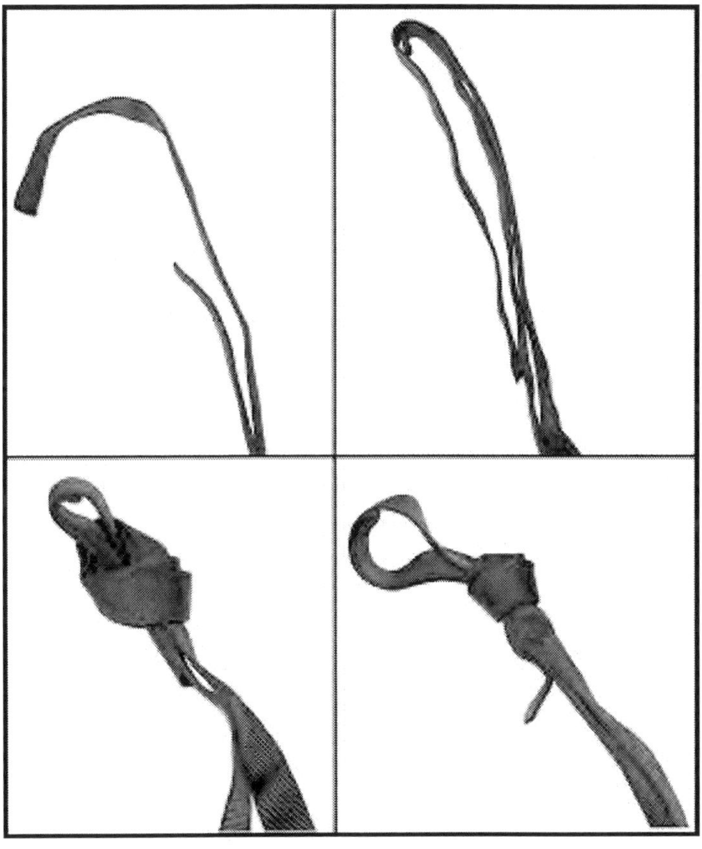

Figure 4-71: Frost knot.

b. **Checkpoints.**
  (1) The tails of the webbing run in opposite directions.
  (2) Three strands of webbing are formed into a tight overhand knot.
  (3) There is a bight and tail exiting the top of the overhand knot.

4-71. GIRTH HITCH

The girth hitch is used to attach a runner to an anchor or piece of equipment (Figure 4-72). It is a special-purpose knot.

a. **Tying the Knot.**
  STEP 1: Form a bight.
  STEP 2: Bring the runner back through the bight.
  STEP 3: Cinch the knot tightly.

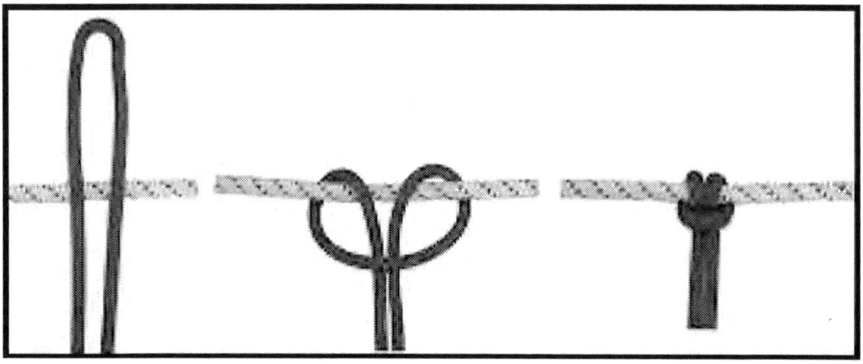

Figure 4-72: Girth hitch.

b. **Checkpoint.**
  (1) Two wraps exist with a locking bar running across the wraps.
  (2) The knot is dressed tightly.

4-72. MUNTER HITCH

The munter hitch, when used in conjunction with a pear-shaped locking carabiner, is used to form a mechanical belay (Figure 4-73).

a. **Tying the Knot.**
  STEP 1. Hold the rope in both hands, palms down about 12 inches apart.

STEP 2. With the right hand, form a loop away from the body toward the left hand. Hold the loop with the left hand.
STEP 3. With the right hand, place the rope that comes from the bottom of the loop over the top of the loop.
STEP 4. Place the bight that has just been formed around the rope into the pear shaped carabiner. Lock the locking mechanism.

b. **Check Points.**
   (1) A bight passes through the carabiner, with the closed end around the standing or running part of the rope.
   (2) The carabiner is locked.

Figure 4-73. Munter hitch.

4-73. RAPPEL SEAT

The rappel seat is an improvised seat rappel harness made of rope (Figure 4-74). It usually requires a sling rope 14 feet or longer.

a. **Tying the Knot.**
STEP 1. Find the middle of the sling rope and make a bight.
STEP 2. Decide which hand will be used as the brake hand and place the bight on the opposite hip.

STEP 3. Reach around behind and grab a single strand of rope. Bring it around the waist to the front and tie two overhands on the other strand of rope, thus creating a loop around the waist.
STEP 4. Pass the two ends between the legs, ensuring they do not cross.
STEP 5. Pass the two ends up under the loop around the waist, bisecting the pocket flaps on the trousers. Pull up on the ropes, tightening the seat.
STEP 6. From rear to front, pass the two ends through the leg loops creating a half hitch on both hips.
STEP 7. Bring the longer of the two ends across the front to the nonbrake hand hip and secure the two ends with a square knot safe tied with overhand knots. Tuck any excess rope in the pocket below the square knot.

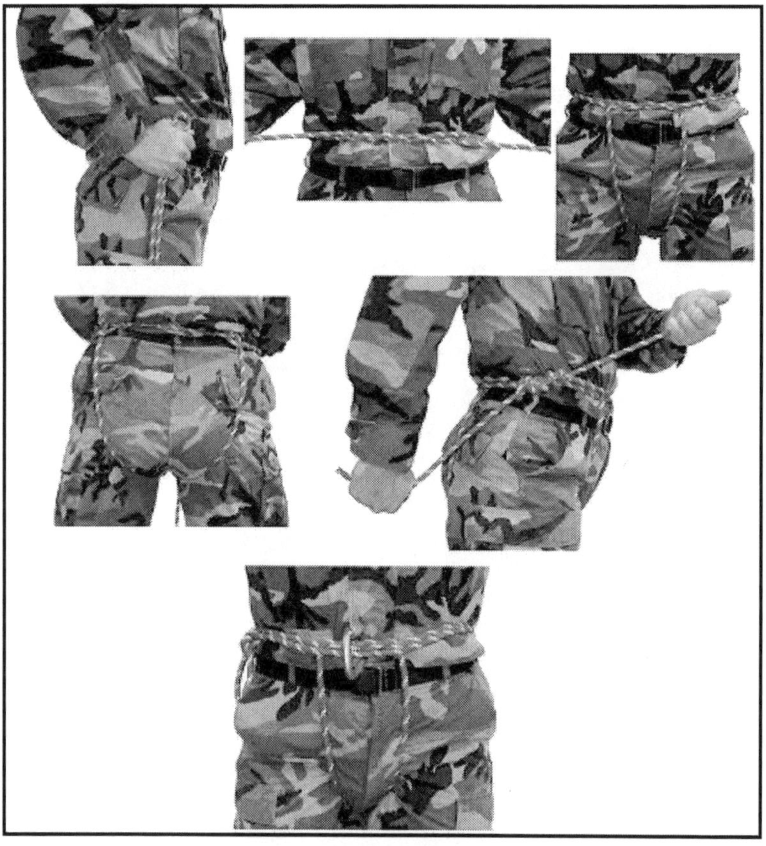

Figure 4-74: Rappel seat.

b. **Check Points.**

(1) There are two overhand knots in the front.

(2) The ropes are not crossed between the legs.

(3) A half hitch is formed on each hip.

(4) Seat is secured with a square knot with overhand safeties on the non-brake hand side.

(5) There is a minimum 4-inch pigtail after the overhand safeties are tied.

4-74. GUARDE KNOT

The guarde knot (ratchet knot, alpine clutch) is a special purpose knot primarily used for hauling systems or rescue (Figure 4-75). The knot works in only one direction and cannot be reversed while under load.

a. **Tying the Knot.**

STEP 1. Place a bight of rope into the two anchored carabiners (works best with two like carabiners, preferably ovals).

STEP 2. Take a loop of rope from the non-load side and place it down into the opposite cararabiner so that the rope comes out between the two carabiners.

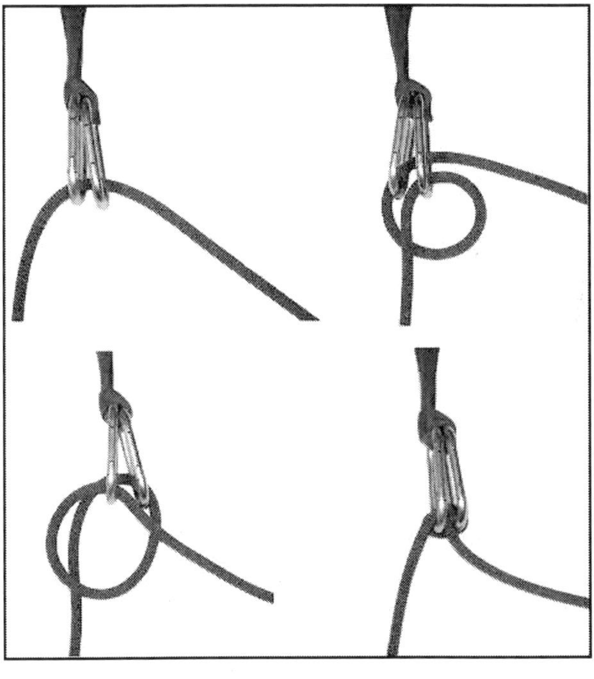

Figure 4-75. Guarde knot.

b. **Check Points.**
   (1) When properly dressed, rope can only be pulled in one direction.
   (2) The knot will not fail when placed under load.

SECTION 5: ANCHORS

This section discusses different types of anchors and their application in rope systems and climbing. Proper selection and placement of anchors is a critical skill that requires a great deal of practice. Failure of any system is most likely to occur at the anchor point. If the anchor is not strong enough to support the intended load, it will fail. Failure is usually the result of poor terrain features selected for the anchor point, or the equipment used in rigging the anchor was placed improperly or in insufficient amounts. When selecting or constructing anchors, always try to make sure the anchor is "bombproof." A bombproof anchor is stronger than any possible load that could be placed on it. An anchor that has more strength than the climbing rope is considered bombproof.

NATURAL ANCHORS

Natural anchors should be considered for use first. They are usually strong and often simple to construct with minimal use of equipment. Trees, boulders, and other terrain irregularities are already in place and simply require a method of attaching the rope. However, natural anchors should be carefully studied and evaluated for stability and strength before use. Sometimes the climbing rope is tied directly to the anchor, but under most circumstances a sling is attached to the anchor and then the climbing rope is attached to the sling with a carabiner(s). (See paragraph 4-81 for slinging techniques.)

4-75. TREES

Trees are probably the most widely used of all natural anchors depending on the terrain and geographical region (Figure 4-76). However, trees must be carefully checked for suitability.

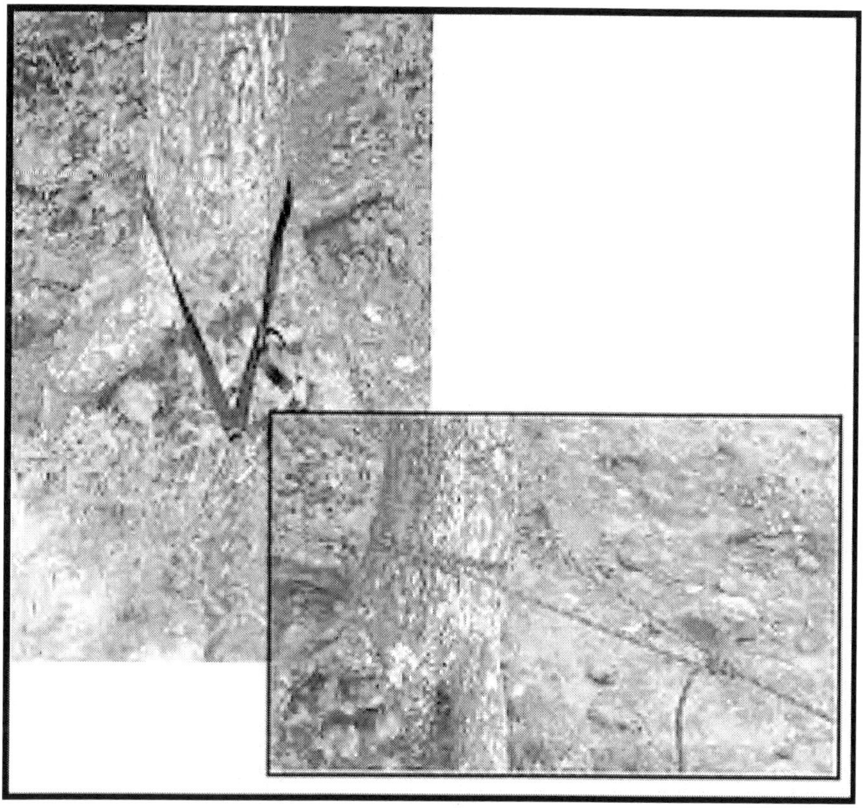

Figure 4-76: Trees used as anchors.

a. In rocky terrain, trees usually have a shallow root system. This can be checked by pushing or tugging on the tree to see how well it is rooted. Anchoring as low as possible to prevent excess leverage on the tree may be necessary.
b. Use padding on soft, sap producing trees to keep sap off ropes and slings.

4-76. BOULDERS

Boulders and rock nubbins make ideal anchors (Figure 4-77). The rock can be firmly tapped with a piton hammer to ensure it is solid. Sedimentary and other loose rock formations are not stable. Talus and scree fields are an indicator that the rock in the area is not solid. All areas around the rock formation that could cut the rope or sling should be padded.

Figure 4-77: Boulders used as anchors.

4-77. CHOCKSTONES

A chockstone is a rock that is wedged in a crack because the crack narrows downward (Figure 4-78). Chockstones should be checked for strength, security, and crumbling and should always be tested before use. All chockstones must be solid and strong enough to support the load. They must have maximum surface contact and be well tapered with the surrounding rock to remain in position.

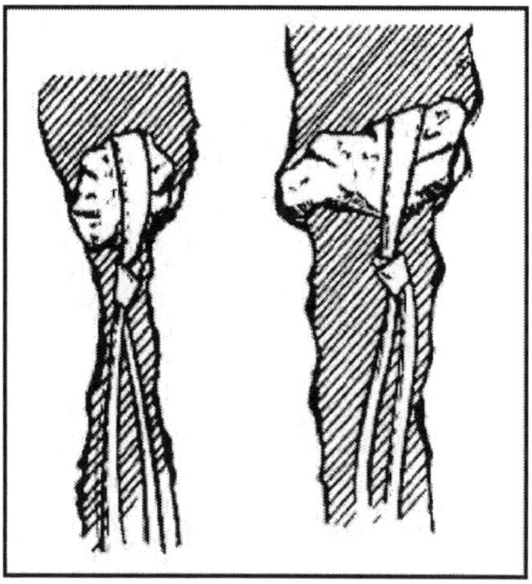

Figure 4-78. Chockstones.

a. Chockstones are often directional they are secure when pulled in one direction but may pop out if pulled in another direction.
b. A creative climber can often make his own chockstone by wedging a rock into position, tying a rope to it, and clipping on a carabiner.
c. Slings should not be wedged between the chockstone and the rock wall since a fall could cut the webbing runner.

4-79. ROCK PROJECTIONS

Rock projections (sometimes called nubbins) often provide suitable protection (Figure 4-80). These include blocks, flakes, horns, and spikes. If rock projections are used, their firmness is important. They should be checked for cracks or weathering that may impair their firmness. If any of these signs exist, the projection should be avoided.

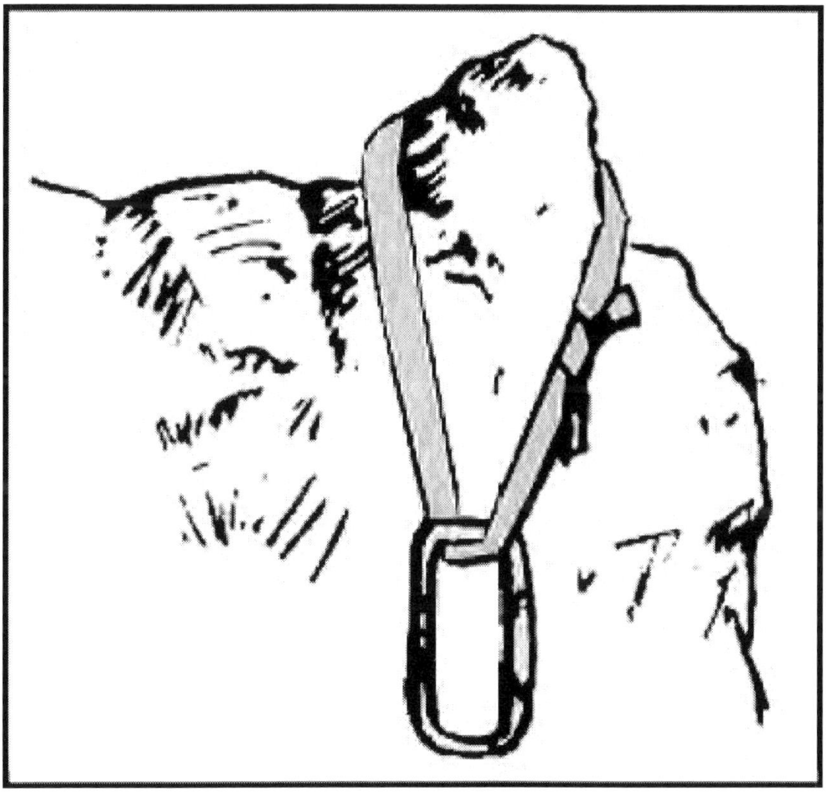

Figure 4-80: Rock projections.

4-80. TUNNELS AND ARCHES

Tunnels and arches are holes formed in solid rock and provide one of the more secure anchor points because they can be pulled in any direction. A sling is threaded through the opening hole and secured with a joining knot or girth hitch. The load-bearing hole must be strong and free of sharp edges (pad if necessary).

4-81. BUSHES AND SHRUBS

If no other suitable anchor is available, the roots of bushes can be used by routing a rope around the bases of several bushes (Figure 4-81). As with trees, the anchoring rope is placed as low as possible to reduce leverage on the anchor. All vegetation should be healthy and well rooted to the ground.

Figure 4-81: Bushes and shrubs.

4-82. SLINGING TECHNIQUES

Three methods are used to attach a sling to a natural anchor drape, wrap, and girth. Whichever method is used, the knot is set off to the side where it will not interfere with normal carabiner movement. The carabiner gate should face away from the ground and open away from the anchor for easy insertion of the rope. When a locking carabiner cannot be used, two carabiners are used with gates opposed. Correctly opposed gates should open on opposite sides and form an "X" when opened (Figure 4-82).

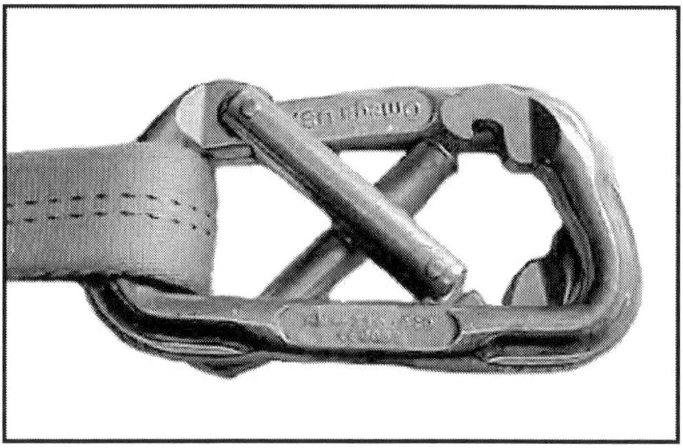

Figure 4-82: Correctly opposed carabiners.

a. **Drape.** Drape the sling over the anchor (Figure 4-83). Untying the sling and routing it around the anchor and then retying is still considered a drape.

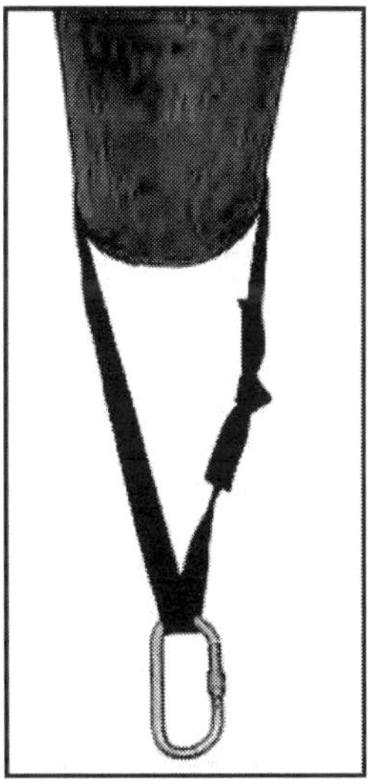

Figure 4-83: Drape.

b. **Wrap.** Wrap the sling around the anchor and connect the two ends together with a carabiner(s) or knot (Figure 4-84).

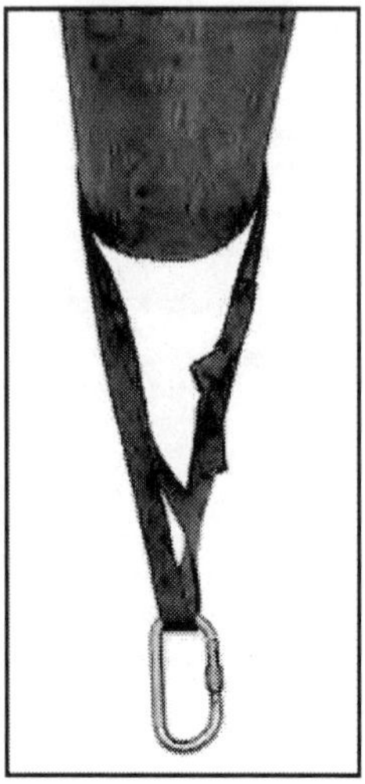

Figure 4-84: Wrap.

c. **Girth.** Tie the sling around the anchor with a girth hitch (Figure 4-85). Although a girth hitch reduces the strength of the sling, it allows the sling to remain in position and not slide on the anchor.

Figure 4-85: Girth.

ANCHORING WITH THE ROPE

The climbing or installation rope can be tied directly to the anchor using several different techniques. This requires less equipment, but also sacrifices some rope length to tie the anchor. The rope can be tied to the anchor using an appropriate anchor knot such as a bowline or a rerouted figure eight. Round turns can be used to help keep the rope in position on the anchor. A tensionless anchor can be used in high-load installations where tension on the attachment point and knot is undesirable.

4-83. ROPE ANCHOR

When tying the climbing or installation rope around an anchor, the knot should be placed approximately the same distance away from the anchor as the diameter of the anchor (Figure 4-86). The knot shouldn't be placed up against the anchor because this can stress and distort the knot under tension.

Figure 4-86: Rope tied to anchor with anchor knot.

4-84. TENSIONLESS ANCHOR

The tensionless anchor is used to anchor the rope on high-load installations such as bridging and traversing (Figure 4-87). The wraps of the rope around the anchor absorb the tension of the installation and keep the tension off the knot and carabiner. The anchor is usually tied with a minimum of four wraps, more if necessary, to absorb the tension. A smooth anchor may require several wraps, whereas a rough barked tree might only require a few. The rope is wrapped from top to bottom. A fixed loop is placed into the end of the rope and attached loosely back onto the rope with a carabiner.

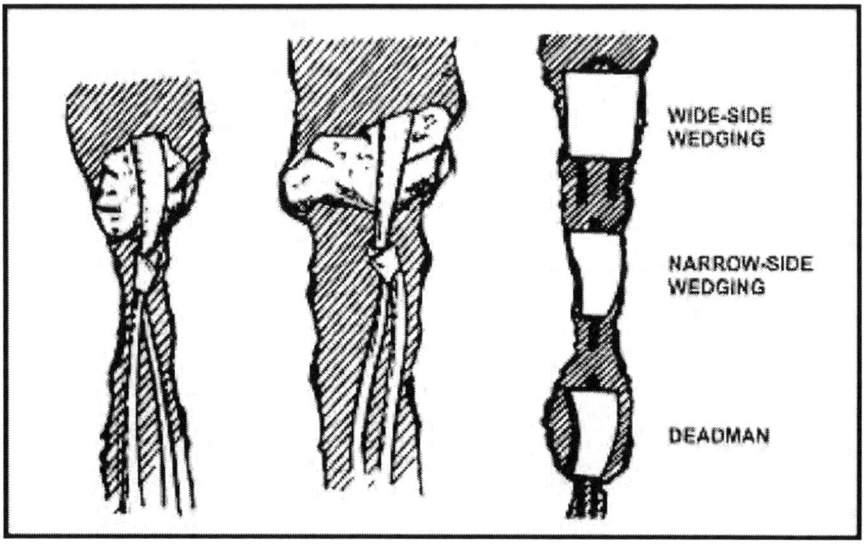

Figure 4-87: Tensionless anchor.

ARTIFICIAL ANCHORS

Using artificial anchors becomes necessary when natural anchors are unavailable. The art of choosing and placing good anchors requires a great deal of practice and experience. Artificial anchors are available in many different types such as pitons, chocks, hexcentrics, and SLCDs. Anchor strength varies greatly; the type used depends on the terrain, equipment, and the load to be placed on it.

4-85. DEADMAN

A "deadman" anchor is any solid object buried in the ground and used as an anchor.

a. An object that has a large surface area and some length to it works best. (A hefty timber, such as a railroad tie, would be ideal.) Large boulders can be used, as well as a bundle of smaller tree limbs or poles. As with natural anchors, ensure timbers and tree limbs are not dead or rotting and that boulders are solid. Equipment, such as skis, ice axes, snowshoes, and ruck sacks, can also be used if necessary.
b. In extremely hard, rocky terrain (where digging a trench would be impractical, if not impossible) a variation of the deadman anchor can be constructed by building above the ground. The sling is attached to the anchor, which is set into the ground as deeply as possible. Boulders are then stacked on top of it until the anchor is strong enough for the load. Though normally not as strong as when buried, this method can work well for light-load installations as in anchoring a hand line for a stream crossing.

> ✍ **NOTE**
> Artificial anchors, such as pitons and bolts, are not widely accepted for use in all areas because of the scars they leave on the rock and the environment. Often they are left in place and become unnatural, unsightly fixtures in the natural environment. For training planning, local laws and courtesies should be taken into consideration for each area of operation.

4-86. PITONS

Pitons have been in use for over 100 years. Although still available, pitons are not used as often as other types of artificial anchors due primarily to their impact on the environment. Most climbers prefer to use chocks, SLCDs and other artificial anchors rather than pitons because they do not scar the rock and are easier to remove. Eye protection should always be worn when driving a piton into rock.

> ✍ **NOTE**
> The proper use and placement of pitons, as with any artificial anchor, should be studied, practiced, and tested while both feet are firmly on the ground and there is no danger of a fall.

a. **Advantages.** Some advantages in using pitons are:
   - Depending on type and placement, pitons can support multiple directions of pull.
   - Pitons are less complex than other types of artificial anchors.
   - Pitons work well in thin cracks where other types of artificial anchors do not.

b. **Disadvantages.** Some disadvantages in using pitons are:
   - During military operations, the distinct sound created when hammering pitons is a tactical disadvantage.
   - Due to the expansion force of emplacing a piton, the rock could spread apart or break causing an unsafe condition.
   - Pitons are more difficult to remove than other types of artificial anchors.
   - Pitons leave noticeable scars on the rock.
   - Pitons are easily dropped if not tied off when being used.

c. **Piton Placement**. The proper positioning or placement of pitons is critical. (Figure 5-12 shows examples of piton placement.) Usually a properly sized piton for a rock crack will fit one half to two thirds into the crack before being driven with the piton hammer. This helps ensure the depth of the crack is adequate for the size piton selected. As pitons are driven into the rock the pitch or sound that is made will change with each hammer blow, becoming higher pitched as the piton is driven in.

   (1) Test the rock for soundness by tapping with the hammer. Driving pitons in soft or rotten rock is not recommended. When this type of rock must be used, clear the loose rock, dirt, and debris from the crack before driving the piton completely in.

   (2) While it is being driven, attach the piton to a sling with a carabiner (an old carabiner should be used, if available) so that if the piton is knocked out of the crack, it will not be lost. The greater the resistance overcome while driving the piton, the firmer the anchor will be. The holding power depends on the climber placing the piton in a sound crack, and on the type of rock. The piton should not spread the rock, thereby loosening the emplacement.

✍ **NOTE**

Pitons that have rings as attachment points might not display much change in sound as they are driven in as long as the ring moves freely.

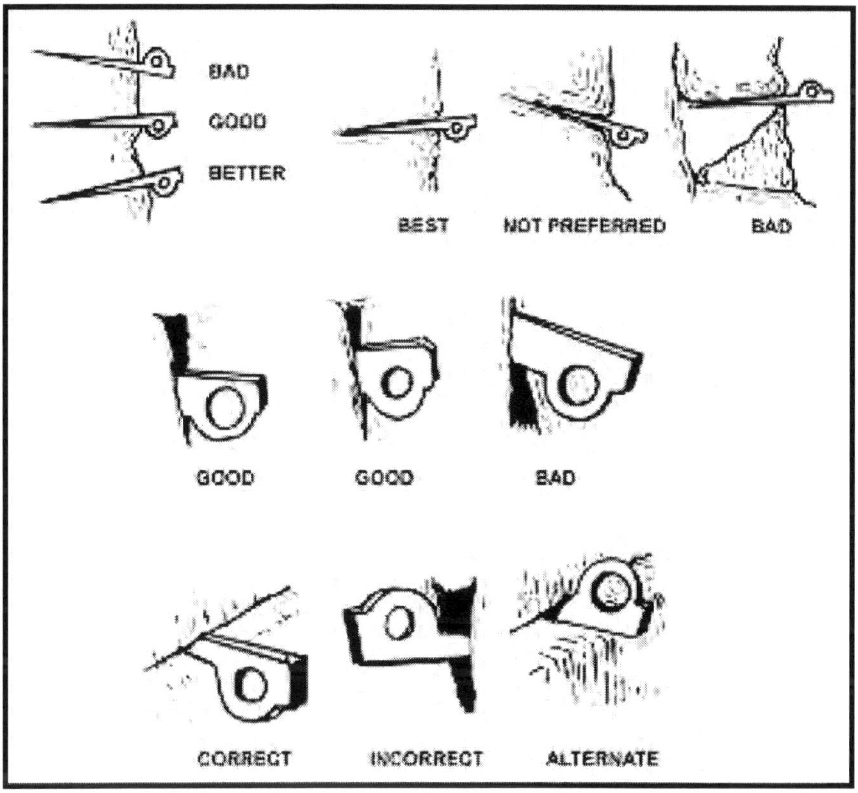

Figure 4-88: Examples of piton placements.

(3) Military mountaineers should practice emplacing pitons using either hand. Sometimes a piton cannot be driven completely into a crack, because the piton is too long. Therefore, it should be tied off using a hero-loop (an endless piece of webbing) (Figure 4-89). Attach this loop to the piton using a girth hitch at the point where the piton enters the rock so that the girth hitch is snug against the rock. Clip a carabiner into the loop.

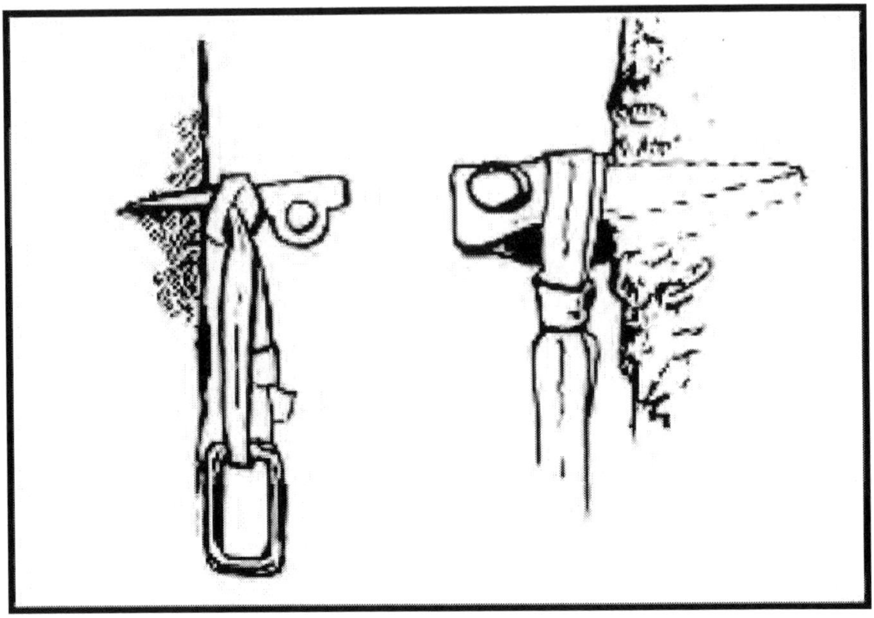

Figure 4-89: Hero-loop.

d. Testing. To test pitons pull up about 1 meter of slack in the climbing rope or use a sling. Insert this rope into a carabiner attached to the piton, then grasp the rope at least 1/2 meter from the carabiner. Jerk vigorously upward, downward, to each side, and then outward while observing the piton for movement. Repeat these actions as many times as necessary. Tap the piton to determine if the pitch has changed. If the pitch has changed greatly, drive the piton in as far as possible. If the sound regains its original pitch, the emplacement is probably safe. If the piton shows any sign of moving or if, upon driving it, there is any question of its soundness, drive it into another place. Try to be in a secure position before testing. This procedure is intended for use in testing an omni-directional anchor (one that withstands a pull in any direction). When a directional anchor (pull in one direction) is used, as in most free and direct-aid climbing situations, and when using chocks, concentrate the test in the direction that force will be applied to the anchor.

e. Removing Pitons. Attach a carabiner and sling to the piton before removal to eliminate the chance of dropping and losing it. Tap the piton firmly along the axis of the crack in which it is located. Alternate tapping from both sides while applying steady pressure. Pulling out on the attached carabiner eventually removes the piton (Figure 4-90).

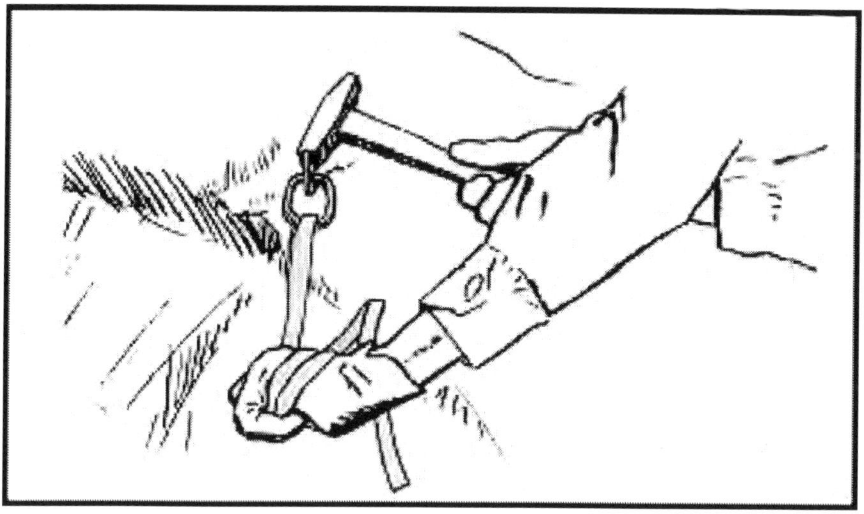

Figure 4-90: Piton removal.

f. Reusing Pitons. Soft iron pitons that have been used, removed, and straightened may be reused, but they must be checked for strength. In training areas, pitons already in place should not be trusted since weather loosens them in time. Also, they may have been driven poorly the first time. Before use, test them as described above and drive them again until certain of their soundness.

4-87. CHOCKS

Chock craft has been in use for many decades. A natural chockstone, having fallen and wedged in a crack, provides an excellent anchor point. Sometimes these chockstones are in unstable positions, but can be made into excellent anchors with little adjustment. Chock craft is an art that requires time and technique to master simple in theory, but complex in practice. Imagination and resourcefulness are key principles to chock craft. The skilled climber must understand the application of mechanical advantage, vectors, and other forces that affect the belay chain in a fall.

a. **Advantages.** The advantages of using chocks are:
- Tactically quiet installation and recovery.
- Usually easy to retrieve and, unless severely damaged, are reusable.
- Light to carry.
- Easy to insert and remove.
- Minimal rock scarring as opposed to pitons.
- Sometimes can be placed where pitons cannot (expanding rock flakes where pitons would further weaken the rock).

b. **Disadvantages.** The disadvantages of using chocks are:
- May not fit in thin cracks, which may accept pitons.
- Often provide only one direction of pull.
- Practice and experience necessary to become proficient in proper placement.

c. **Placement.** The principles of placing chocks are to find a crack with a constriction at some point, place a chock of appropriate size above and behind the constriction, and set the chock by jerking down on the chock loop (Figure 4-91). Maximum surface contact with a tight fit is critical. Chocks are usually good for a single direction of pull.

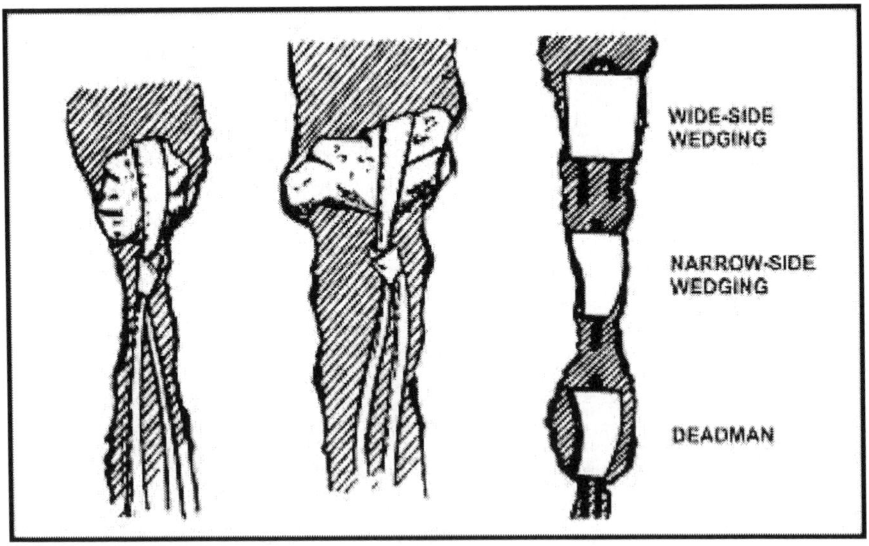

Figure 4-91: Chock placements.

(1) Avoid cracks that have crumbly (soft) or deteriorating rock, if possible. Some cracks may have loose rock, grass, and dirt, which should be removed before placing the chock. Look for a constriction point in the crack, then select a chock to fit it.

(2) When selecting a chock, choose one that has as much surface area as possible in contact with the rock. A chock resting on one small crystal or point of rock is likely to be unsafe. A chock that sticks partly out of the crack is avoided. Avoid poor protection. Ensure that the chock has a wire or runner long enough; extra ropes, cord, or webbing may be needed to extend the length of the runner.

(3) End weighting of the placement helps to keep the protection in position. A carabiner often provides enough weight

(4) Parallel-sided cracks without constrictions are a problem. Chocks designed to be used in this situation rely on camming principles to remain emplaced. Weighting the emplacement with extra hardware is often necessary to keep the chocks from dropping out.

   (a) Emplace the wedge-shaped chock above and behind the constriction; seat it with a sharp downward tug.

   (b) Place a camming chock with its narrow side into the crack, then rotate it to the attitude it will assume under load; seat it with a sharp downward tug.

d. **Testing.** After seating a chock, test it to ensure it remains in place. A chock that falls out when the climber moves past it is unsafe and offers no protection. To test it, firmly pull the chock in every anticipated direction of pull. Some chock placements fail in one or more directions; therefore, use pairs of chocks in opposition.

4-88. SPRING-LOADED CAMMING DEVICE

The SLCD offers quick and easy placement of artificial protection. It is well suited in awkward positions and difficult placements, since it can be emplaced with one hand. It can usually be placed quickly and retrieved easily (Figure 4-92).

a. To emplace an SLCD hold the device in either hand like a syringe, pull the retractor bar back, place the device into a crack, and release the retractor bar. The SLCD holds well in parallel-sided hand- and fist-sized cracks. Smaller variations are available for finger-sized cracks.

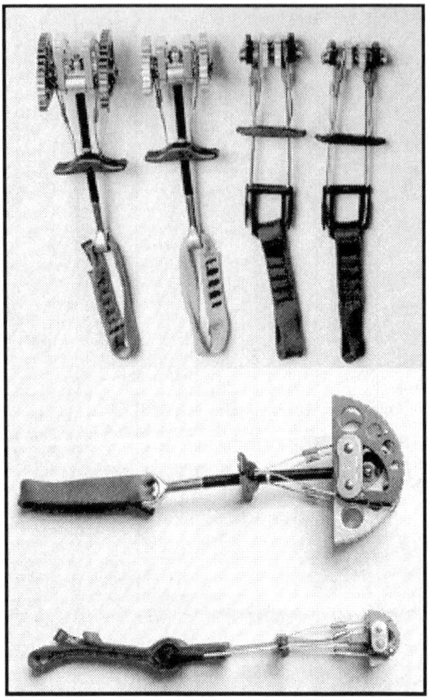

Figure 4-92: SLCD placements.

b. Careful study of the crack should be made before selecting the device for emplacement. It should be placed so that it is aligned in the direction of force applied to it. It should not be placed any deeper than is needed for secure placement, since it may be impossible to reach the extractor bar for removal. An SLCD should be extended with a runner and placed so that the direction of pull is parallel to the shaft; otherwise, it may rotate and pull out. The versions that have a semi-rigid wire cable shaft allow for greater flexibility and usage, without the danger of the shaft snapping off in a fall.

4-89. BOLTS

Bolts are often used in fixed-rope installations and in aid climbing where cracks are not available.

a. Bolts provide one of the most secure means of establishing protection. The rock should be inspected for evidence of crumbling, flaking, or cracking, and should be tested with a hammer. Emplacing a bolt with a hammer and a hand drill is a time-consuming and difficult process that requires drilling a hole in the rock deeper than the length of the bolt. This normally takes more than 20 minutes for one hole. Electric or even gas-powered drills can be used to greatly shorten drilling time. However, their size and weight can make them difficult to carry on the climbing route.

b. A hanger (carrier) and nut are placed on the bolt, and the bolt is inserted and then driven into the hole. A climber should never hammer on a bolt to test or "improve" it, since this permanently weakens it. Bolts should be used with carriers, carabiners, and runners.

c. When using bolts, the climber uses a piton hammer and hand drill with a masonry bit for drilling holes. Some versions are available in which the sleeve is hammered and turned into the rock (self-drilling), which bores the hole. Split bolts and expanding sleeves are common bolts used to secure hangers and carriers (Figure 4-94). Surgical tubing is useful in blowing dust out of the holes. Nail type bolts are emplaced by driving the nail with a hammer to expand the sleeve against the wall of the drilled hole. Safety glasses should always be worn when emplacing bolts.

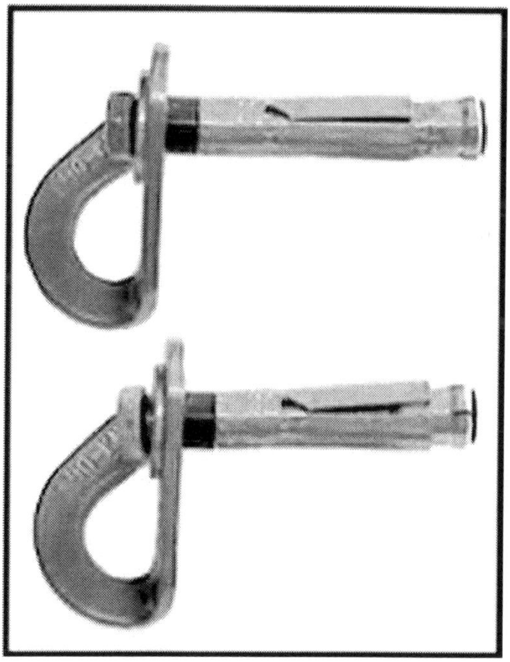

Figure 4-94: Bolt with expanding sleeve.

4-90. EQUALIZING ANCHORS

Equalizing anchors are made up of more than one anchor point joined together so that the intended load is shared equally. This not only provides greater anchor strength, but also adds redundancy or backup because of the multiple points.

a. **Self-equalizing Anchor.** A self-equalizing anchor will maintain an equal load on each individual point as the direction of pull changes (Figure 4-95). This is sometimes used in rappelling when the route must change left or right in the middle of the rappel. A self-equalizing anchor should only be used when necessary because if any one of the individual points fail, the anchor will extend and shock-load the remaining points or even cause complete anchor failure.

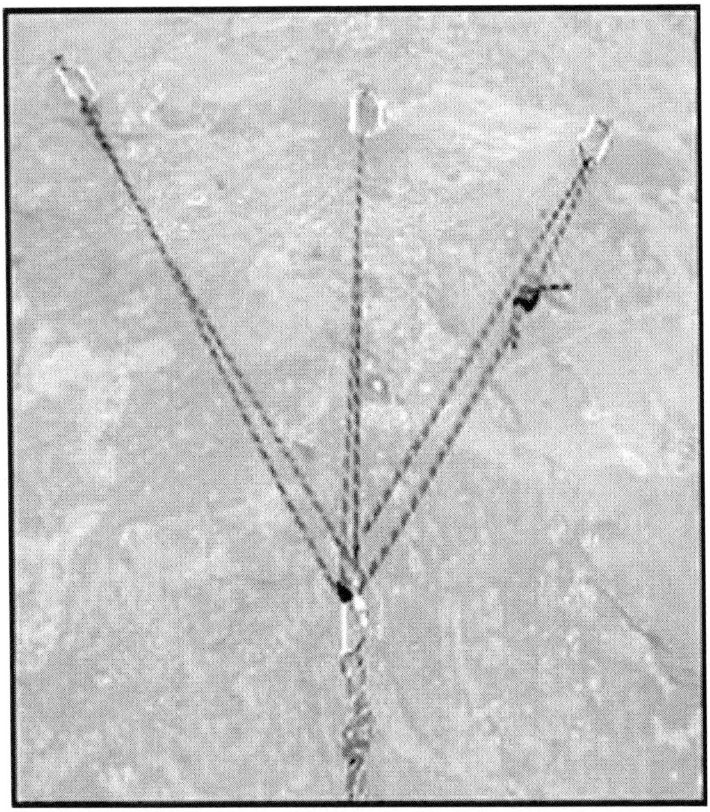

Figure 4-95: Self-equalizing anchors.

b. Pre-equalized Anchor. A pre-equalized anchor distributes the load equally to each individual point (Figure 4-96). It is aimed in the direction of the load. A pre-equalized anchor prevents extension and shock-loading of the anchor if an individual point fails. An anchor is pre-equalized by tying an overhand or figure-eight knot in the webbing or sling.

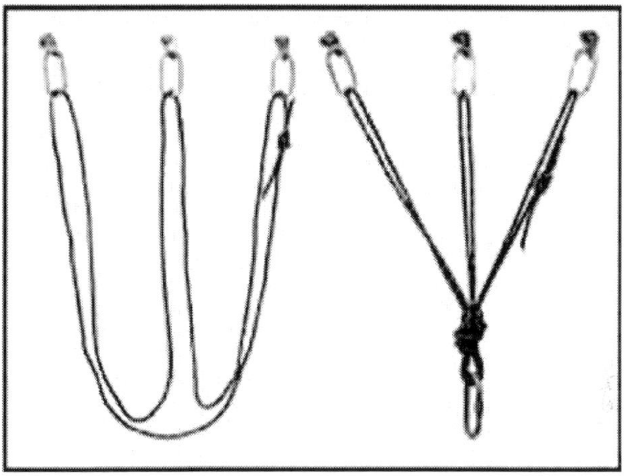

Figure 4-96: Pre-equalized anchor.

> ✍ **NOTE**
>
> When using webbing or slings, the angles of the webbing or slings directly affect the load placed on an anchor. An angle greater than 90 degrees can result in anchor failure (Figure 4-97).

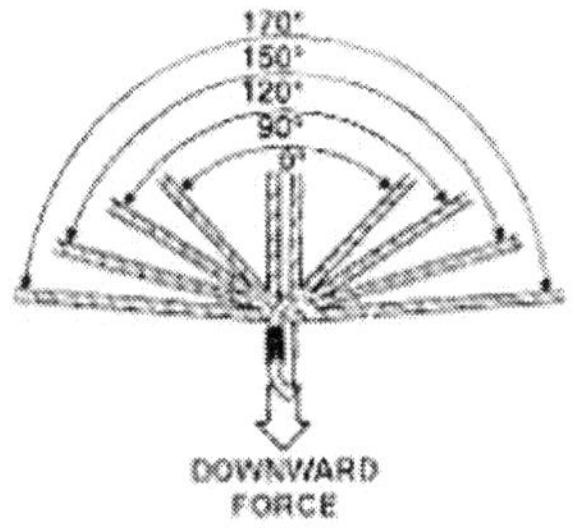

TENSION IN MULTIPLE ANCHOR RIGGING	
ANGLE	RESULTING LEG TENSION*
170°	1150%
150°	200%
120°	100%
90°	70%
0°	50%

*On each leg relative to downward force.

Figure 4-97 : Effects of angles on an anchor.

SECTION 6: CLIMBING

A steep rock face is a terrain feature that can be avoided most of the time through prior planning and good route selection. Rock climbing can be time consuming, especially for a larger unit with a heavy combat load. It can leave the climbing party totally exposed to weather, terrain hazards, and the enemy for the length of the climb.

Sometimes steep rock cannot be avoided. Climbing relatively short sections of steep rock (one or two pitches) may prove quicker and safer than using alternate routes. A steep rock route would normally be considered an unlikely avenue of approach and, therefore, might be weakly defended or not defended at all.

All personnel in a unit preparing for deployment to mountainous terrain should be trained in the basics of climbing. Forward observers, reconnaissance personnel, and security teams are a few examples of small units who may require rock climbing skills to gain their vantage points in mountainous terrain. Select personnel demonstrating the highest degree of skill and experience should be trained in roped climbing techniques. These personnel will have the job of picking and "fixing" the route for the rest of the unit.

Rock climbing has evolved into a specialized "sport" with a wide range of varying techniques and styles. This chapter focuses on the basics most applicable to military operations.

CLIMBING FUNDAMENTALS

A variety of refined techniques are used to climb different types of rock formations. The foundation for all of these styles is the art of climbing. Climbing technique stresses climbing with the weight centered over the feet, using the hands primarily for balance. It can be thought of as a combination of the balanced movement required to walk a tightrope and the technique used to ascend a ladder. No mountaineering equipment is required; however, the climbing technique is also used in roped climbing.

4-91. ROUTE SELECTION

The experienced climber has learned to climb with the "eyes." Even before getting on the rock, the climber studies all possible routes, or "lines," to the top looking for cracks, ledges, nubbins, and other irregularities in the rock that will be used for footholds and handholds, taking note of any larger ledges or benches for resting places. When picking the line, he mentally climbs the route, rehearsing the step-by-step sequence of movements that will be required to do the climb, ensuring himself that the route has an adequate number of holds and the difficulty of the climb will be well within the limit of his ability.

4-92. TERRAIN SELECTION FOR TRAINING

Route selection for military climbing involves picking the easiest and quickest possible line for all personnel to follow. However, climbing skill and experience can only be developed by increasing the length and difficulty of routes as training progresses. In the training environment, beginning lessons in climbing should be performed CLOSE to the ground on lower-angled rock with plenty of holds for the hands and feet. Personnel not climbing can act as "otters" for those climbing. In later lessons, a "top-rope" belay can be used for safety, allowing e individual to increase the length and difficulty of the climb under the protection of the climbing rope.

4-93. PREPARATION

In preparation for climbing, the boot soles should be dry and clean. A small stick can be used to clean out dirt and small rocks that might be caught between the lugs of the boot sole. If the soles are wet or damp, dry them off by stomping and rubbing the soles on clean, dry rock. All jewelry should be removed from the fingers. Watches and bracelets can interfere with hand placements and may become damaged if worn while climbing. Helmets should be worn to protect the head from injury if an object, such as a rock or climbing gear, falls from climbers above. Most climbing helmets are not designed to provide protection from impact to the head if the wearer falls, but will provide a minimal amount of protection if a climber comes in contact with the rock during climbing.

> **⚠ CAUTION**
>
> Rings can become caught on rock facial features and or lodged into cracks, which could cause injuries during a slip or fall.

4-94. SPOTTING

Spotting is a technique used to add a level of safety to climbing without a rope. A second man stands below and just outside of the climbers fall path and helps (spots) the climber to land safely if he should fall. Spotting is only applicable if the climber is not going above the spotters head on the rock. Beyond that height a roped climbing should be conducted. If an individual climbs beyond the effective range of the spotter(s), he has climbed TOO HIGH for his own safety. The duties of the spotter are to help prevent the falling climber from impacting the head and or spine, help the climber land feet first, and reduce the impact of a fall.

> ⚠ **CAUTION**
>
> The spotter should not catch the climber against the rock because additional injuries could result. If the spotter pushes the falling climber into the rock, deep abrasions of the skin or knee may occur. Ankle joints could be twisted by the fall if the climber's foot remained high on the rock. The spotter might be required to fully support the weight of the climber causing injury to the spotter.

4-95. CLIMBING TECHNIQUE

Climbing involves linking together a series of movements based on foot and hand placement, weight shift, and movement. When this series of movements is combined correctly, a smooth climbing technique results. This technique reduces excess force on the limbs, helping to minimize fatigue. The basic principle is based on the five body parts described here.

a. **Five Body Parts.** The five body parts used for climbing are the right hand, left hand, right foot, left foot, and body (trunk). The basic principle to achieve smooth climbing is to move only one of the five body parts at a time. The trunk is not moved in conjunction with a foot or in conjunction with a hand, a hand is not moved in conjunction with a foot, and so on. Following this simple technique forces both legs to do all the lifting simultaneously.

b. **Stance or Body Position.** Body position is probably the single most important element to good technique. A relaxed, comfortable stance is essential. (Figure 4-98 shows a correct climbing stance, and Figure 4-99 shows an incorrect stance.) The body should be in a near vertical or erect stance with the weight centered over the feet. Leaning in towards the rock will cause the feet to push outward, away from the rock, resulting in a loss of friction between the boot sole and rock surface. The legs are straight and the heels are kept low to reduce fatigue. Bent legs and tense muscles tire quickly. If strained for too long, tense muscles may vibrate uncontrollably. This vibration, known as "Elvis-ing" or "sewing-machine leg" can be cured by straightening the leg, lowering the heel, or moving on to a more restful position. The hands are used to maintain balance. Keeping the hands between waist and shoulder level will reduce arm fatigue.

Figure 4-98: Correct climbing stance-balanced over both feet.

Figure 4-99: Incorrect stance-stretched out.

(1) Whenever possible, three points of contact are maintained with the rock. Proper positioning of the hips and shoulders is critical. When using two footholds and one handhold, the hips and shoulders should be centered over both feet. In most cases, as the climbing progresses, the body is resting on one foot with two handholds for balance. The hips and shoulders must be centered over the support foot to maintain balance, allowing the "free" foot to maneuver.

(2) The angle or steepness of the rock also determines how far away from the rock the hips and shoulders should be. On low-angle slopes, the hips are moved out away from the rock to keep the body in balance with the weight over the feet. The shoulders can be moved closer to the rock to reach handholds. On steep rock, the hips are pushed closer to the rock. The shoulders are moved away from the rock by arching the back. The body is still in balance over the feet and the eyes can see where the hands need to go. Sometimes, when footholds are small, the hips are moved back to increase friction between the foot and the

rock. This is normally done on quick, intermediate holds. It should be avoided in the rest position as it places more weight on the arms and hands. When weight must be placed on handholds, the arms should be kept straight to reduce fatigue. Again, flexed muscles tire quickly.

c. **Climbing Sequence.** The steps defined below provide a complete sequence of events to move the entire body on the rock. These are the basic steps to follow for a smooth climbing technique. Performing these steps in this exact order will not always be necessary because the nature of the route will dictate the availability of hand and foot placements. The basic steps are weight, shift, and movement (movement being either the foot, hand, or body). (A typical climbing sequence is shown in Figure 4-100.)

STEP ONE: Shift the weight from both feet to one foot. This will allow lifting of one foot with no effect on the stance.
STEP TWO: Lift the unweighted foot and place it in a new location, within one to two feet of the starting position, with no effect on body position or balance (higher placement will result in a potentially higher lift for the legs to make, creating more stress, and is called a high step) The trunk does not move during foot movement.
STEP THREE: Shift the weight onto both feet. (Repeat steps 1 through 3 for remaining foot.)
STEP FOUR: Lift the body into a new stance with both legs.
STEP FIVE: Move one hand to a new position between waist and head height. During this movement, the trunk should be completely balanced in position and the removed hand should have no effect on stability.
STEP SIX: Move the remaining hand as in Step 5.

Now the entire body is in a new position and ready to start the process again. Following these steps will prevent lifting with the hands and arms, which are used to maintain stance and balance. If both legs are bent, leg extension can be performed as soon as one foot has been moved. Hand movements can be delayed until numerous foot movements have been made, which not only creates shorter lifts with the legs, but may allow a better choice for the next hand movements because the reach will have increased.

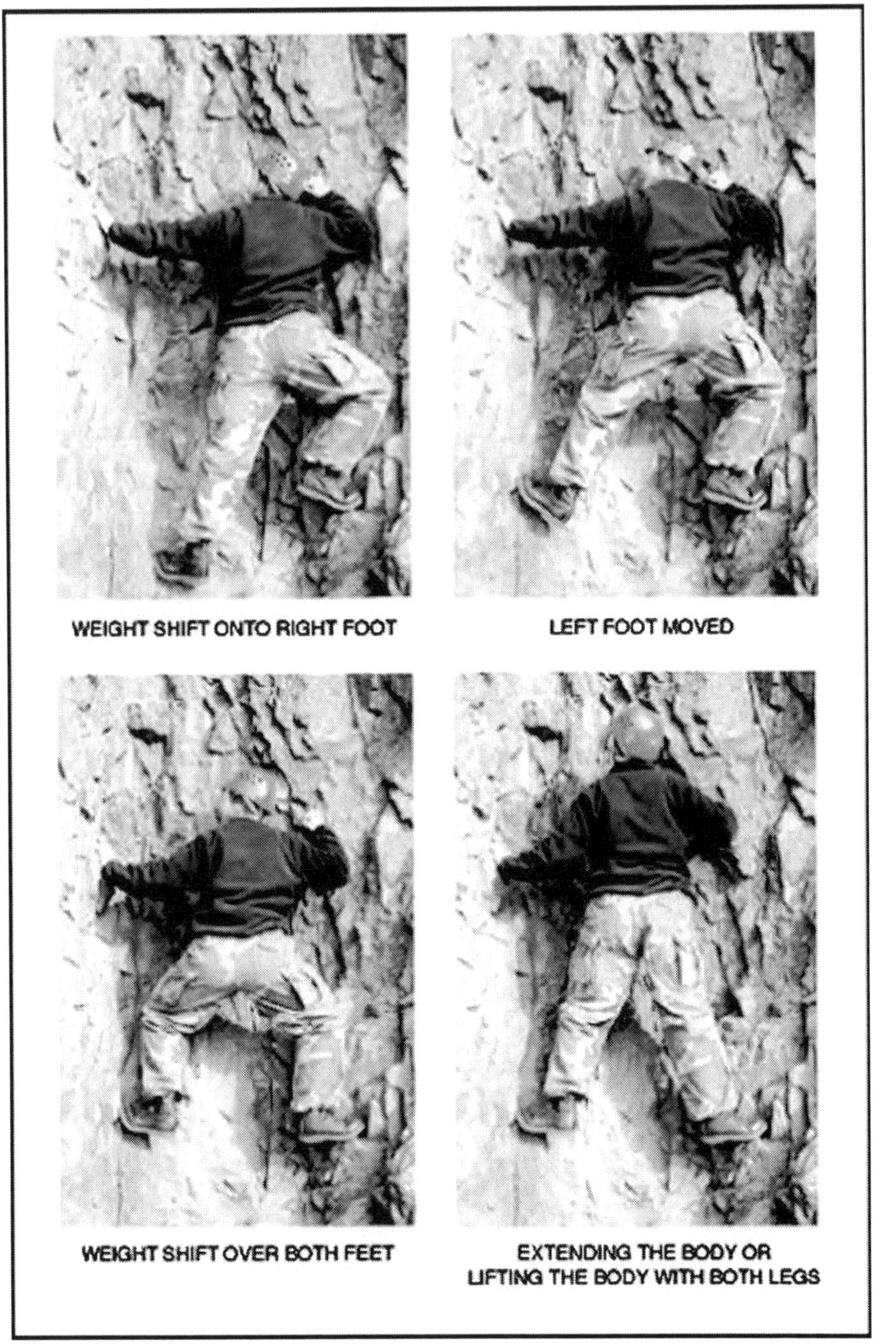

Figure 4-100: Typical climbing sequence.

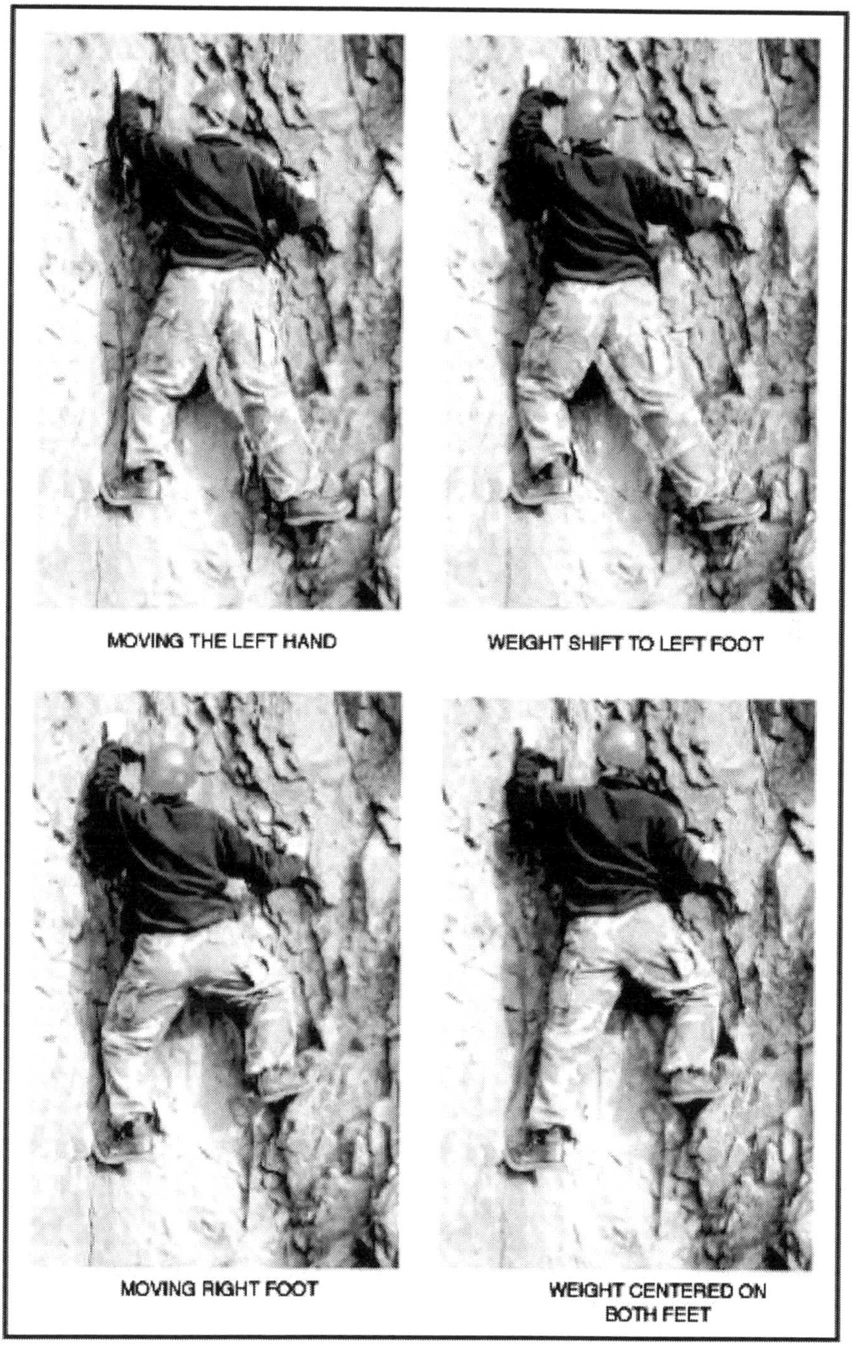

Figure 4-100B: Typical climbing sequence (continued).

Figure 4-100: Typical climbing sequence (continued).

(1) Many climbers will move more than one body part at a time, usually resulting in lifting the body with one leg or one leg and both arms. This type of lifting is inefficient, requiring one leg to perform the work of two or using the arms to lift the body. Proper climbing technique is lifting the body with the legs, not the arms, because the legs are much stronger.

(2) When the angle of the rock increases, these movements become more critical. Holding or pulling the body into the rock with the arms and hands may be necessary as the angle increases (this is still not lifting with the arms). Many climbing routes have angles greater than ninety degrees (overhanging) and the arms are used to support partial body weight. The same technique applies even at those angles.

(3) The climber should avoid moving on the knees and elbows. Other than being uncomfortable, even painful, to rest on, these bony portions of the limbs offer little friction and "feel" on the rock.

4-96. SAFETY PRECAUTIONS

The following safety precautions should be observed when rock climbing.

a. While ascending a seldom or never traveled route, you may encounter precariously perched rocks. If the rock will endanger your second, it may be possible to remove it from the route and trundle it, tossing it down. This is extremely dangerous to climbers below and should not be attempted unless you are absolutely sure no men are below. If you are not sure that the flight path is clear, do not do it. Never dislodge loose rocks carelessly. Should a rock become loose accidentally, immediately shout the warning "ROCK" to alert climbers below. Upon hearing the warning, personnel should seek immediate cover behind any rock bulges or overhangs available, or flatten themselves against the rock to minimize exposure.

b. Should a climber fall, he should do his utmost to maintain control and not panic. If on a low-angle climb, he may be able to arrest his own fall by staying in contact with the rock, grasping for any possible hold available. He should shout the warning "FALLING" to alert personnel below.

> **⚠ CAUTION**
> Grasping at the rock in a fall can result in serious injuries to the upper body. If conducting a roped climb, let the rope provide protection.

c. When climbing close to the ground and without a rope, a spotter can be used for safety. The duties of the spotter are to ensure the falling climber does not impact the head or spine, and to reduce the impact of a fall.
d. Avoid climbing directly above or below other climbers (with the exception of spotters). When personnel must climb at the same time, following the same line, a fixed rope should be installed.
e. Avoid climbing with gloves on because of the decreased "feel" for the rock. The use of gloves in the training environment is especially discouraged, while their use in the mountains is often mandatory when it is cold. A thin polypropylene or wool glove is best for rock climbing, although heavier cotton or leather work gloves are often used for belaying.
f. Be extremely careful when climbing on wet or moss-covered rock; friction on holds is greatly reduced.
g. Avoid grasping small vegetation for handholds; the root systems can be shallow and will usually not support much weight.

4-97. MARGIN OF SAFETY

Besides observing the standard safety precautions, the climber can avoid catastrophe by climbing with a wide margin of safety. The margin of safety is a protective buffer the climber places between himself and potential climbing hazards. Both subjective (personnel-related) and objective (environmental) hazards must be considered when applying the margin of safety. The leader must apply the margin of safety taking into account the strengths and weaknesses of the entire team or unit.

a. When climbing, the climber increases his margin of safety by selecting routes that are well within the limit of his ability. When leading a group of climbers, he selects a route well within the ability of the weakest member.
b. When the rock is wet, or when climbing in other adverse weather conditions, the climber's ability is reduced and routes are selected accordingly. When the climbing becomes difficult

or exposed, the climber knows to use the protection of the climbing rope and belays. A lead climber increases his margin of safety by placing protection along the route to limit the length of a potential fall.

USE OF HOLDS

The climber should check each hold before use. This may simply be a quick, visual inspection if he knows the rock to be solid. When in doubt, he should grab and tug on the hold to test it for soundness BEFORE depending on it. Sometimes, a hold that appears weak can actually be solid as long as minimal force is applied to it, or the force is applied in a direction that strengthens it. A loose nubbin might not be strong enough to support the climber's weight, but it may serve as an adequate handhold. Be especially careful when climbing on weathered, sedimentary-type rock.

4-98. CLIMBING WITH THE FEET

"Climb with the feet and use the hands for balance" is extremely important to remember. In the early learning stages of climbing, most individuals will rely heavily on the arms, forgetting to use the feet properly. It is true that solid handholds and a firm grip are needed in some combination techniques; however, even the most strenuous techniques require good footwork and a quick return to a balanced position over one or both feet. Failure to climb any route, easy or difficult, is usually the result of poor footwork.

a. The beginning climber will have a natural tendency to look up for handholds. Try to keep the hands low and train your eyes to look down for footholds. Even the smallest irregularity in the rock can support the climber once the foot is positioned properly and weight is committed to it.

b. The foot remains on the rock as a result of friction. Maximum friction is obtained from a correct stance over a properly positioned foot. The following describes a few ways the foot can be positioned on the rock to maximize friction.

(1) ***Maximum Sole Contact***. The principle of using full sole contact, as in mountain walking, also applies in climbing. Maximum friction is obtained by placing as much of the boot sole on the rock as possible. Also, the leg muscles can relax the most when the entire foot is placed on the rock. (Figure 4-101 shows examples of maximum and minimum sole contact.)

(a) Smooth, low-angled rock (slab) and rock containing large "bucket" holds and ledges are typical formations where the entire boot sole should be used.

(b) On some large holds, like bucket holds that extend deep into the rock, the entire foot cannot be used. The climber may not be able to achieve a balanced position if the foot is stuck too far underneath a bulge in the rock. In this case, placing only part of the foot on the hold may allow the climber to achieve a balanced stance. The key is to use as much of the boot sole as possible. Remember to keep the heels low to reduce strain on the lower leg muscles.

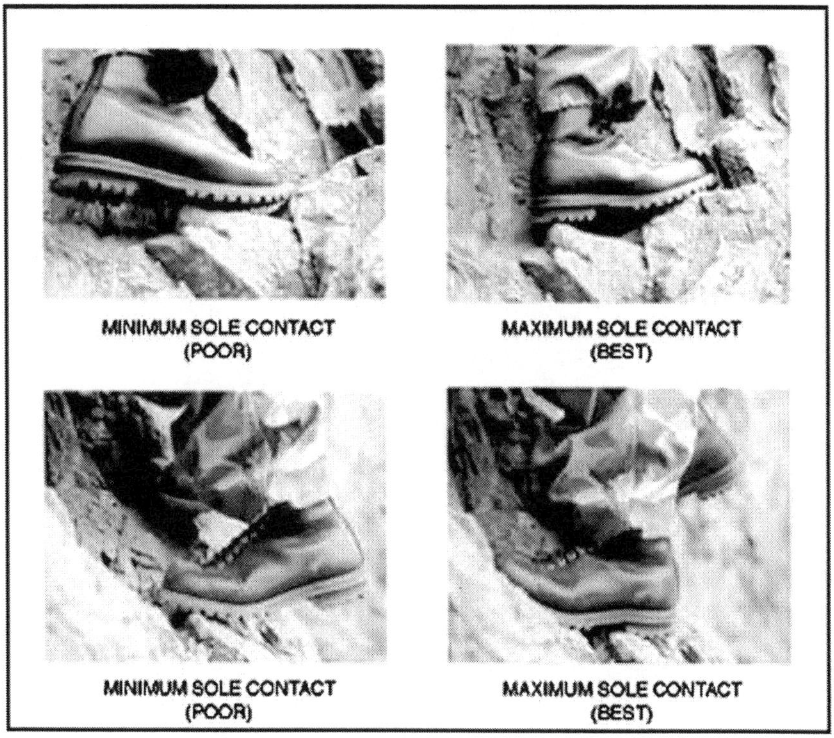

Figure 4-101: Examples of maximum and minimum sole contact.

(2) *Edging*. The edging technique is used where horizontal crack systems and other irregularities in the rock form

small, well-defined ledges. The edge of the boot sole is placed on the ledge for the foothold. Usually, the inside edge of the boot or the edge area around the toes is used. Whenever possible, turn the foot sideways and use the entire inside edge of the boot. Again, more sole contact equals more friction and the legs can rest more when the heel is on the rock. (Figure 4-102 shows examples of the edging technique.)

(a) On smaller holds, edging with the front of the boot, or toe, may be used. Use of the toe is most tiring because the heel is off the rock and the toes support the climber's weight. Remember to keep the heel low to reduce fatigue. Curling and stiffening the toes in the boot increases support on the hold. A stronger position is usually obtained on small ledges by turning the foot at about a 45-degree angle, using the strength of the big toe and the ball of the foot.

(b) Effective edging on small ledges requires stiff-soled footwear. The stiffer the sole, the better the edging capability. Typical mountain boots worn by the US military have a relatively flexible lugged sole and, therefore, edging ability on smaller holds will be somewhat limited.

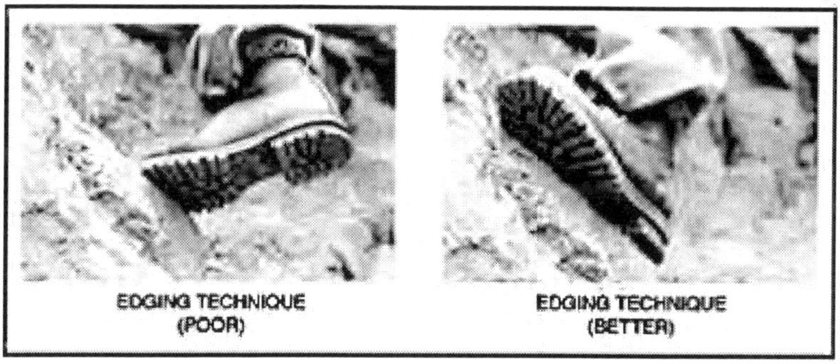

Figure 4-102: Examples of edging technique.

(3) ***Smearing.*** When footholds are too small to use a good edging technique, the ball of the foot can be "smeared" over the hold. The smearing technique requires the boot

to adhere to the rock by deformation of the sole and by friction. Rock climbing shoes are specifically designed to maximize friction for smearing; some athletic shoes also work well. The Army mountain boot, with its softer sole, usually works better for smearing than for edging. Rounded, down-sloping ledges and low-angled slab rock often require good smearing technique. (Figure 4-103 shows examples of the smearing technique.)

(a) Effective smearing requires maximum friction between the foot and the rock. Cover as much of the hold as possible with the ball of the foot. Keeping the heel low will not only reduce muscle strain, but will increase the amount of surface contact between the foot and the rock.

(b) Sometimes flexing the ankles and knees slightly will place the climber's weight more directly over the ball of the foot and increase friction; however, this is more tiring and should only be used for quick, intermediate holds. The leg should be kept straight whenever possible.

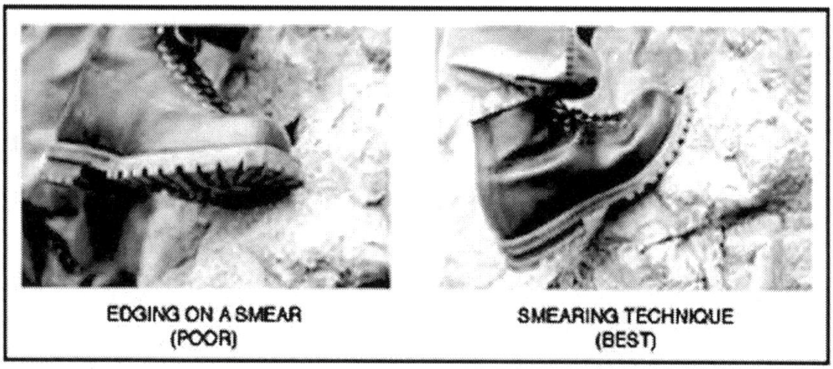

Figure 4-103: Examples of the smearing technique.

(4) ***Jamming.*** The jamming technique works on the same principal as chock placement. The foot is set into a crack in such a way that it "jams" into place, resisting a downward pull. The jamming technique is a specialized skill used to climb vertical or near vertical cracks when no other holds are available on the rock face. The technique is

not limited to just wedging the feet; fingers, hands, arms, even the entire leg or body are all used in the jamming technique, depending on the size of the crack. Jam holds are described in this text to broaden the range of climbing skills. Jamming holds can be used in a crack while other hand/foot holds are used on the face of the rock. Many cracks will have facial features, such as edges, pockets, and soon, inside and within reach. Always look or feel for easier to use features. (Figure 4-104 shows examples of jamming.)

(a) The foot can be jammed in a crack in different ways. It can be inserted above a constriction and set into the narrow portion, or it can be placed in the crack and turned, like a camming device, until it locks in place tight enough to support the climber's weight. Aside from these two basic ideas, the possibilities are endless. The toes, ball of the foot, or the entire foot can be used. Try to use as much of the foot as possible for maximum surface contact. Some positions are more tiring, and even more painful on the foot, than others. Practice jamming the foot in various ways to see what offers the most secure, restful position.

(b) Some foot jams may be difficult to remove once weight has been committed to them, especially if a stiffer sole boot is used. The foot is less likely to get stuck when it is twisted or "cammed" into position. When removing the boot from a crack, reverse the way it was placed to prevent further constriction.

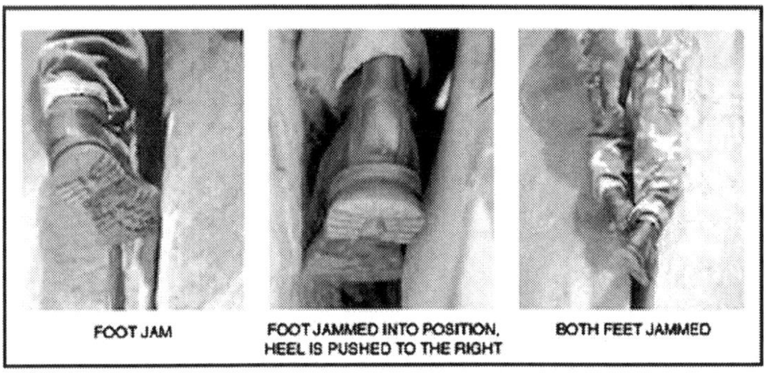

Figure 4-104: Examples of jamming.

4-98. USING THE HANDS

The hands can be placed on the rock in many ways. Exactly how and where to position the hands and arms depends on what holds are available, and what configuration will best support the current stance as well as the movement to the next stance. Selecting handholds between waist and shoulder level helps in different ways. Circulation in the arms and hands is best when the arms are kept low. Secondly, the climber has less tendency to "hang" on his arms when the handholds are at shoulder level and below. Both of these contribute to a relaxed stance and reduce fatigue in the hands and arms.

a. As the individual climbs, he continually repositions his hands and arms to keep the body in balance, with the weight centered over the feet. On lower-angled rock, he may simply need to place the hands up against the rock and extend the arm to maintain balance; just like using an ice ax as a third point of contact in mountain walking. Sometimes, he will be able to push directly down on a large hold with the palm of the hand. More often though, he will need to "grip" the rock in some fashion and then push or pull against the hold to maintain balance.

b. As stated earlier, the beginner will undoubtedly place too much weight on the hands and arms. If we think of ourselves climbing a ladder, our body weight is on our legs. Our hands grip, and our arms pull on each rung only enough to maintain our balance and footing on the ladder. Ideally, this is the amount of grip and pull that should be used in climbing. Of course, as the size and availability of holds decreases, and the steepness of the rock approaches the vertical, the grip must be stronger and more weight might be placed on the arms and handholds for brief moments. The key is to move quickly from the smaller, intermediate holds to the larger holds where the weight can be placed back on the feet allowing the hands and arms to relax. The following describes some of the basic handholds and how the hand can be positioned to maximize grip on smaller holds.

(1) ***Push Holds.*** Push holds rely on the friction created when the hand is pushed against the rock. Most often a climber will use a push hold by applying "downward pressure" on a ledge or nubbin. This is fine, and works well; however, the climber should not limit his use of push holds to the application of down pressure. Pushing

sideways, and on occasion, even upward on less obvious holds can prove quite secure. Push holds often work best when used in combination with other holds. Pushing in opposite directions and "push-pull" combinations are excellent techniques. (Figure 4-105 shows examples of push holds.)

(a) An effective push hold does not necessarily require the use of the entire hand. On smaller holds, the side of the palm, the fingers, or the thumb may be all that is needed to support the stance. Some holds may not feel secure when the hand is initially placed on them. The hold may improve or weaken during the movement. The key is to try and select a hold that will improve as the climber moves past it.

(b) Most push holds do not require much grip; however, friction might be increased by taking advantage of any rough surfaces or irregularities in the rock. Sometimes the strength of the hold can be increased by squeezing, or "pinching," the rock between the thumb and fingers (see paragraph on pinch holds).

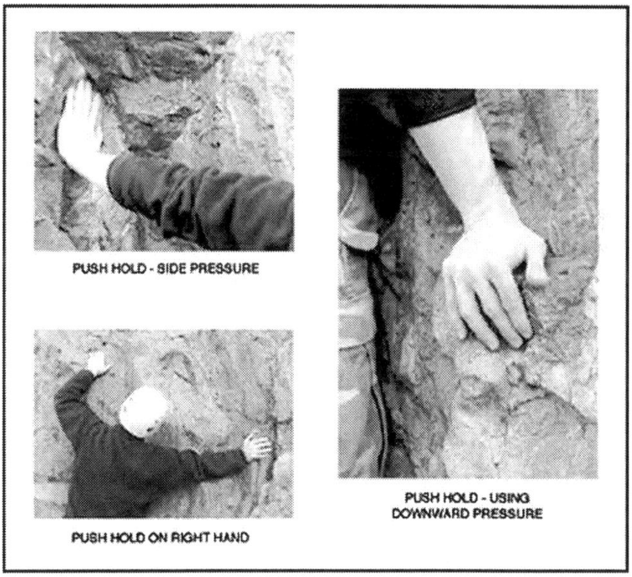

Figure 4-105. Examples of push holds.

(2) ***Pull Holds.*** Pull holds, also called "cling holds," which are grasped and pulled upon, are probably the most widely used holds in climbing. Grip plays more of a role in a pull hold, and, therefore, it normally feels more secure to the climber than a push hold. Because of this increased feeling of security, pull holds are often overworked. These are the holds the climber has a tendency to hang from. Most pull holds do not require great strength, just good technique. Avoid the "death grip" syndrome by climbing with the feet. (Figure 4-106 shows examples of pull holds.)

(a) Like push holds, pressure on a pull hold can be applied straight down, sideways, or upward. Again, these are the holds the climber tends to stretch and

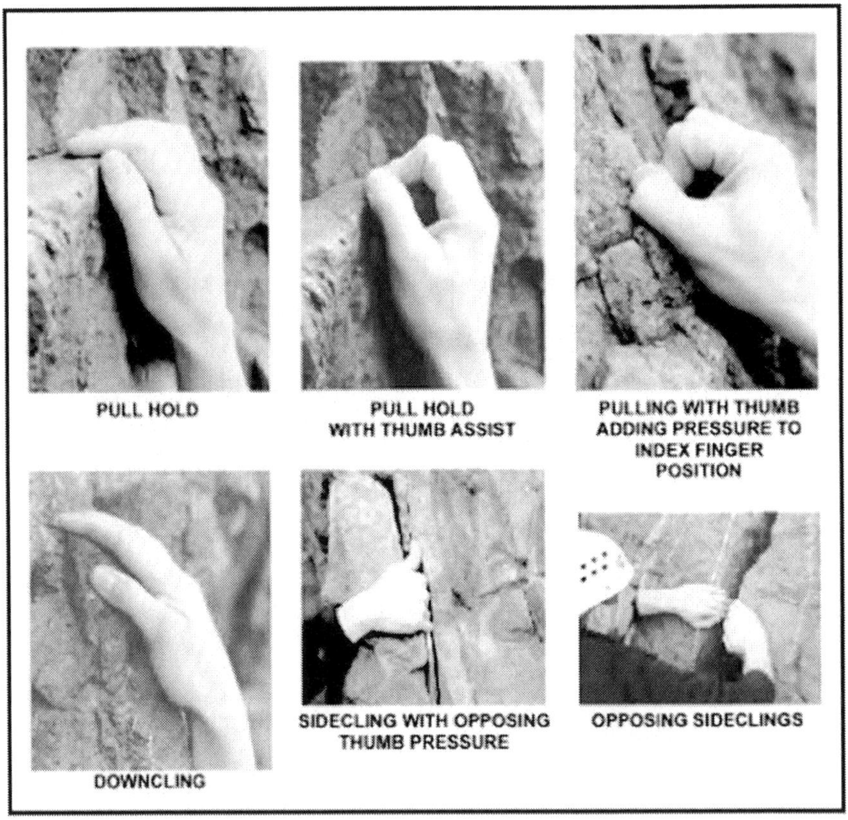

Figure 4-106. Examples of pull holds.

reach for, creating an unbalanced stance. Remember to try and keep the hands between waist and shoulder level, making use of intermediate holds instead of reaching for those above the head.

(b) Pulling sideways on vertical cracks can be very secure. There is less tendency to hang from "side-clings" and the hands naturally remain lower. The thumb can often push against one side of the crack, in opposition to the pull by the fingers, creating a stronger hold. Both hands can also be placed in the same crack, with the hands pulling in opposite directions. The number of possible combinations is limited only by the imagination and experience of the climber.

(c) Friction and strength of a pull hold can be increased by the way the hand grips the rock. Normally, the grip is stronger when the fingers are closed together; however, sometimes more friction is obtained by spreading the fingers apart and placing them between irregularities on the rock surface. On small holds, grip can often be improved by bending the fingers upward, forcing the palm of the hand to push against the rock. This helps to hold the finger tips in place and reduces muscle strain in the hand. Keeping the forearm up against the rock also allows the arm and hand muscles to relax more.

(d) Another technique that helps to strengthen a cling hold for a downward pull is to press the thumb against the side of the index finger, or place it on top of the index finger and press down. This hand configuration, known as a "ring grip," works well on smaller holds.

(3) ***Pinch Holds.*** Sometimes a small nubbin or protrusion in the rock can be "squeezed" between the thumb and fingers. This technique is called a pinch hold. Friction is applied by increasing the grip on the rock. Pinch holds are often overlooked by the novice climber because they feel insecure at first and cannot be relied upon to support much body weight. If the climber has his weight over his feet properly, the pinch hold will work well in providing balance. The pinch hold can also be used as a gripping

technique for push holds and pull holds. (Figure 4-107 shows examples of pinch holds.)

Figure 4-107: Examples of pinch holds.

(4) ***Jam Holds.*** Like foot jams, the fingers and hands can be wedged or cammed into a crack so they resist a downward or outward pull. Jamming with the fingers and hands can be painful and may cause minor cuts and abrasions to tender skin. Cotton tape can be used to protect the fingertips, knuckles, and the back of the hand; however, prolonged jamming technique requiring hand taping should be avoided. Tape also adds friction to the hand in jammed position. (Figure 4-108 shows examples of jam holds.)

(a) The hand can be placed in a crack a number of ways. Sometimes an open hand can be inserted and wedged into a narrower portion of the crack. Other times a clenched fist will provide the necessary grip. Friction can be created by applying cross pressure between the fingers and the back of the hand. Another technique for vertical cracks is to place the hand in the crack with the thumb pointed either up or down. The hand is then clenched as much as possible. When the arm is straightened, it will twist the hand and tend to cam it into place. This combination of clenching and camming usually produces the most friction, and the most secure hand jam in vertical cracks.

(b) In smaller cracks, only the fingers will fit. Use as many fingers as the crack will allow. The fingers can sometimes be stacked in some configuration to increase friction. The thumb is usually kept outside the crack in finger jams and pressed against the rock to increase friction or create cross pressure. In

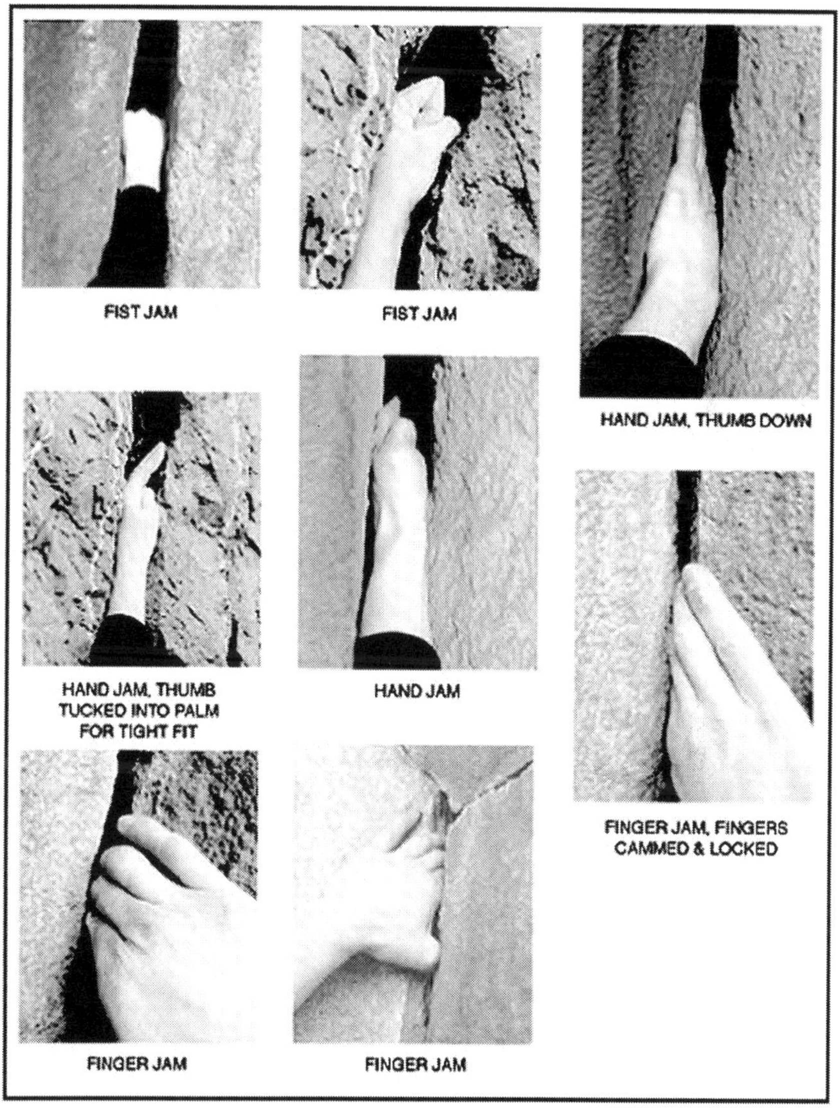

Figure 4-108: Examples of jam holds.

vertical cracks it is best to insert the fingers with the thumb pointing down to make use of the natural camming action of the fingers that occurs when the arm is twisted towards a normal position.

(c) Jamming technique for large cracks, or "off widths," requiring the use of arm, leg, and body jams, is another technique. To jam or cam an arm, leg, or body into an off width, the principle is the same as for fingers, hands, or feet-you are making the jammed appendage "fatter" by folding or twisting it inside the crack. For off widths, you may place your entire arm inside the crack with the arm folded and the palm pointing outward. The leg can be used, from the calf to the thigh, and flexed to fit the crack. Routes requiring this type of climbing should be avoided as the equipment normally used for protection might not be large enough to protect larger cracks and openings. However, sometimes a narrower section may be deeper in the crack allowing the use of "normal" size protection.

4-99. COMBINATION TECHNIQUES

The positions and holds previously discussed are the basics and the ones most common to climbing. From these fundamentals, numerous combination techniques are possible. As the climber gains experience, he will learn more ways to position the hands, feet, and body in relation to the holds available; however, he should always strive to climb with his weight on his feet from a balanced stance.

a. Sometimes, even on an easy route, the climber may come upon a section of the rock that defies the basic principles of climbing. Short of turning back, the only alternative is to figure out some combination technique that will work. Many of these type problems require the hands and feet to work in opposition to one another. Most will place more weight on the hands and arms than is desirable, and some will put the climber in an "out of balance" position. To make the move, the climber may have to "break the rules" momentarily. This is not a problem and is done quite frequently by experienced climbers. The key to using these type of combination techniques is to plan and execute them deliberately, without lunging or groping for holds, yet quickly, before the hands, arms, or other body parts tire. Still, most of these maneuvers require good technique more than great strength, though a certain degree of hand and arm strength certainly helps.

b. Combination possibilities are endless. The following is a brief description of some of the more common techniques.

(1) ***Change Step.*** The change step, or hop step, can be used when the climber needs to change position of the feet. It is commonly used when traversing to avoid crossing the feet, which might put the climber in an awkward position. To prevent an off balance situation, two solid handholds should be used. The climber simply places his weight on his handholds while here positions the feet. He often does this with a quick "hop," replacing the lead foot with the trail foot on the same hold. Keeping the forearms against the rock during the maneuver takes some of the strain off the hands, while at the same time strengthening the grip on the holds.

(2) ***Mantling.*** Mantling is a technique that can be used when the distance between the holds increases and there are no immediate places to move the hands or feet. It does require a ledge (mantle) or projection in the rock that the climber can press straight down upon. (Figure 4-109 shows the mantling sequence.)

(a) When the ledge is above head height, mantling begins with pull holds, usually "hooking" both hands over the ledge. The climber pulls himself up until his head is above the hands, where the pull holds become push holds. He elevates himself until the arms are straight and he can lock the elbows to relax the muscles. Rotating the hands inward during the transition to push holds helps to place the palms more securely on the ledge. Once the arms are locked, a foot can be raised and placed on the ledge. The climber may have to remove one hand to make room for the foot. Mantling can be fairly strenuous; however, most individuals should be able to support their weight, momentarily, on one arm if they keep it straight and locked. With the foot on the ledge, weight can be taken off the arms and the climber can grasp the holds that were previously out of reach.

Once balanced over the foot, he can stand up on the ledge and plan his next move.

(b) Pure mantling uses arm strength to raise the body; however, the climber can often smear the balls of the feet against the rock and "walk" the feet up during the maneuver to take some of the weight off the arms. Sometimes edges will be available for short steps in the process.

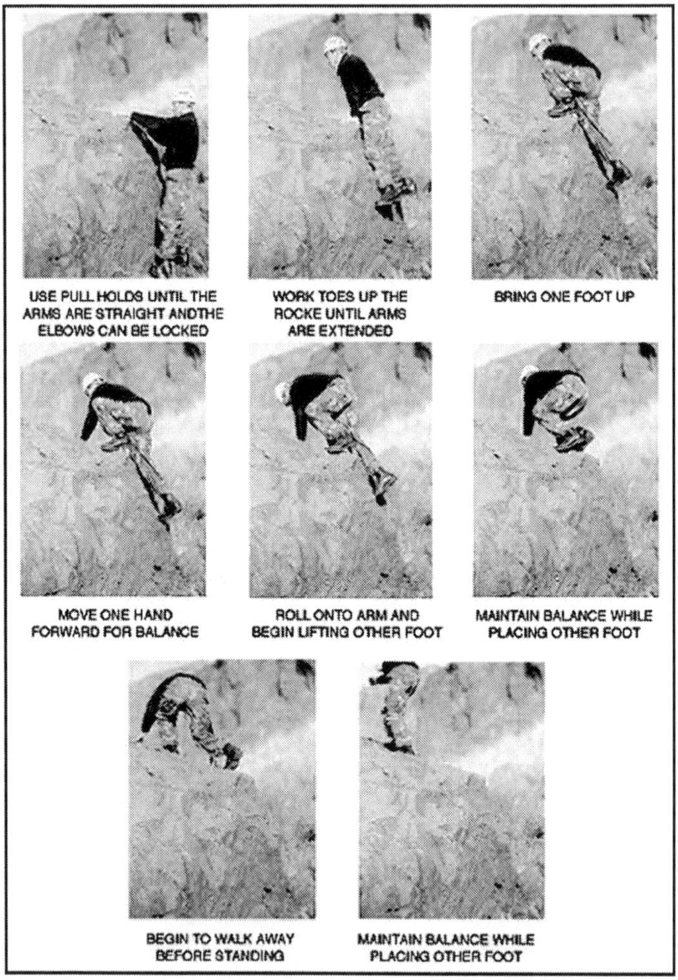

Figure 4-109. Mantling sequence.

(3) ***Undercling***. An "undercling" is a classic example of handholds and footholds working in opposition (Figure 4-110). It is commonly used in places where the rock projects outward, forming a bulge or small overhang. Underclings can be used in the tops of buckets, also. The hands are placed "palms-up" underneath the bulge, applying an upward pull. Increasing this upward pull creates a counterforce, or body tension, which applies more weight and friction to the footholds. The arms and legs should be kept as straight as possible to reduce fatigue. The climber can often lean back slightly in the undercling position, enabling him to see above the overhang better and search for the next hold.

Figure 4-110: Undercling.

(4) ***Lieback***. The "lieback" is another good example of the hands working in opposition to the feet. The technique is often used in a vertical or diagonal crack separating two rock faces that come together at, more or less, a right angle (commonly referred to as a dihedral). The crack edge closest to the body is used for handholds while the feet are pressed against the other edge. The climber bends at the waist, putting the body into an L-shaped position. Leaning away from the crack on two pull holds, body tension creates friction between the feet and the hands. The feet must be kept relatively high to maintain weight, creating maximum friction between the sole and the rock surface. Either full sole contact or the smearing technique can be used, whichever seems to produce the most friction.

(a) The climber ascends a dihedral by alternately shuffling the hands and feet upward. The lieback technique can be extremely tiring, especially when the dihedral is near vertical. If the hands and arms tire out before completing the sequence, the climber will likely fall. The arms should be kept straight throughout the entire maneuver so the climber's weight is pulling against bones and ligaments, rather than muscle. The legs should be straightened whenever possible.

(b) Placing protection in a lieback is especially tiring. Look for edges or pockets for the feet in the crack or on the face for a better position to place protection from, or for a rest position. Often, a lieback can be avoided with closer examination of the available face features. The lieback can be used alternately with the jamming technique, or vice versa, for variation or to get past a section of a crack with difficult or nonexistent jam possibilities. The lieback can sometimes be used as a face maneuver (Figure 4-111).

Figure 4-111: Lieback on a face.

(5) ***Stemming.*** When the feet work in opposition from a relatively wide stance, the maneuver is known as stemming. The stemming technique can sometimes be used on faces, as well as in a dihedral in the absence of solid handholds for the lieback (Figure 4-112).

Figure 4-112: Stemming on a face.

(a) The classic example of stemming is when used in combination with two opposing push holds in wide, parallel cracks, known as chimneys. Chimneys are cracks in which the walls are at least 1 foot apart and just big enough to squeeze the body into. Friction is created by pushing outward with the hands and feet on each side of the crack. The climber ascends the chimney by alternately moving the hands and feet up the crack (Figure 4-113). Applying pressure with the back and bottom is usually necessary in wider chimneys. Usually, full sole contact of the shoes will provide the most friction, although smearing may work best in some instances. Chimneys that do not allow a full stemming position can be negotiated using the arms, legs, or body as an integral contact point. This technique will often feel more secure since there is more body to rock contact.

Figure 4-113: Chimney sequence.

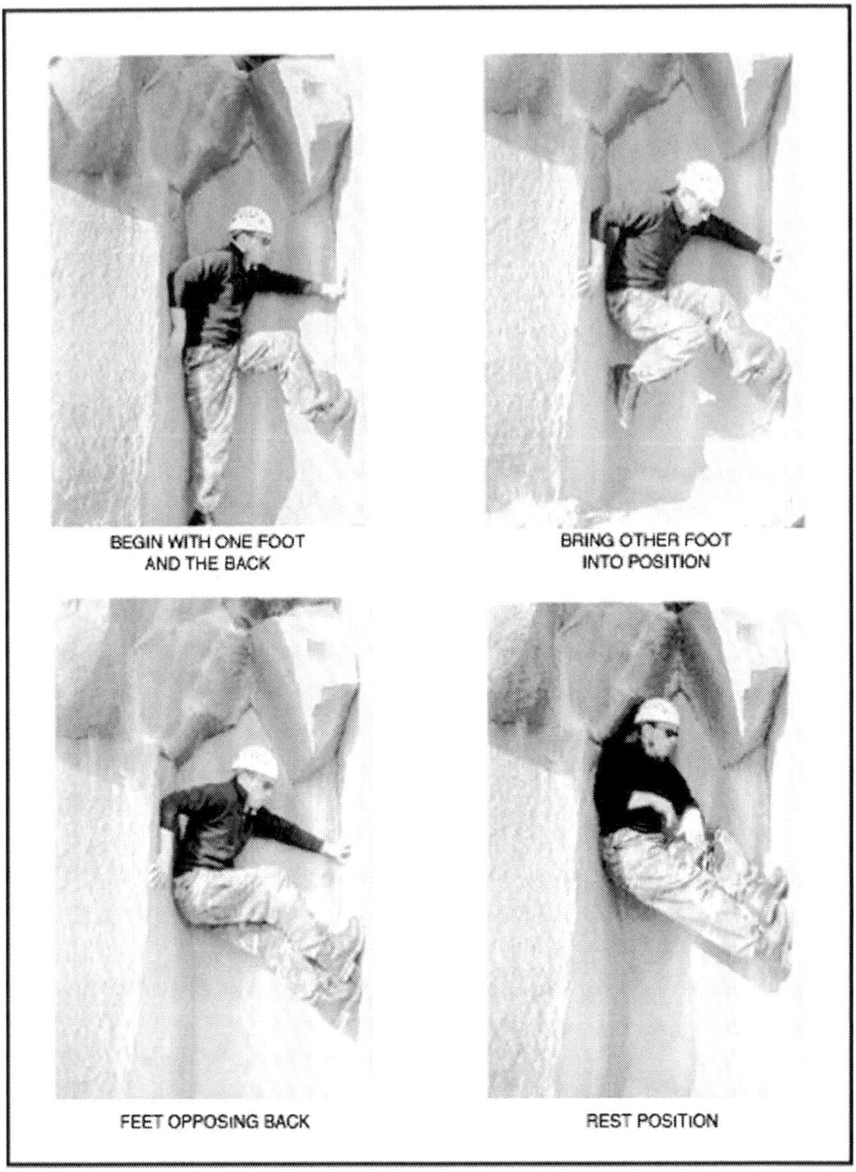

Figure 4-114: Chimney sequence (continued).

(b) The climber can sometimes rest by placing both feet on the same side of the crack, forcing the body against the opposing wall. The feet must be kept relatively high up under the body so the force is directed sideways against the walls of the crack. The arms should be straightened with the elbows locked whenever possible to reduce muscle strain. The climber must ensure that the crack does not widen beyond the climbable width before committing to the maneuver. Remember to look for face features inside chimneys for more security in the climb.

(c) Routes requiring this type of climbing should be avoided as the equipment normally used for protection might not be large enough to protect chimneys. However, face features, or a much narrower crack irotection.

(6) ***Slab Technique***. A slab is a relatively smooth, low-angled rock formation that requires a slightly modified climbing technique (Figure 4-115). Since slab rock normally contains few, if any holds, the technique requires maximum friction and perfect balance over the feet.

(a) On lower-angled slab, the climber can often stand erect and climb using full sole contact and other mountain walking techniques. On steeper slab, the climber will need to apply good smearing technique. Often, maximum friction cannot be attained on steeper slab from an erect stance. The climber will have to flex the ankles and knees so his weight is placed more directly over the balls of the feet. He may then have to bend at the waist to place the hands on the rock, while keeping the hips over his feet.

(b) The climber must pay attention to any changes in slope angle and adjust his body accordingly. Even the slightest change in the position of the hips over the feet can mean the difference between a good grip or a quick slip. The climber should also take advantage of any rough surfaces, or other irregularities in the rock he can place his hands or feet on, to increase friction.

Figure 4-115: Slab technique.

(7) ***Down Climbing.*** Descending steep rock is normally performed using a roped method; however, the climber may at some point be required to down climb a route. Even if climbing ropes and related equipment are on hand, down climbing easier terrain is often quicker than taking the time to rig a rappel point. Also, a climber might find himself confronted with difficulties part way up a route that exceed his climbing ability, or the abilities of others to follow. Whatever the case may be, down climbing is a skill well worth practicing.

⚠ CAUTIONS

1. Down climbing can inadvertently lead into an unforeseen dangerous position on a descent. When in doubt, use a roped descent.
2. Down climbing is accomplished at a difficulty level well below the ability of the climber. When in doubt, use a roped descent.

(a) On easier terrain, the climber can face outward, away from the rock, enabling him to see the route better and descend quickly. As the steepness and difficulty increase, he can often turn sideways, still having a good view of the descent route, but being better able to use the hands and feet on the holds available. On the steepest terrain, the climber will have to face the rock and down climb using good climbing techniques.

(b) Down climbing is usually more difficult than ascending a given route. Some holds will be less visible when down climbing, and slips are more likely to occur. The climber must often lean well away from the rock to look for holds and plan his movements. More weight is placed on the arms and handholds at times to accomplish this, as well as to help lower the climber to the next foothold. Hands should be moved to holds as low as waist level to give the climber more range of movement with each step. If the handholds are too high, he may have trouble reaching the next foothold. The climber must be careful not to overextend himself, forcing a release of his handholds before reaching the next foothold.

⚠ CAUTION

Do not drop from good handholds to a standing position. A bad landing could lead to injured ankles or a fall beyond the planned landing area.

(c) Descending slab formations can be especially tricky. The generally lower angle of slab rock may give the climber a false sense of security, and a tendency to move too quickly. Down climbing must be slow and deliberate, as in ascending, to maintain perfect balance and weight distribution over the feet. On lower-angle slab the climber may be able to stand more or less erect, facing outward or sideways, and descend using good flat foot technique. The climber should avoid the tendency to move faster, which can lead to uncontrollable speed.

(d) On steeper slab, the climber will normally face the rock and down climb, using the same smearing technique as for ascending. An alternate method for descending slab is to face away from the rock in a "crab" position (Figure 4-116). Weight is still concentrated over the feet, but may be shifted partly onto the hands to increase overall friction. The climber is able to maintain full sole contact with the rock and see the entire descent route. Allowing the buttocks to "drag behind" on the rock will decrease the actual weight on the footholds, reducing friction, and leading to the likelihood of a slip. Facing the rock, and down-climbing with good smearing technique, is usually best on steeper slab.

Figure 4-116: Descending slab in the crab position.

ROPED CLIMBING

When the angle, length, and difficulty of the proposed climbing route surpasses the ability of the climbers' safety margin (possibly on class 4 and usually on class 5 terrain), ropes must be used to proceed. Roped climbing is only safe if accomplished correctly. Reading this manual does not constitute skill with ropes-much training and practice is necessary. Many aspects of roped climbing take time to understand and learn. Ropes are normally not used in training until the basic principles of climbing are covered.

> ✍ **NOTE**
> A rope is completely useless for climbing unless the climber knows how to use it safely.

4-99. TYING-IN TO THE CLIMBING ROPE

Over the years, climbers have developed many different knots and procedures for tying-in to the climbing rope. Some of the older methods of tying directly into the rope require minimal equipment and are relatively easy to inspect; however, they offer little support to the climber, may induce further injuries, and may even lead to strangulation in a severe fall. A severe fall, where the climber might fall 20 feet or more and be left dangling on the end of the rope, is highly unlikely in most instances, especially for most personnel involved in military climbing. Tying directly into the rope is perfectly safe for many roped party climbs used in training on lower-angled rock. All climbers should know how to properly tie the rope around the waist in case a climbing harness is unavailable.

4-100. PRESEWN HARNESSES

Although improvised harnesses are made from readily available materials and take little space in the pack or pocket, presewn harnesses provide other aspects that should be considered. No assembly is required, which reduces preparation time for roped movement. All presewn harnesses provide a range of adjustability. These harnesses have a fixed buckle that, when used correctly, will not fail before the nylon materials connected to it. However, specialized equipment, such as a presewn harness, reduce the flexibility of gear. Presewn harnesses are bulky, also.

a. **Seat Harness**. Many presewn seat harnesses are available with many different qualities separating them, including cost.
   (1) The most notable difference will be the amount and placement of padding. The more padding the higher the price and the more comfort. Gear loops sewn into the waist belt on the sides and in the back are a common feature and are usually strong enough to hold quite a few carabiners and or protection. The gear loops will vary in number from one model/manufacturer to another.
   (2) Although most presewn seat harnesses have a permanently attached belay loop connecting the waist belt and the leg loops, the climbing rope should be run around the waist belt and leg loop connector. The presewn belay loop adds another link to the chain of possible failure points and only gives one point of security whereas running the rope through the waist belt and leg loop connector provides two points of contact.
   (3) If more than two men will be on the rope, connect the middle position(s) to the rope with a carabiner routed the same as stated in the previous paragraph.
   (4) Many manufactured seat harnesses will have a presewn loop of webbing on the rear. Although this loop is much stronger than the gear loops, it is not for a belay anchor. It is a quick attachment point to haul an additional rope.

b. **Chest Harness.** The chest harness will provide an additional connecting point for the rope, usually in the form of a carabiner loop to attach a carabiner and rope to. This type of additional connection will provide a comfortable hanging position on the rope, but otherwise provides no additional protection from injury during a fall (if the seat harness is fitted correctly).
   (1) A chest harness will help the climber remain upright on the rope during rappelling or ascending a fixed rope, especially while wearing a heavy pack. (If rappelling or ascending long or multiple pitches, let the pack hang on a drop cord below the feet and attached to the harness tie-in point.)
   (2) The presewn chest harnesses available commercially will invariably offer more comfort or performance features, such as padding, gear loops, or ease of adjustment, than an improvised chest harness.

c. **Full-Body Harness.** Full-body harnesses incorporate a chest and seat harness into one assembly. This is the safest harness to use as it relocates the tie-in point higher, at the chest, reducing the chance of an inverted position when hanging on the rope. This is especially helpful when moving on ropes with heavy packs. A full-body harness only affects the body position when hanging on the rope and will not prevent head injury in a fall.

⚠ CAUTION

This type of harness does not prevent the climber from falling headfirst. Body position during a fall is affected only by the forces that generated the fall, and this type of harness promotes an upright position only when hanging on the rope from the attachment point.

4-101. IMPROVISED HARNESSES

Without the use of a manufactured harness, many methods are still available for attaching oneself to a rope. Harnesses can be improvised using rope or webbing and knots.

a. **Swami Belt.** The swami belt is a simple, belt-only harness created by wrapping rope or webbing around the waistline and securing the ends. One-inch webbing will provide more comfort. Although an effective swami belt can be assembled with a minimum of one wrap, at least two wraps are recommended for comfort, usually with approximately ten feet of material. The ends are secured with an appropriate knot.

b. **Bowline-on-a-Coil.** Traditionally, the standard method for attaching oneself to the climbing rope was with a bowline-on-a-coil around the waist. The extra wraps distribute the force of a fall over a larger area of the torso than a single bowline would, and help prevent the rope from riding up over the rib cage and under the armpits. The knot must be tied snugly around the narrow part of the waist, just above the bony portions of the hips (pelvis). Avoid crossing the wraps by keeping them spread over the waist area. "Sucking in the gut" a bit when making the wraps will ensure a snug fit.

(1) The bowline-on-a-coil can be used to tie-in to the end of the rope (Figure 4-117). The end man should have a

minimum of four wraps around the waist before completing the knot.

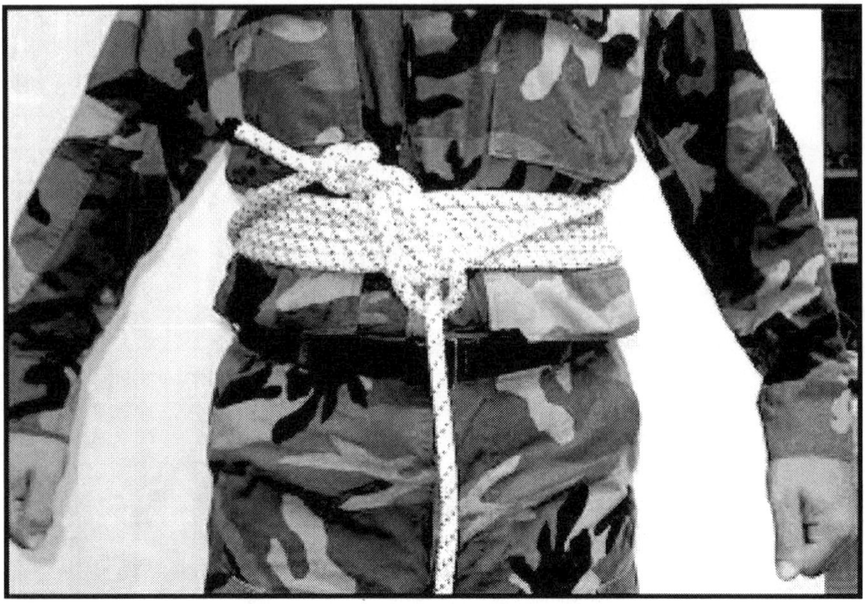

Figure 4-117: Tying-in with a bowline-on-a-coil.

(2) The bowline-on-a-coil is a safe and effective method for attaching to the rope when the terrain is low-angled, WITHOUT THE POSSIBILITY OF A SEVERE FALL. When the terrain becomes steeper, a fall will generate more force on the climber and this will be felt through the coils of this type of attachment. A hard fall will cause the coils to ride up against the ribs. In a severe fall, any tie-in around the waist only could place a "shock load" on the climber's lower back. Even in a relatively short fall, if the climber ends up suspended in mid-air and unable to regain footing on the rock, the rope around the waist can easily cut off circulation and breathing in a relatively short time.

(3) The climbing harness distributes the force of a fall over the entire pelvic region, like a parachute harness. Every climber should know how to tie some sort of improvised climbing harness from sling material. A safe, and

comfortable, seat/chest combination harness can be tied from one-inch tubular nylon.

c. **Improvised Seat Harness.** A seat harness can be tied from a length of webbing approximately 25 feet long (Figure 4-118).

(1) Locate the center of the rope. Off to one side, tie two fixed loops approximately 6 inches apart (overhand loops). Adjust the size of the loops so they fit snugly around the thigh. The loops are tied into the sling "off center" so the remaining ends are different lengths. The short end should be approximately 4 feet long (4 to 5 feet for larger individuals).

(2) Slip the leg loops over the feet and up to the crotch, with the knots to the front. Make one complete wrap around the waist with the short end, wrapping to the outside, and hold it in place on the hip. Keep the webbing flat and free of twists when wrapping.

(3) Make two to three wraps around the waist with the long end in the opposite direction (wrapping to the outside), binding down on the short end to hold it in place. Grasping both ends, adjust the waist wraps to a snug fit. Connect the ends with the appropriate knot between the front and one side so you will be able to see what you are doing.

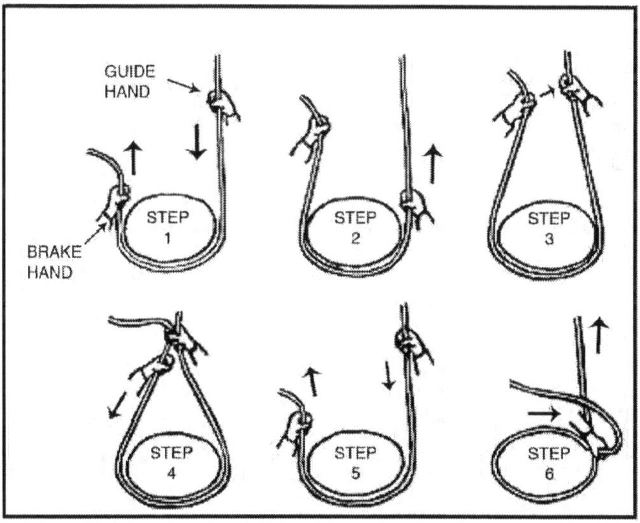

Figure 4-118: Improvised seat and chest harness.

d. **Improvised Chest Harness.** The chest harness can be tied from rope or webbing, but remember that with webbing, wider is better and will be more comfortable when you load this harness. Remember as you tie this harness that the remaining ends will need to be secured so choose the best length. Approximately 6 to 10 feet usually works.
   (1) Tie the ends of the webbing together with the appropriate knot, making a sling 3 to 4 feet long.
   (2) Put a single twist into the sling, forming two loops.
   (3) Place an arm through each loop formed by the twist, just as you would put on a jacket, and drape the sling over the shoulders. The twist, or cross, in the sling should be in the middle of the back.
   (4) Join the two loops at the chest with a carabiner. The water knot should be set off to either side for easy inspection (if a pack is to be worn, the knot will be uncomfortable if it gets between the body and the pack). The chest harness should fit just loose enough to allow necessary clothing and not to restrict breathing or circulation. Adjust the size of the sling if necessary.

e. **Improvised Full-Body Harness.** Full-body harnesses incorporate a chest and seat harness into one assembly.
   (1) The full-body harness is the safest harness because it relocates the tie-in point higher, at the chest, reducing the chance of an inverted hanging position on the rope. This is especially helpful when moving on ropes with heavy packs. A full-body harness affects the body position only when hanging on the rope.

> ⚠ **CAUTION**
> A full-body harness does not prevent falling head first; body position in a fall is caused by the forces that caused the fall.

   (2) Although running the rope through the carabiner of the chest harness does, in effect, create a type of full-body harness, it is not a true full-body harness until the chest harness and the seat harness are connected as one piece. A true full-body harness can be improvised by connecting

the chest harness to the seat harness, but not by just tying the rope into both the two harnesses must be "fixed" as one harness. Fix them together with a short loop of webbing or rope so that the climbing rope can be connected directly to the chest harness and your weight is supported by the seat harness through the connecting material.

f. Attaching the Rope to the Improvised Harness. The attachment of the climbing rope to the harness is a CRITICAL LINK. The strength of the rope means nothing if it is attached poorly, or incorrectly, and comes off the harness in a fall. The climber ties the end of the climbing rope to the seat harness with an appropriate knot. If using a chest harness, the standing part of the rope is then clipped into the chest harness carabiner. The seat harness absorbs the main force of the fall, and the chest harness helps keep the body upright.

⚠ CAUTION

The knot must be tied around all the waist wraps and the 6-inch length of webbing between the leg loops.

(1) A middleman must create a fixed loop to tie in to. A rethreaded figure-eight loop tied on a doubled rope or the three loop bowline can be used. If using the three loop bowline, ensure the end, or third loop formed in the knot, is secured around the tie-in loops with an overhand knot. The standing part of the rope going to the lead climber is clipped into the chest harness carabiner.

✍ NOTE

The climbing rope is not clipped into the chest harness when belaying.

(2) The choice of whether to tie-in with a bowline-on-a-coil or into a climbing harness depends entirely on the climber's judgment, and possibly the equipment available. A good rule of thumb is: "Wear a climbing harness when

the potential for severe falls exists and for all travel over snow-covered glaciers because of the crevasse fall hazard."

(3) Under certain conditions many climbers prefer to attach the rope to the seat harness with a locking carabiner, rather than tying the rope to it. This is a common practice for moderate snow and ice climbing, and especially for glacier travel where wet and frozen knots become difficult to untie.

> **⚠ CAUTION**
> Because the carabiner gate may be broken or opened by protruding rocks during a fall, tie the rope directly to the harness for maximum safety.

BELAY TECHNIQUES

Tying-in to the climbing rope and moving as a member of a rope team increases the climber's margin of safety on difficult, exposed terrain. In some instances, such as when traveling over snow-covered glaciers, rope team members can often move at the same time, relying on the security of a tight rope and "team arrest" techniques to halt a fall by any one member. On steep terrain, however, simultaneous movement only helps to ensure that if one climber falls, he will jerk the other rope team members off the slope. For the climbing rope to be of any value on steep rock climbs, the rope team must incorporate "belays" into the movement.

Belaying is a method of managing the rope in such a way that, if one person falls, the fall can be halted or "arrested" by another rope team member (belayer). One person climbs at a time, while being belayed from above or below by another. The belayer manipulates the rope so that friction, or a "brake," can be applied to halt a fall. Belay techniques are also used to control the descent of personnel and equipment on fixed rope installations, and for additional safety on rappels and stream crossings.

Belaying is a skill that requires practice to develop proficiency. Setting up a belay may at first appear confusing to the beginner, but with practice, the procedure should become "second nature." If confronted with a peculiar problem during the setup of a belay, try to use common sense and apply the basic principles stressed throughout this text. Remember the following key points:

- Select the best possible terrain features for the position and use terrain to your advantage.
- Use a well braced, sitting position whenever possible.
- Aim and anchor the belay for all possible load directions.
- Follow the "minimum" rule for belay anchors-2 for a downward pull, 1 for an upward pull.
- Ensure anchor attachments are aligned, independent, and snug.
- Stack the rope properly.
- Choose a belay technique appropriate for the climbing.
- Use a guide carabiner for rope control in all body belays.
- Ensure anchor attachments, guide carabiner (if applicable), and rope running to the climber are all on the guide hand side.

> ⚠ **CAUTION**
> Never remove the brake hand from the rope while belaying. If the brake hand is removed, there is no belay.

- The brake hand remains on the rope when belaying.
- Ensure you are satisfied with your position before giving the command "BELAY ON."
- The belay remains in place until the climber gives the command "OFF BELAY."

> ⚠ **CAUTION**
> The belay remains in place the from the time the belayer commands *"BELAY ON"* until the climber commands *"OFF BELAY."*

4-102. PROCEDURE FOR MANAGING THE ROPE

A number of different belay techniques are used in modern climbing, ranging from the basic "body belays" to the various "mechanical belays," which incorporate some type of friction device.

a. Whether the rope is wrapped around the body, or run through a friction device, the rope management procedure is basically the same. The belayer must be able to perform three basic functions: manipulate the rope to give the climber slack during movement, take up rope to remove excess slack, and apply the brake to halt a fall.

b. The belayer must be able to perform all three functions while maintaining "total control" of the rope at all times. Total control means the brake hand is NEVER removed from the rope. When giving slack, the rope simply slides through the grasp of the brake hand, at times being fed to the climber with the other "feeling" or guide hand. Taking up rope, however, requires a certain technique to ensure the brake hand remains on the rope at all times. The following procedure describes how to take up excess rope and apply the brake in a basic body belay.

(1) Grasping the rope with both hands, place it behind the back and around the hips. The hand on the section of rope between the belayer and the climber would be the guide hand. The other hand is the brake hand.

(2) Take in rope with the brake hand until the arm is fully extended. The guide hand can also help to pull in the rope (Figure 4-118, step 1).

(3) Holding the rope in the brake hand, slide the guide hand out, extending the arm so the guide hand is father away from the body than the brake hand (Figure 4-118, step 2).

(4) Grasp both parts of the rope, to the front of the brake hand, with the guide hand (Figure 4-118, step 3).

(5) Slide the brake hand back towards the body (Figure 4-118, step 4).

(6) Repeat step 5 of Figure 4-118. The brake can be applied at any moment during the procedure. It is applied by wrapping the rope around the front of the hips while increasing grip with the brake hand (Figure 4-118, step 6).

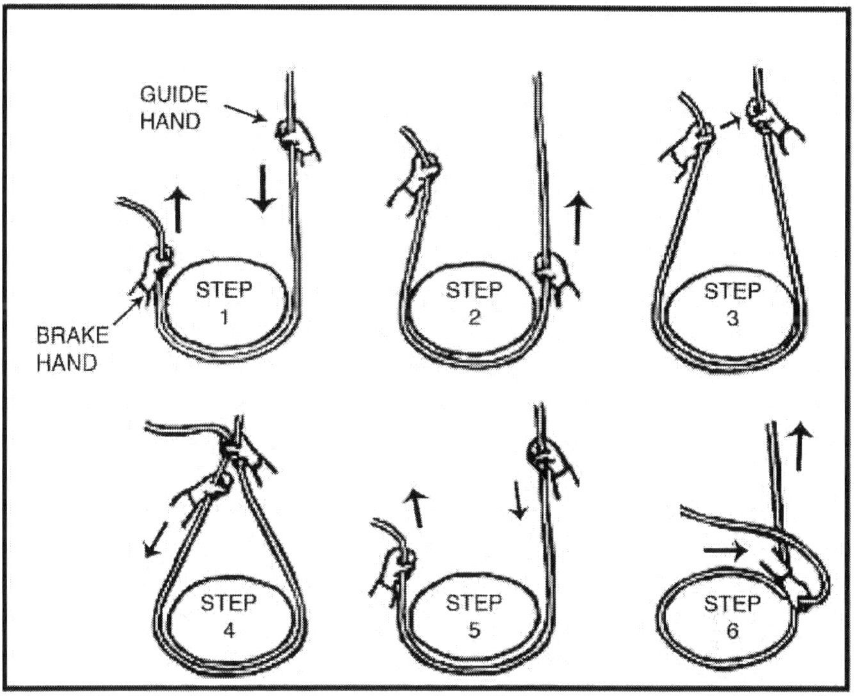

Figure 4-118: Managing the rope.

4-103. CHOOSING A BELAY TECHNIQUE

The climber may choose from a variety of belay techniques. A method that works well in one situation may not be the best choice in another. The choice between body belays and mechanical belays depends largely on equipment available, what the climber feels most comfortable with, and the amount of load, or fall force, the belay may have to absorb. The following describes a few of the more widely used techniques, and the ones most applicable to military mountaineering.

a. **Body Belay**. The basic body belay is the most widely used technique on moderate terrain. It uses friction between the rope and the clothed body as the rope is pressured across the clothing. It is the simplest belay, requiring no special equipment, and should be the first technique learned by all climbers. A body belay gives the belayer the greatest "feel" for the climber, letting him know when to give slack or take up rope. Rope management in a body belay is quick and easy, especially for beginners, and is effective in snow and ice climbing when ropes often become wet, stiff, and frozen. The body belay, in its various forms, will hold low to moderate impact falls well. It has been known to arrest some severe falls, although probably not without inflicting great pain on the belayer.

⚠ CAUTION

The belayer must ensure he is wearing adequate clothing to protect his body from rope burns when using a body belay. Heavy duty cotton or leather work gloves can also be worn to protect the hands.

(1) ***Sitting Body Belay***. The sitting body belay is the preferred position and is usually the most secure (Figure 4-119). The belayer sits facing the direction where the force of a fall will likely come from, using terrain to his advantage, and attempts to brace both feet against the rock to support his position. It is best to sit in a slight depression, placing the buttocks lower than the feet, and straightening the

Figure 4-119: Sitting body belay.

legs for maximum support. When perfectly aligned, the rope running to the climber will pass between the belayer's feet, and both legs will equally absorb the force of a fall. Sometimes, the belayer may not be able to sit facing the direction he would like, or both feet cannot be braced well. The leg on the "guide hand" side should then point towards the load, bracing the foot on the rock when possible. The belayer can also "straddle" a large tree or rock nubbin for support, as long as the object is solid enough to sustain the possible load.

(2) ***Standing Body Belay***. The standing body belay is used on smaller ledges where there is no room for the belayer to sit (Figure 4-120). What appears at first to be a fairly unstable position can actually be quite secure when belay anchors are placed at or above shoulder height to support the stance when the force will be downward.

Figure 4-120. Standing body belay.

(a) For a body belay to work effectively, the belayer must ensure that the rope runs around the hips properly, and remains there under load when applying the brake. The rope should run around the narrow portion of the pelvic girdle, just below the bony high points of the hips. If the rope runs too high, the force of a fall could injure the belayer's midsection and lower rib cage. If the rope runs too low, the load may pull the rope below the buttocks, dumping the belayer out of position. It is also possible for a strong upward or downward pull to strip the rope away from the belayer, rendering the belay useless.

(b) To prevent any of these possibilities from happening, the belay rope is clipped into a carabiner attached to the guide hand side of the seat harness (or bowline-on-a-coil). This "guide carabiner" helps keep the rope in place around the hips and prevents loss of control in upward or downward loads (Figure 4-121).

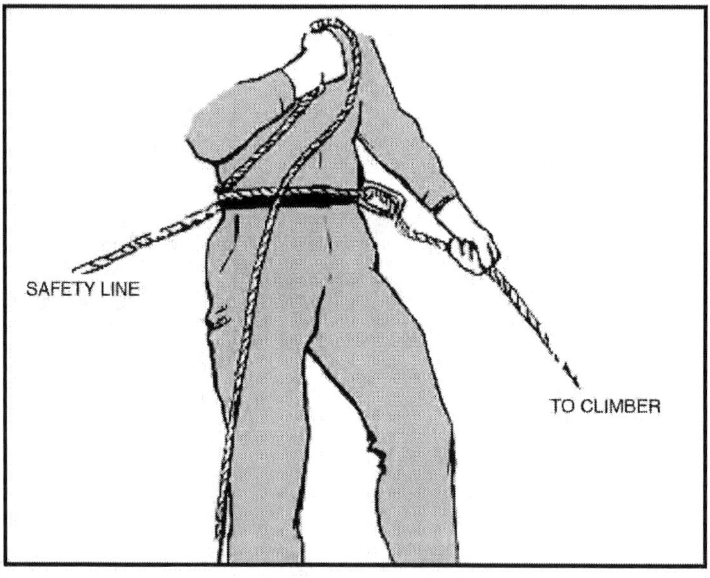

Figure 4-121: Guide carabiner for rope control in a body belay.

b. **Mechanical Belay.** A mechanical belay must be used whenever there is potential for the lead climber to take a severe fall. The holding power of a belay device is vastly superior to any body belay under high loads. However, rope management in a mechanical belay is more difficult to master and requires more practice. For the most part, the basic body belay should be totally adequate on a typical military route, as routes used during military operations should be the easiest to negotiate.

(1) ***Munter Hitch***. The Munter hitch is an excellent mechanical belay technique and requires only a rope and a carabiner (Figure 4-122). The Munter is actually a two-way friction hitch. The Munter hitch will flip back and forth through the carabiner as the belayer switches from giving slack to taking up rope. The carabiner must be large enough, and of the proper design, to allow this function. The locking pear-shaped carabiner, or pearabiner, is designed for the Munter hitch.

(a) The Munter hitch works exceptionally well as a lowering belay off the anchor. As a climbing belay, the carabiner should be attached to the front of the belayer's seat harness. The hitch is tied by forming a loop and a bight in the rope, attaching both to the carabiner. It's fairly easy to place the bight on the carabiner backwards, which forms an obvious, useless hitch. Put some tension on the Munter to ensure it is formed correctly, as depicted in the following illustrations.

(b) The Munter hitch will automatically "lock-up" under load as the brake hand grips the rope. The brake is

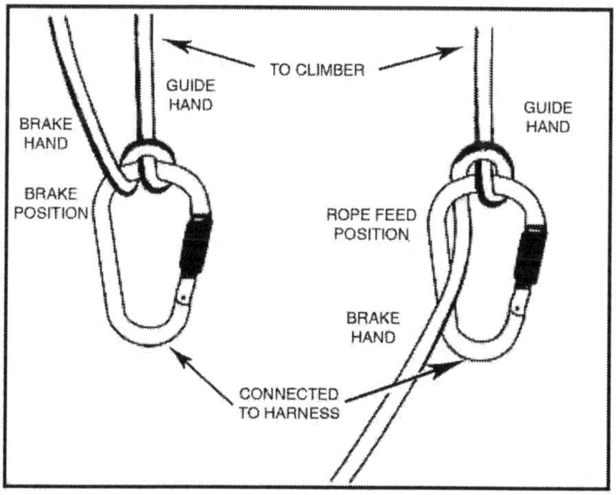

Figure 4-122: Munter hitch.

increased by pulling the slack rope away from the body, towards the load. The belayer must be aware that flipping the hitch DOES NOT change the function of the hands. The hand on the rope running to the climber, or load, is always the guide hand.

(2) ***Figure-Eight Device.*** The figure-eight device is a versatile piece of equipment and, though developed as a rappel device, has become widely accepted as an effective mechanical belay device (Figure 4-123). The advantage of any mechanical belay is friction required to halt a fall is applied on the rope through the device, rather than around the belayer's body. The device itself provides rope control for upward and downward pulls and excellent friction for halting severe falls. The main principle behind the figure-eight device in belay mode is the friction developing on the rope as it reaches and exceeds the 90-degree angle between the rope entering the device and leaving the device. As a belay device, the figure-eight works well for both belayed climbing and for lowering personnel and equipment on fixed-rope installations.

(a) As a climbing belay, a bight placed into the climbing rope is run through the "small eye" of the device and attached to a locking carabiner at the front of the belayer's seat harness. A short, small diameter

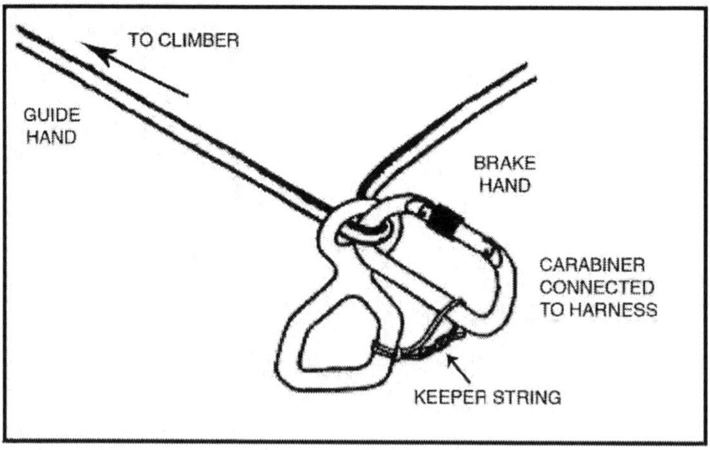

Figure 4-123: Figure-eight device.

safety rope is used to connect the "large eye" of the figure eight to the locking carabiner for control of the device. The guide hand is placed on the rope running to the climber. Rope management is performed as in a body belay. The brake is applied by pulling the slack rope in the brake hand towards the body, locking the rope between the device and the carabiner.

(b) As a lowering belay, the device is normally attached directly to the anchor with the rope routed as in rappelling.

> ✍ **NOTE**
> Some figure-eight descenders should not be used as belay devices due to their construction and design. Always refer to manufacturer's specifications and directions before use.

(3) ***Mechanical Camming Device.*** The mechanical camming device has an internal camming action that begins locking the rope in place as friction is increased. Unlike the other devices, the mechanical camming device can stop a falling climber without any input from the belayer. A few other devices perform similarly to this, but have no moving parts. Some limitations to these type devices are minimum and maximum rope diameters.

(4) ***Other Mechanical Belay Devices.*** There are many other commercially available mechanical belay devices. Most of these work with the same rope movement direction and the same braking principle. The air traffic controller (ATC), slotted plate, and other tube devices are made in many different shapes. These all work on the same principle as the figure-eight device friction increases on the rope as it reaches and exceeds the 90-degreeangle between the rope entering the device and leaving the device.

4-105. ESTABLISHING A BELAY

A belay can be established using either a direct or indirect connection. Each type has advantages and disadvantages. The choice will depend on the intended use of the belay.

a. **Direct Belay**. The direct belay removes any possible forces from the belayer and places this force completely on the anchor. Used often for rescue installations or to bring a second climber up to anew belay position in conjunction with the Munter hitch, the belay can be placed above the belayer's stance, creating a comfortable position and ease of applying the brake. Also, if the second falls or weights the rope, the belayer is not locked into a position. Direct belays provide no shock-absorbing properties from the belayer's attachment to the system as does the indirect belay; therefore, the belayer is apt to pay closer attention to the belaying process.
b. **Indirect Belay**. An indirect belay, the most commonly used, uses a belay device attached to the belayer's harness. This type of belay provides dynamic shock or weight absorption by the belayer if the climber falls or weights the rope, which reduces the direct force on the anchor and prevents a severe shock load to the anchor.

4-106. SETTING UP A BELAY

In rock climbing, climbers must sometimes make do with marginal protection placements along a route, but belay positions must be made as "bombproof" as possible. Additionally, the belayer must set up the belay in relation to where the fall force will come from and pay strict attention to proper rope management for the belay to be effective. All belay positions are established with the anchor connection to the front of the harness. If the belay is correctly established, the belayer will feel little or no force if the climber falls or has to rest on the rope. Regardless of the actual belay technique used, five basic steps are required to set up a sound belay.

a. **Select Position and Stance.** Once the climbing line is picked, the belayer selects his position. It's best if the position is off to the side of the actual line, putting the belayer out of the direct path of a potential fall or any rocks kicked loose by the climber. The position should allow the belayer to maintain a comfortable, relaxed stance, as he could be in the position for a fairly long time. Large ledges that allow a well braced, sitting stance are preferred. Look for belay positions close to bombproof natural anchors. The position must at least allow for solid artificial placements.
b. **Aim the Belay.** With the belay position selected, the belay must now be "aimed." The belayer determines where the

rope leading to the climber will run and the direction the force of a fall will likely come from. When a lead climber begins placing protection, the fall force on the belayer will be in some upward direction, and in line with the first protection placement. If this placement fails under load, the force on the belay could be straight down again. The belayer must aim his belay for all possible load directions, adjusting his position or stance when necessary. The belay can be aimed through an anchor placement to immediately establish an upward pull; however, the belayer must always be prepared for the more severe downward fall force in the event intermediate protection placements fail.

c. **Anchor the Belay.** For a climbing belay to be considered bombproof, the belayer must be attached to a solid anchor capable of withstanding the highest possible fall force. A solid natural anchor would be ideal, but more often the belayer will have to place pitons or chocks. A single artificial placement should never be considered adequate for anchoring a belay (except at ground level). Multiple anchor points capable of supporting both upward and downward pulls should be placed. The rule of thumb is to place two anchors for a downward pull and one anchor for an upward pull as a MINIMUM. The following key points also apply to anchoring belays.
   (1) Each anchor must be placed in line with the direction of pull it is intended to support.
   (2) Each anchor attachment must be rigged "independently" so a failure of one will not shock load remaining placements or cause the belayer to be pulled out of position.
   (3) The attachment between the anchor and the belayer must be snug to support the stance. Both belayer's stance and belay anchors should absorb the force of a fall.
   (4) It is best for the anchors to be placed relatively close to the belayer with short attachments. If the climber has to be tied-off in an emergency, say after a severe fall, the belayer can attach a Prusik sling to the climbing rope, reach back, and connect the sling to one of the anchors. The load can be placed on the Prusik and the belayer can come out of the system to render help.
   (5) The belayer can use either a portion of the climbing rope or slings of the appropriate length to connect himself to the anchors. It's best to use the climbing rope whenever

possible, saving the slings for the climb. The rope is attached using either figure eight loops or clove hitches. Clove hitches have the advantage of being easily adjusted. If the belayer has to change his stance at some point, he can reach back with the guide hand and adjust the length of the attachment through the clove hitch as needed.

(6) The anchor attachments should also help prevent the force of a fall from "rotating" the belayer out of position. To accomplish this, the climbing rope must pass around the "guide-hand side" of the body to the anchors. Sling attachments are connected to the belayer's seat harness (or bowline-on-a-coil) on the guide-hand side.

(7) Arrangement of rope and sling attachments may vary according to the number and location of placements. Follow the guidelines set forth and remember the key points for belay anchors; "in line", "independent", and "snug". Figure 6-27 shows an example of a common

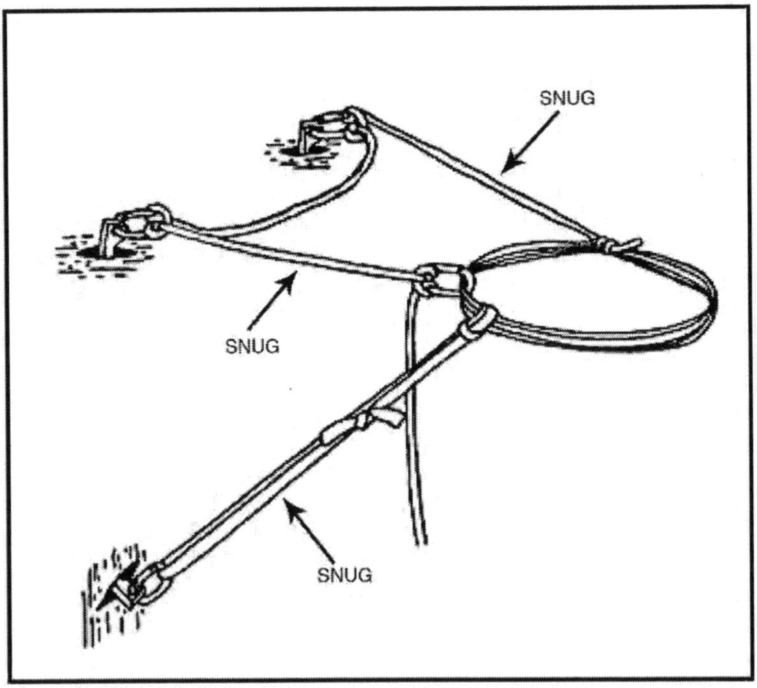

Figure 4-124: Anchoring a belay.

arrangement, attaching the rope to the two "downward" anchors and a sling to the "upward" anchor. Note how the rope is connected from one of the anchors back to the belayer. This is not mandatory, but often helps "line-up" the second attachment.

d. **Stack the Rope**. Once the belayer is anchored into position, he must stack the rope to ensure it is free of twists and tangles that might hinder rope management in the belay. The rope should be stacked on the ground, or on the ledge, where it will not get caught in cracks or nubbins as it is fed out to the climber.
   (1) On small ledges, the rope can be stacked on top of the anchor attachments if there is no other place to lay it, but make sure to stack it carefully so it won't tangle with the anchored portion of the rope or other slings. The belayer must also ensure that the rope will not get tangled around his legs or other body parts as it "feeds" out.
   (2) The rope should never be allowed to hang down over the ledge. If it gets caught in the rock below the position, the belayer may have to tie-off the climber and come out of the belay to free the rope; a time-consuming and unnecessary task. The final point to remember is the rope must be stacked "from the belayer's end" so the rope running to the climber comes off the "top" of the stacked pile.

e. Attach the Belay. The final step of the procedure is to attach the belay. With the rope properly stacked, the belayer takes the rope coming off the top of the pile, removes any slack between himself and the climber, and applies the actual belay technique. If using a body belay, ensure the rope is clipped into the guide carabiner.
   (1) The belayer should make one quick, final inspection of his belay. If the belay is set up correctly, the anchor attachments, guide carabiner if applicable, and the rope running to the climber will all be on the "guide hand" side, which is normally closest to the rock (Figure 4-125). If the climber takes a fall, the force, if any, should not have any negative effect on the belayer's involvement in the system. The brake hand is out away from the slope where it won't be jammed between the body and the rock. The guide hand can be placed on the rock to help support the stance when applying the brake.

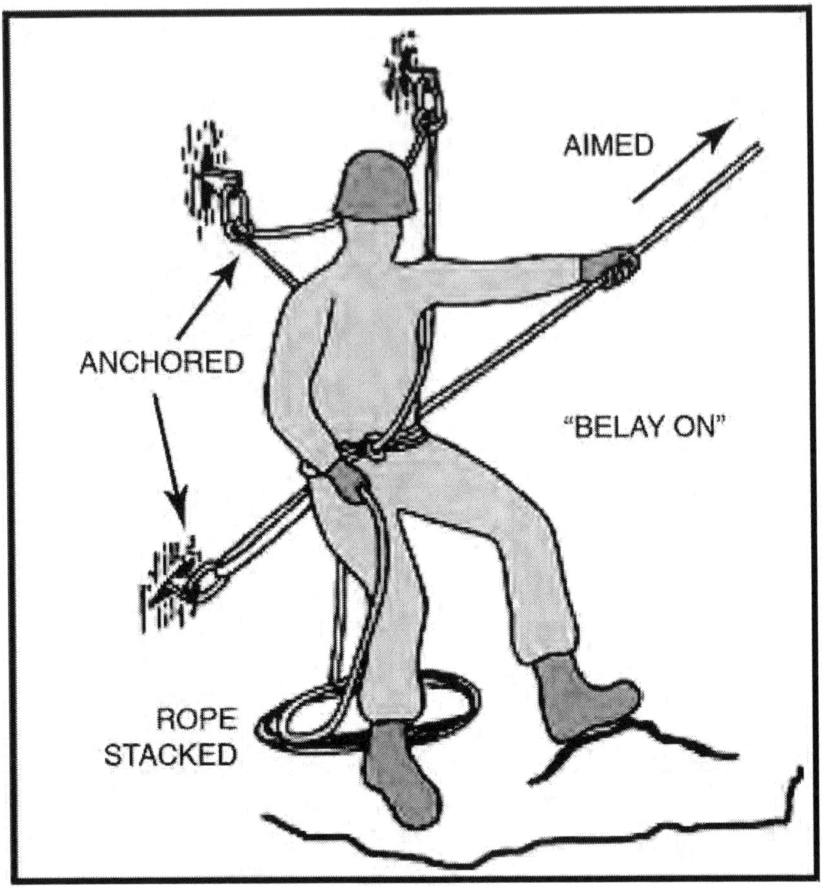

Figure 4-125: Belay setup.

(2) When the belayer is satisfied with his position, he gives the signal, "BELAY ON!". When belaying the "second", the same procedure is used to set up the belay. Unless the belay is aimed for an upward pull, the fall force is of course downward and the belayer is usually facing away from the rock, the exception being a hanging belay on a vertical face. If the rope runs straight down to the climber and the anchors are directly behind the position, the belayer may choose to brake with the hand he feels most comfortable with. Anchor attachments, guide carabiner, and rope running to the climber through the guide

hand must still be aligned on the same side to prevent the belayer from being rotated out of position, unless the belayer is using an improvised harness and the anchor attachment is at the rear.

4-107. TOP-ROPE BELAY

A "top-rope" is a belay setup used in training to protect a climber while climbing on longer, exposed routes. A solid, bombproof anchor is required at the top of the pitch. The belayer is positioned either on the ground with the rope running through the top anchor and back to the climber, or at the top at the anchor. The belayer takes in rope as the climber proceeds up the rock. If this is accomplished with the belayer at the bottom, the instructor can watch the belayer while he coaches the climber through the movements.

> **⚠ CAUTION**
>
> Do not use a body belay for top-rope climbing. The rope will burn the belayer if the climber has to be lowered.

CLIMBING COMMANDS

Communication is often difficult during a climb. As the distance between climber and belayer increases, it becomes harder to distinguish one word from another and the shortest sentence may be heard as nothing more than jumbled syllables. A series of standard voice commands were developed over the years to signal the essential rope management functions in a belayed climb. Each command is concise and sounds a bit different from another to reduce the risk of a misunderstanding between climber and belayer. They must be pronounced clearly and loudly so they can be heard and understood in the worst conditions.

4-108. VERBAL COMMANDS

Verbal Commands Chart lists standard rope commands and their meanings in sequence as they would normally be used on a typical climb. (Note how the critical "BELAY" commands are reversed so they sound different and will not be confused.)

BELAYER	CLIMBER	MEANING/ACTION TAKEN
"BELAY ON"		The belay is on; you may climb when ready; the rope will be managed as needed.
	"CLIMBING" (as a courtesy)	I am ready to climb.
"CLIMB" (as a courtesy)		Proceed, and again, the rope will be managed as necessary.
"ROCK"	"ROCK"	**PROTECT YOURSELF FROM FALLING OBJECTS.** Signal will be echoed by all climbers in the area. If multipitch climbing, ensure climbers below hear.
	"TAKE ROPE"	Take in excess rope between us without pulling me off the route. Belayer takes in rope.
	"SLACK"	Release all braking/tension on the rope so I can have slack without pulling the rope. Belayer removes brake/tension.
	"TENSION"	Take all the slack, apply brake, and hold me. My weight will be on the rope. Belayer removes slack and applies brake.
	"FALLING"	I am falling. Belayer applies brake to arrest the fall.
"TWEN-TY-FIVE"		You have approximately 25 feet of rope left. Start looking for the next belay position. Climber selects a belay position.
"FIF-TEEN"		You have approximately 15 feet of rope left. Start looking for the next belay position. Climber selects a belay position within the next few feet.
"FIVE"	Set up the belay.	You have 5 feet of rope left. Set up the belay position. You have no more rope. Climber sets up the belay.
Removes the belay, remains anchored. Prepares to climb.	"OFF BELAY"	I have finished climbing and I am anchored. You may remove the belay. Belayer removes the belay and, remaining anchored, prepares to climb.

4-109. ROPE TUG COMMANDS

Sometimes the loudest scream cannot be heard when the climber and belayer are far apart. This is especially true in windy conditions, or when the climber is around a corner, above an overhang, or at the back of a ledge. It may be necessary to use a series of "tugs" on the rope in place of the standard voice commands. To avoid any possible confusion with interpretation of multiple rope tug commands, use only one.

a. While a lead climb is in progress, the most important command is "BELAY ON." This command is given only by the climber when the climber is anchored and is prepared for the second to begin climbing. With the issue of this command, the second knows the climber is anchored and the second prepares to climb.
b. For a rope tug command, the leader issues three distinct tugs on the rope AFTER anchoring and putting the second on belay. This is the signal for "BELAY ON" and signals the second to climb when ready. The new belayer keeps slack out of the rope.

ROPED CLIMBING METHODS

In military mountaineering, the primary mission of a roped climbing team is to "fix" a route with some type of rope installation to assist movement of less trained personnel in the unit. This duty falls upon the most experienced climbers in the unit, usually working in two- or three-man groups or teams called assault climbing teams. Even if the climbing is for another purpose, roped climbing should be performed whenever the terrain becomes difficult and exposed.

4-110. TOP-ROPED CLIMBING

Top-roped climbing is used for training purposes only. This method of climbing is not used for movement due to the necessity of pre-placing anchors at the top of a climb. If you can easily access the top of a climb, you can easily avoid the climb itself.

a. For training, top-roped climbing is valuable because it allows climbers to attempt climbs above their skill level and or to hone present skills without the risk of a fall. Top-roped climbing may be used to increase the stamina of a climber training to climb longer routes as well as for a climber practicing protection placements.

b. The belayer is positioned either at the base of a climb with the rope running through the top anchor and back to the climber or at the top at the anchor. The belayer takes in rope as the climber moves up the rock, giving the climber the same protection as a belay from above. If this is accomplished with the belayer at the bottom, the instructor is able to keep an eye on the belayer while he coaches the climber through the movements.

4-111. LEAD CLIMBING

A lead climb consists of a belayer, a leader or climber, rope(s), and webbing or hardware used to establish anchors or protect the climb. As he climbs the route, the leader emplaces "intermediate" anchors, and the climbing rope is connected to these anchors with a carabiner. These "intermediate" anchors protect the climber against a fall-thus the term "protecting the climb."

> ✍ **NOTE**
>
> Intermediate anchors are commonly referred to as "protection," "pro," "pieces," "pieces of pro," "pro placements," and so on. For standardization within this publication, these specific anchors will be referred to as "protection;" anchors established for other purposes, such as rappel points, belays, or other rope installations, will be referred to as "anchors."

> ⚠ **CAUTION**
>
> During all lead climbing, each climber in the team is either anchored or being belayed.

a. Lead climbing with two climbers is the preferred combination for movement on technically difficult terrain. Two climbers are at least twice as fast as three climbers, and are efficient for installing a "fixed rope," probably the most widely used rope installation in the mountains. A group of three climbers are typically used on moderate snow, ice, and snow-covered glaciers where the rope team can often move at the same time, stopping occasionally to set up belays on particularly difficult sections. A group or team of three climbers is sometimes used in rock climbing because of an odd number of personnel, a shortage of ropes (such as six climbers and only two ropes), or to protect and assist an individual who has little or no experience in climbing and belaying. Whichever technique is chosen, a standard roped climbing procedure is used for maximum speed and safety.

b. When the difficulty of the climbing is within the "leading ability" of both climbers, valuable time can be saved by "swinging leads." This is normally the most efficient method for climbing multipitch routes. The second finishes cleaning the first pitch and continues climbing, taking on the role of lead climber. Unless he requires equipment from the other rack or desires a break, he can climb past the belay and immediately begin leading. The belayer simply adjusts his position, re-aiming the belay once the new leader begins placing protection. Swinging leads, or "leapfrogging," should be planned before starting the climb so the leader knows to anchor the upper belay for both upward and downward pulls during the setup.

c. The procedures for conducting a lead climb with a group of two are relatively simple. The most experienced individual is the "lead" climber or leader, and is responsible for selecting the route. The leader must ensure the route is well within his ability and the ability of the second. The lead climber carries most of the climbing equipment in order to place protection along the route and set up the next belay. The leader must also ensure that the second has the necessary equipment, such as a piton hammer, nut tool, etc., to remove any protection that the leader may place.

(1) The leader is responsible for emplacing protection frequently enough and in such a manner that, in the event that either the leader or the second should fall, the fall will be neither long enough nor hard enough to result in injury. The leader must also ensure that the rope is routed in a way that will allow it to run freely through the protection placements, thus minimizing friction, or "rope drag".

(2) The other member of the climbing team, the belayer (sometimes referred to as the "second"), is responsible for belaying the leader, removing the belay anchor, and

retrieving the protection placed by the leader between belay positions (also called "cleaning the pitch").

(3) Before the climb starts, the second will normally set up the first belay while the leader is arranging his rack. When the belay is ready, the belayer signals, "BELAY ON", affirming that the belay is "on" and the rope will be managed as necessary. When the leader is ready, he double checks the belay. The leader can then signal, "CLIMBING", only as a courtesy, to let the belayer know he is ready to move. The belayer can reply with "CLIMB", again, only as a courtesy, reaffirming that the belay is "on" and the rope will be managed as necessary. The leader then begins climbing.

(4) While belaying, the second must pay close attention to the climber's every move, ensuring that the rope runs free and does not inhibit the climber's movements. If he cannot see the climber, he must "feel" the climber through the rope. Unless told otherwise by the climber, the belayer can slowly give slack on the rope as the climber proceeds on the route. The belayer should keep just enough slack in the rope so the climber does not have to pull it through the belay. If the climber wants a tighter rope, it can be called for. If the belayer notices too much slack developing in the rope, the excess rope should be taken in quickly. It is the belayer's responsibility to manage the rope, whether by sight or feel, until the climber tells him otherwise.

(5) As the leader protects the climb, slack will sometimes be needed to place the rope through the carabiner (clipping), in a piece of protection above the tie-in point on the leaders harness. In this situation, the leader gives the command "SLACK" and the belayer gives slack, (if more slack is needed the command will be repeated). The leader is able to pull a bight of rope above the tie-in point and clip it into the carabiner in the protection above. When the leader has completed the connection, or the clip, the command "TAKE ROPE" is given by the leader and the belayer takes in the remaining slack.

(6) The leader continues on the route until either a designated belay location is reached or he is at the end of or near the end of the rope. At this position, the leader

sets an anchor, connects to the anchor and signals "OFF BELAY". The belayer prepares to climb by removing all but at least one of his anchors and secures the remaining equipment. The belayer remains attached to at least one anchor until the command "BELAY ON" is given.

d. When the leader selects a particular route, he must also determine how much, and what types, of equipment might be required to safely negotiate the route. The selected equipment must be carried by the leader. The leader must carry enough equipment to safely protect the route, additional anchors for the next belay, and any other items to be carried individually such as rucksacks or individual weapons.

(1) The leader will assemble, or "rack," the necessary equipment onto his harness or onto slings around the head and shoulder. A typical leader "rack" consists of:

- Six to eight small wired stoppers on a carabiner.
- Four to six medium to large wired stoppers on a carabiner.
- Assorted hexentrics, each on a separate carabiner.
- SLCDs of required size, each on a separate carabiner.
- Five to ten standard length runners, with two carabiners on each.
- Two to three double length runners, with two carabiners on each.
- Extra carabiners.
- Nut tool.

> ✍ **NOTE**
> The route chosen will dictate, to some degree, the necessary equipment. Members of a climbing team may need to consolidate gear to climb a particular route.

(2) The belayer and the leader both should carry many duplicate items while climbing.

- Short Prusik sling.
- Long Prusik sling.
- Cordellette.
- 10 feet of 1-inch webbing.

- 20 feet of 1-inch webbing.
- Belay device (a combination belay/rappel device is multifunctional).
- Rappel device (a combination belay/rappel device is multifunctional).
- Large locking carabiner (pear shape carabiners are multifunctional).
- Extra carabiners.
- Nut tool (if stoppers are carried).

> ✍ **NOTE**
> If using an over the shoulder gear sling, place the items in order from smallest to the front and largest to the rear.

e. Leading a difficult pitch is the most hazardous task in roped climbing. The lead climber may be exposed to potentially long, hard falls and must exercise keen judgment in route selection, placement of protection, and routing of the climbing rope through the protection. The leader should try to keep the climbing line as direct as possible to the next belay to allow the rope to run smoothly through the protection with minimal friction. Protection should be placed whenever the leader feels he needs it, and BEFORE moving past a difficult section.

> ⚠ **CAUTION**
> The climber must remember he will fall twice the distance from his last piece of protection before the rope can even begin to stop him.

(1) ***Placing Protection***. Generally, protection is placed from one stable position to the next. The anchor should be placed as high as possible to reduce the potential fall distance between placements. If the climbing is difficult, protection should be placed more frequently. If the climbing becomes easier, protection can be placed farther apart, saving hardware for difficult sections. On some routes an extended diagonal or horizontal movement, known as

a traverse, is required. As the leader begins this type of move, he must consider the second's safety as well as his own. The potential fall of the second will result in a pendulum swing if protection is not adequate to prevent this. The danger comes from any objects in the swinging path of the second.

⚠ CAUTION

Leader should place protection prior to, during, and upon completion of any traverse. This will minimize the potential swing, or pendulum, for both the leader and second if either should fall.

(2) ***Correct Clipping Technique***. Once an anchor is placed, the climber "clips" the rope into the carabiner (Figure 4-127). As a carabiner hangs from the protection, the rope can be routed through the carabiner in two possible ways. One way will allow the rope to run smoothly as the climber moves past the placement; the other way will often create a dangerous situation in which the rope could become "unclipped" from the carabiner if the leader were to fall on this piece of protection. In addition, a series of incorrectly clipped carabiners may contribute to rope drag. When placing protection, the leader must ensure the carabiner on the protection does not hang with the carabiner gate facing the rock; when placing protection in a crack ensure the carabiner gate is not facing into the crack.

- Grasp the rope with either hand with the thumb pointing down the rope towards the belayer
- Pull enough rope to reach the carabiner with a bight.
- Note the direction the carabiner is hanging from the protection.
- Place the bight into the carabiner so that, when released, the rope does not cause the carabiner to twist.
  (a) If the route changes direction, clipping the carabiner will require a little more thought. Once leaving that piece of protection, the rope may force the carabiner to twist if not correctly clipped. If the clip is made correctly, a rotation of the clipped

carabiner to ensure that the gate is not resting against the rock may be all that is necessary.

⚠ CAUTION

Ensure the carabiner gate is not resting against a protrusion or crack edge in the rock surface; the rock may cause the gate to open.

(b) Once the rope is clipped into the carabiner, the climber should check to see that it is routed correctly by pulling on the rope in the direction it will travel when the climber moves past that position.

(c) Another potential hazard peculiar to leading should be eliminated before the climber continues. The carabiner is attached to the anchor or runner with the gate facing away from the rock and opening down for easy insertion of the rope. However, in a leader fall, it is possible for the rope to run back over the carabiner as the climber falls below the placement. If the carabiner is left with the gate facing the direction of the route there is a chance that the rope will open the gate and unclip itself entirely from the placement. To prevent this possibility, the climber should ensure that after the clip has been made, the gate is facing away from the direction of the route. There are two ways to accomplish this: determine which direction the gate will face before the protection or runner is placed or once clipped, rotate the carabiner upwards 180 degrees. This problem is more apt to occur if bent gate carabiners are used. Straight gate ovals or "Ds" are less likely to have this problem and are stronger and are highly recommended. Bent gate carabiners are easier to clip the rope into and are used mostly on routes with bolts preplaced for protection. Bent gate carabiners are not recommended for many climbing situations.

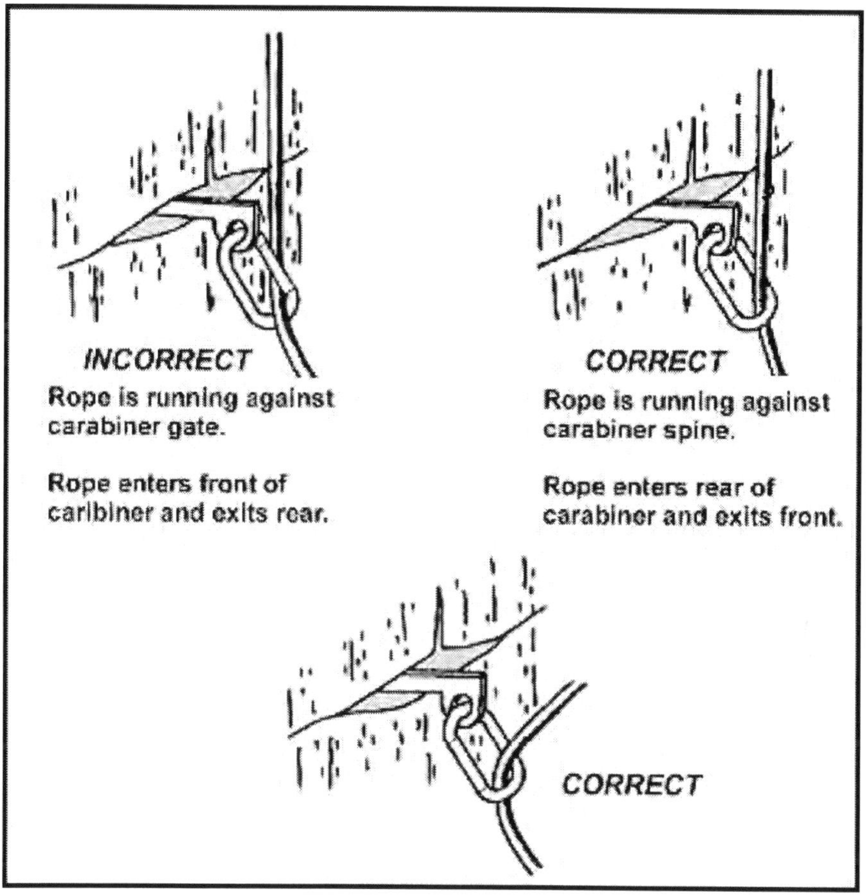

Figure 4-127: Clipping on to protection.

(3) ***Reducing Rope Drag; Using Runners.*** No matter how direct the route, the climber will often encounter problems with "rope drag" through the protection positions. The friction created by rope drag will increase to some degree every time the rope passes through a carabiner, or anchor. It will increase dramatically if the rope begins to "zigzag" as it travels through the carabiners. To prevent this, the placements should be positioned so the rope creates a smooth, almost straight line as it passes through the carabiners (Figure 4-128). Minimal rope drag is an inconvenience; severe rope drag may actually pull the climber off balance, inducing a fall.

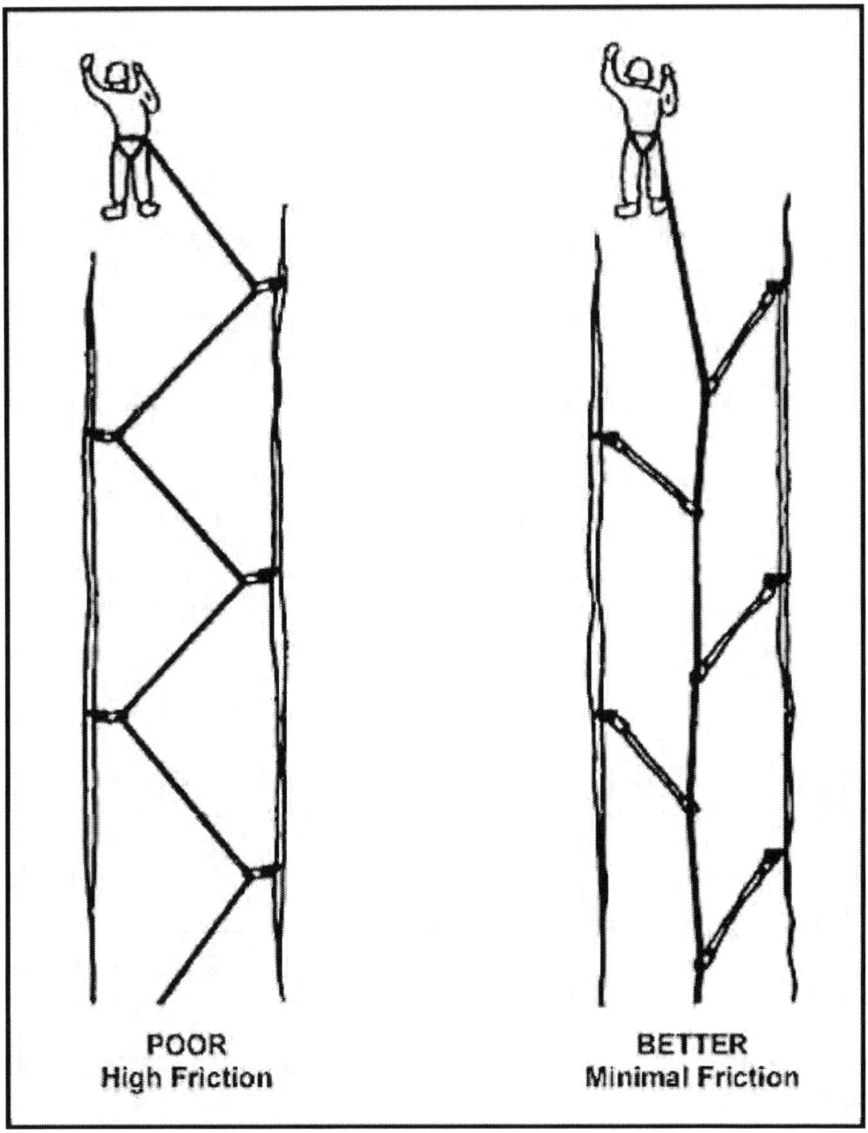

Figure 4-128: Use of slings on protection.

⚠ CAUTION

Rope drag can cause confusion when belaying the second or follower up to a new belay position. Rope drag can be mistaken for the climber, causing the belayer to not take in the necessary slack in the rope and possibly resulting in a serious fall.

(a) If it is not possible to place all the protection so the carabiners form a straight line as the rope moves through, you should "extend" the protection (Figure 4-129). Do this by attaching an appropriate length sling, or runner, to the protection to extend the rope connection in the necessary direction. The runner is attached to the protection's carabiner while the rope is clipped into a carabiner at the other end of the runner. Extending placements with runners will allow the climber to vary the route slightly while the rope continues to run in a relatively straight line.

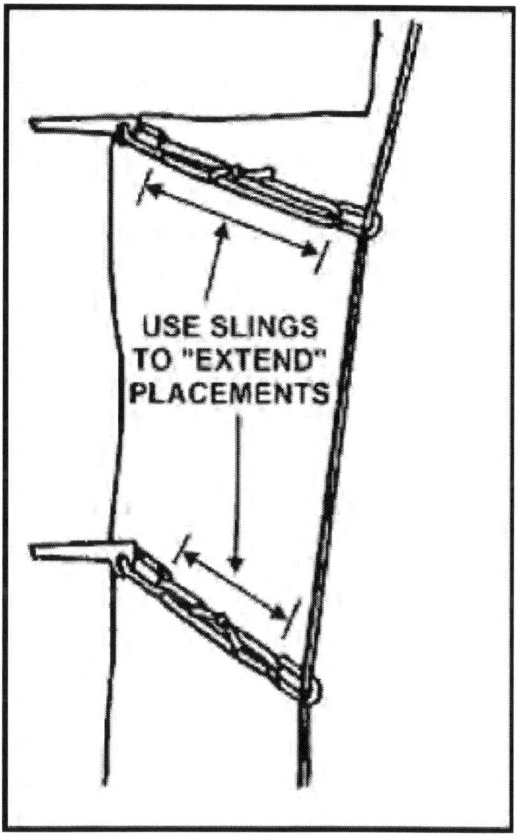

Figure 4-129: Use of slings to extend placement positions.

(b) Not only is rope drag a hindrance, it can cause undue movement of protection as the rope tightens between any "out of line" placements. Rope drag through chock placements can be dangerous. As the climber moves above the placements, an outward or upward pull from rope drag may cause correctly set chocks to pop out, even when used "actively". Most all chocks placed for leader protection should be extended with a runner, even if the line is direct to eliminate the possibility of movement.

(c) Wired chocks are especially prone to wiggling loose as the rope pulls on the stiff cable attachment. All wired chocks used for leader protection should be extended to reduce the chance of the rope pulling them out (Figure 4-130). Some of the larger chocks, such as roped Hexentrics and Tri-Cams, have longer slings pre-attached that will normally serve as an adequate runner for the placement. Chocks with smaller sling attachments must often be extended with a runner. Many of today's chocks are manufactured with pre-sewn webbing installed instead of cable.

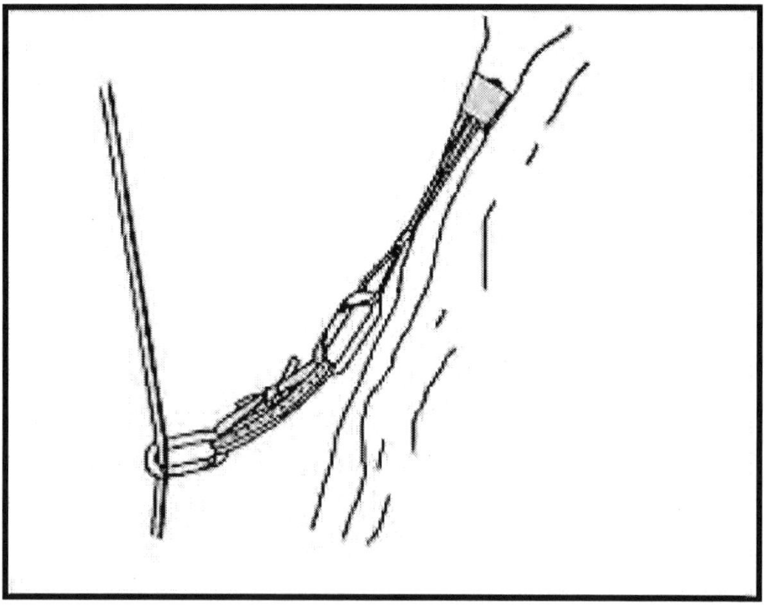

Figure 4-130: Use of sling on a wired stopper.

(d) When a correctly placed piton is used for protection, it will normally not be affected by rope drag. A correctly placed piton is generally a multi-directional anchor, therefore, rope drag through pitons will usually only affect the leader's movements but will continue to protect as expected.

(e) Rope drag will quite often move SLCDs out of position, or "walk" them deeper into the crack than initially placed, resulting in difficult removal or inability to remove the mat all. Furthermore, most cases of SLCD movement result in the SLCD moving to a position that does not provide protection in the correct direction or no protection at all due to the lobes being at different angles from those at the original position.

> ✍ **NOTE**
>
> Any placement extended with a runner will increase the distance of a potential fall by the actual length of the sling. Try to use the shortest runners possible, ensuring they are long enough to function properly.

f. Belaying the follower is similar to belaying a top-roped climb in that the follower is not able to fall any farther than rope stretch will allow. This does not imply there is no danger in following. Sharp rocks, rock fall, and inadequately protected traverses can result in damage to equipment or injury to the second.

g. Following, or seconding, a leader has a variety of responsibilities. The second has to issue commands to the leader, as well as follow the leader's commands. Once the lead climber reaches a good belay position, he immediately establishes an anchor and connects to it. When this is completed he can signal "OFF BELAY" to the belayer. The second can now remove the leader's belay and prepare to climb. The second must remain attached to at least one of the original anchors while the leader is preparing the next belay position. The removed materials and hardware can be organized and secured on the second's rack in preparation to climb.

(1) When the leader has established the new belay position and is ready to belay the follower, the "new" belayer signals "BELAY ON." The second, now the climber, removes any remaining anchor hardware/materials and completes any final preparations. The belayer maintains tension on the rope, unless otherwise directed, while the final preparations are taking place, since removal of these remaining anchors can introduce slack into the rope. When the second is ready, he can, as a courtesy, signal "CLIMBING," and the leader can, again as a courtesy, reply with "CLIMB."

(2) Upon signaling "BELAY ON," the belayer must remove and keep all slack from the rope. (This is especially important as in many situations the belayer cannot see the follower. A long pitch induces weight and sometimes "drag" on the rope and the belayer above will have difficulty distinguishing these from a rope with no slack.)

h. When removing protection, the man cleaning the pitch should rack it properly to facilitate the exchange and or arrangement of equipment at the end of the pitch. When removing the protection, or "cleaning the pitch", SLCDs or chocks may be left attached to the rope to prevent loss if they are accidentally dropped during removal. If necessary, the hardware can remain on the rope until the second reaches a more secure stance. If removing a piton, the rope should be unclipped from the piton to avoid the possibility of damaging the rope with a hammer strike.

(1) The second may need to place full body weight on the rope to facilitate use of both hands for protection removal by giving the command "TENSION." The second must also ensure that he does not climb faster than the rope is being taken in by the belayer. If too much slack develops, he should signal "TAKE ROPE" and wait until the excess is removed before continuing the climb. Once the second completes the pitch, he should immediately connect to the anchor. Once secured, he can signal "OFF BELAY." The leader removes the belay, while remaining attached to an anchor. The equipment is exchanged or organized in preparation for the next pitch or climb.

(2) When the difficulty of the climbing is within the "leading ability" of both climbers, valuable time can be saved

by "swinging leads." This is normally the most efficient method for climbing multi-pitch routes. The second finishes cleaning the first pitch and continues climbing, taking on the role of lead climber. Unless he requires equipment from the belayer or desires a break, he can climb past the belay and immediately begin leading. The belayer simply adjusts his position, re-aiming the belay once the new leader begins placing protection. Swinging leads, or "leap frogging," should be planned before starting the climb so the leader knows to anchor the upper belay for both upward and downward pulls during the setup.

4-112. AID CLIMBING

When a route is too difficult to free climb and is unavoidable, if the correct equipment is available you might aid climb the route. Aid climbing consists of placing protection and putting full body weight on the piece. This allows you to hang solely on the protection you place, giving you the ability to ascend more difficult routes than you can free climb. Clean aid consists of using SLCDs and chocks, and is the simplest form of aid climbing.

a. **Equipment.** Aid climbing can be accomplished with various types of protection. Regardless of the type of protection used, the method of aid climbing is the same. In addition to the equipment for free climbing, other specialized equipment will be needed.

(1) ***Pitons.*** Pitons are used the same as for free climbing. Most piton placements will require the use of both hands. Piton usage will usually leave a scar in the rock just by virtue of the hardness of the piton and the force required to set it with a hammer. Swinging a hammer to place pitons will lead to climber fatigue sooner than clean aid. Since pitons are multidirectional, the strength of a well-placed piton is more secure than most clean aid protection. Consider other forms of protection when noise could be hazardous to tactics.

(2) ***Bolts.*** Bolts are used when no other protection will work. They are a more permanent form of protection and more time is needed to place them. Placing bolts creates more noise whether drilled by hand or by motorized drill. Bolts used in climbing are a multi-part expanding system

pounded into predrilled holes and then tightened to the desired torque with a wrench or other tool. Bolts are used in many ways in climbing today. The most common use is with a hanger attached and placed for anchors in face climbing. However, bolts can be used for aid climbing, with or without the hanger.

(a) Placing bolts for aid climbing takes much more time than using pitons or clean aid. Bolting for aid climbing consists of consecutive bolts about 2 feet apart. Drilling a deep enough hole takes approximately thirty minutes with a hand drill and up to two minutes with a powered hammer drill. A lot of time and work is expended in a short distance no matter how the hole is drilled. (The weight of a powered hammer drill becomes an issue in itself.) Noise will also be a factor in both applications. A constant pounding with a hammer on the hand drill or the motorized pounding of the powered drill may alert the enemy to the position. The typical climbing bolt/hanger combination normally is left in the hole where it was placed.

(b) Other items that can be used instead of the bolt/hanger combination are the removable and reusable "spring-loaded removable bolts" such as rivets (hex head threaded bolts sized to fit tightly into the hole and pounded in with a hammer), split-shaft rivets, and some piton sizes that can be pounded into the holes. When using rivets or bolts without a hanger, place a loop of cable over the head and onto the shaft of the rivet or bolt and attach a carabiner to the other end of the loop (a stopper with the chock slid back will suffice). Rivet hangers are available that slide onto the rivet or bolt after it is placed and are easily removed for reuse. Easy removal means a slight loss of security while in use.

(3) ***SLCDs.*** SLCDs are used the same as for free climbing, although in aid climbing, full bodyweight is applied to the SLCD as soon as it is placed.

(4) ***Chocks.*** Chocks are used the same as for free climbing, although in aid climbing, weight is applied to the chock as soon as it is placed.

(5) ***Daisy Chains.*** Daisy chains are tied or presewn loops of webbing with small tied or presewn loops approximately every two inches. The small loops are just large enough for two or three carabiners. Two daisy chains should be girth-hitched to the tie-in point in the harness.

(6) ***Etriers (or Aiders).*** Etriers (aiders) are tied or presewn webbing loops with four to six tiedor presewn internal loops, or steps, approximately every 12 inches. The internal loops are large enough to easily place one booted foot into. At least two etriers (aiders) should be connected by carabiner to the free ends of the daisy chains.

(7) ***Fifi Hook.*** A fifi hook is a small, smooth-surfaced hook strong enough for body weight. The fifi hook should be girth-hitched to the tie-in point in the harness and is used in the small loops of the daisy chain. A carabiner can be used in place of the fifi hook, although the fifi hook is simpler and adequate.

(8) ***Ascenders.*** Ascenders are mechanical devices that will move easily in one direction on the rope, but will lock in place if pushed or pulled the other direction. (Prusiks can be used but are more difficult than ascenders.)

b. **Technique.** The belay will be the same as in normal lead climbing and the rope will be routed through the protection the same way also. The big difference is the movement up the rock. With the daisy chains, aiders, and fifi hook attached to the rope tie-in point of the harness as stated above, and secured temporarily to a gear loop or gear sling, the climb continues as follows:

(1) The leader places the first piece of protection as high as can safely be reached and attaches the appropriate sling/ carabiner

(2) Attach one daisy chain/ aider group to the newly placed protection

(3) Clip the rope into the protection, (the same as for normal lead climbing)

(4) Insure the protection is sound by weighting it gradually; place both feet, one at a time, into the steps in the aider, secure your balance by grasping the top of the aider with your hands.

(5) When both feet are in the aider, move up the steps until your waist is no higher than the top of the aider.

(6) Place the fifi hook (or substituted carabiner) into the loop of the daisy chain closest to the daisy chain/aider carabiner, this effectively shortens the daisy chain; maintain tension on the daisy chain as the hook can fall out of the daisy chain loop if it is unweighted.

> ✍ **NOTE**
>
> Moving the waist higher than the top of the aider is possible, but this creates a potential for a fall to occur even though you are on the aider and "hooked" close to the protection with the daisy chain. As the daisy chain tie-in point on the harness moves above the top of the aider, you are no longer supported from above by the daisy chain, you are now standing above your support. From this height, the fifi hook can easily fall out of the daisy chain loop if it is unweighted. If this happens, you could fall the full length of the daisy chain resulting in a static fall on the last piece of protection placed.

(7) Release one hand from the aider and place the next piece of protection, again, as high as you can comfortably reach; if using pitons or bolts you may need both hands free- "lean" backwards slowly, and rest your upper body on the daisy chain that you have "shortened" with the fifi hook

(8) Clip the rope into the protection

(9) Attach the other daisy chain/aider group to the next piece of protection

(10) Repeat entire process until climb is finished

c. **Seconding.** When the pitch is completed, the belayer will need to ascend the route. To ascend the route, use ascenders instead of Prusiks, ascenders are much faster and safer than Prusiks. Attach each ascender to a daisy chain/aider group with carabiners. To adjust the maximum reach/height of the ascenders on the rope, adjust the effective length of the daisy chains with a carabiner the same as with the fifi hook; the typical height will be enough to hold the attached ascender in the hand at nose level. When adjusted to the correct height, the arms need not support much body weight. If the ascender is too high, you will have difficulty reaching and maintaining a grip on the handle.

(1) Unlike lead climbing, there will be a continuous load on the rope during the cleaning of the route, this would normally increase the difficulty of removing protection. To make this easier, as you approach the protection on the ascenders, move the ascenders, one at a time, above the piece. When your weight is on the rope above the piece, you can easily unclip and remove the protection.

⚠ CAUTION

If both ascenders should fail while ascending the pitch, a serious fall could result. To prevent this possibility, tie-in short on the rope every 10-20 feet by tying a figure eight loop and clipping it into the harness with a separate locking carabiner as soon as the ascent is started. After ascending another 20 feet, repeat this procedure. Do not unclip the previous figure eight until the new knot is attached to another locking carabiner. Clear each knot as you unclip it.

✍ NOTES

1. Ensure the loops formed by the short tie-ins do not catch on anything below as you ascend.
2. If the nature of the rock will cause the "hanging loop" of rope, formed by tying in at the end of the rope, to get caught as you move upward, do not tie into the end of the rope.

(2) Seconding an aid pitch can be done in a similar fashion as seconding free-climbed pitches. The second can be belayed from above as the second "climbs" the protection. However, the rope is unclipped from the protection before the aider/daisy chain is attached.

d. Seconding Through a Traverse. While leading an aid traverse, the climber is hanging on the protection placed in front of the current position. If the second were to clean the section by hanging on the rope while cleaning, the protection will be pulled in more than one direction, possibly resulting in the protection failing. To make this safer and easier, the second should hang on the protection just as the leader did. As the second moves to the beginning of the traverse, one ascender/ daisy chain/aider group is removed from the rope and clipped to the protection with a carabiner, (keep the ascenders attached to the daisy chain/aider group for convenience when the traverse ends). The second will negotiate the traverse by leapfrogging the daisy chain/aider groups on the next protection just as the leader did. Cleaning is accomplished by removing the protection a sit is passed when all weight is removed from it. This is in effect a self-belay. The second maintains a shorter safety tie-in on the rope than for vertical movement to reduce the possibility of a lengthy pendulum if the protection should pull before intended.

e. Clean Aid Climbing. Clean aid climbing consists of using protection placed without a hammer or drill involvement: chocks, SLCDs, hooks, and other protection placed easily by hand. This type of aid climbing will normally leave no trace of the climb when completed. When climbing the aiders on clean aid protection, ensure the protection does not "move" from its original position.

(1) Hooks are any device that rests on the rock surface without a camming or gripping action. Hooks are just what the name implies, a curved piece of hard steel with a hole in one end for webbing attachment. The hook blade shape will vary from one model to another, some have curved or notched "blades" to better fit a certain crystal shape on a face placement. These types of devices due to their passive application, are only secure while weighted by the climber.

(2) Some featureless sections of rock can be negotiated with hook use, although bolts can be used. Hook usage is faster and quieter but the margin of safety is not there unless hooks are alternated with more active forms of protection. If the last twenty foot section of a route is negotiated with hooks, a forty foot fall could result.

4-113. THREE-MAN CLIMBING TEAM

Often times a movement on steep terrain will require a team of more than two climbers, which involves more difficulties. A four-man team (or more) more than doubles the difficulty found in three men climbing together. A four-man team should be broken down into two groups of two unless prevented by a severe lack of gear.

a. Given one rope, a three-man team is at a disadvantage on a steep, belayed climb. It takes at least twice as long to climb an average length pitch because of the third climber and the extra belaying required. The distance between belay positions will be halved if only one rope is used because one climber must tie in at the middle of the rope. Two ropes are recommended for a team of three climbers.

> ✍ **NOTE**
>
> Time and complications will increase when a three-man team uses only one rope. For example: a 100-foot climb with a 150-foot rope would normally require two belays for two climbers; a100-foot climb with a 150-foot rope would require six belays for three climbers.

b. At times a three-man climb may be unavoidable and personnel should be familiar with the procedure. Although a team of three may choose from many different methods, only two are described below. If the climb is only one pitch, the methods will vary.

> ⚠ **CAUTION**
>
> When climbing with a team of three, protected traverses will require additional time. The equipment used to protect the traverse must be left in place to protect both the second and third climbers.

(1) The first method can be used when the belay positions are not large enough for three men. If using one rope, two

climbers tie in at each end and the other at the mid-point. When using two ropes, the second will tie in at one end of both ropes, and the other two climbers will each tie in to the other ends. The most experienced individual is the leader, or number 1 climber. The second, or number 2 climber, is the stronger of the remaining two and will be the belayer for both number 1 and number 3. Number 3 will be the last to climb. Although the number 3 climber does no belaying in this method, each climber should be skilled in the belay techniques required. The sequence for this method (in one pitch increments) is as follows (repeated until the climb is complete):

(a) Number 1 ascends belayed by number 2. Number 2 belays the leader up the first pitch while number 3 is simply anchored to the rock for security (unless starting off at ground level) and manages the rope between himself and number 2. When the leader completes the pitch, he sets up the next belay and belays number 2 up.

(b) Number 2 ascends belayed by number 1, and cleans the route (except for traverses). Number 2 returns the hardware to the leader and belays him up the next pitch. When the leader completes this pitch, he again sets up a new belay. When number 2 receives "OFF BELAY" from the leader, he changes ropes and puts number 3 on belay. He should not have to change anchor attachments because the position was already aimed for a downward as well as an upward pull when he belayed the leader.

(c) Number 3 ascends belayed by number 2. When number 3 receives "BELAY ON," he removes his anchor and climbs to number 2's position. When the pitch is completed he secures himself to one of number 2's belay anchors. When number 1's belay is ready, he brings up number 2 while number 3 remains anchored for security. Number 2 again cleans the pitch and the procedure is continued until the climb is completed.

(d) In this method, number 3 performs no belay function. He climbs when told to do so by number 2. When number 3 is not climbing, he remains anchored to the rock for security. The standard rope commands are used; however, the number 2 climber may include the trailing climber's name or number in the commands to avoid confusion as to who should be climbing.

(e) Normally, only one climber would be climbing at a time; however, the number 3climber could ascend a fixed rope to number 2's belay position using proper ascending technique, with no effect on the other two members of the team. This would save time for a team of three, since number 2 would not have to belay number 3 and could be either belaying number 1 to the next belay or climbing to number 1. If number 3 is to ascend a fixed rope to the next belay position, the rope will be loaded with number 3'sweight, and positioned directly off the anchors established for the belay. The rope should be located so it does not contact any sharp edges. The rope to the ascending number 3 could be secured to a separate anchor, but this would require additional time and gear.

(2) The second method uses either two ropes or a doubled rope, and number 2 and number 3climb simultaneously. This requires either a special belay device that accepts two ropes, such as the tuber type, or with two Munter hitches. The ropes must travel through the belay device(s) without affecting each other.

(a) As the leader climbs the pitch, he will trail a second rope or will be tied in with a figure eight in the middle of a doubled rope. The leader reaches the next belay position and establishes the anchor and then places both remaining climbers on belay. One remaining climber will start the ascent toward the leader and the other will start when a gap of at least 10 feet is created between the two climbers. The belayer will have to remain alert for differences in rope movement and the climbers will have to climb at the same speed. One of the "second" climbers also cleans the pitch.

(b) Having at least two experienced climbers in this team will also save time. The belayer will have additional requirements to meet as opposed to having just one second. The possible force on the anchor will be twice that of one second. The second that is not cleaning the pitch can climb off route, but staying on route will usually prevent a possible swing if stance is not maintained.

CHAPTER 5:

SEA SURVIVAL

Perhaps the most difficult survival situation to be in is sea survival. Short- or long-term survival depends upon rations and equipment available and your ingenuity. You must be resourceful to survive.

Water covers about 75 percent of the earth's surface, with about 70 percent being oceans and seas. You can assume that you will sometime cross vast expanses of water. There is always the chance that the plane or ship you are on will become crippled by such hazards as storms, collision, fire, or war.

THE OPEN SEA

As a survivor on the open sea, you will face waves and wind. You may also face extreme heat or cold. To keep these environmental hazards from becoming serious problems, take precautionary measures as soon as possible. Use the available resources to protect yourself from the elements and from heat or extreme cold and humidity.

Protecting yourself from the elements meets only one of your basic needs. You must also be able to obtain water and food. Satisfying these three basic needs will help prevent serious physical and psychological problems. However, you must know how to treat health problems that may result from your situation.

Precautionary Measures

Your survival at sea depends upon—

- Your knowledge of and ability to use the available survival equipment.

- Your special skills and ability to apply them to cope with the hazards you face.
- Your will to live.

When you board a ship or aircraft, find out what survival equipment is on board, where it is stowed, and what it contains. For instance, how many life preservers and lifeboats or rafts are on board? Where are they located? What type of survival equipment do they have? How much food, water, and medicine do they contain? How many people are they designed to support?

If you are responsible for other personnel on board, make sure you know where they are and they know where you are.

DOWN AT SEA

If you are in an aircraft that goes down at sea, take the following actions once you clear the aircraft. Whether you are in the water or in a raft —

- Get clear and upwind of the aircraft as soon as possible, but stay in the vicinity until the aircraft sinks.
- Get clear of fuel-covered water in case the fuel ignites.
- Try to find other survivors.

A search for survivors usually takes place around the entire area of and near the crash site. Missing personnel may be unconscious and floating low in the water. Figure 5-1 illustrates rescue procedures.

The best technique for rescuing personnel from the water is to throw them a life preserver attached to a line. Another is to send a swimmer (rescuer) from the raft with a line attached to a flotation device that will support the rescuer's weight. This device will help conserve a rescuer's energy while recovering the survivor. The least acceptable technique is to send an attached swimmer without flotation devices to retrieve a survivor. In all cases, the rescuer wears a life preserver. A rescuer should not underestimate the strength of a panic-stricken person in the water. A careful approach can prevent injury to the rescuer.

When the rescuer approaches a survivor in trouble from behind, there is little danger the survivor will kick, scratch, or grab him. The rescuer swims to a point directly behind the survivor and grasps the

FIGURE 5-1: RESCUE FROM WATER

life preserver's backstrap. The rescuer uses the sidestroke to drag the survivor to the raft.

If you are in the water, make your way to a raft. If no rafts are available, try to find a large piece of floating debris to cling to. Relax; a person who knows how to relax in ocean water is in very little danger of drowning. The body's natural buoyancy will keep at least the top of the head above water, but some movement is needed to keep the face above water.

Floating on your back takes the least energy. Lie on your back in the water, spread your arms and legs, and arch your back. By controlling your breathing in and out, your face will always be out of the water and you may even sleep in this position for short periods. Your head will be partially submerged, but your face will be above water. If you cannot float on your back or if the sea is too rough, float face down in the water as shown in Figure 5-2.

The following are the best swimming strokes during a survival situation:

- *Dog paddle.* This stroke is excellent when clothed or wearing a life jacket. Although slow in speed, it requires very little energy.
- *Breaststroke.* Use this stroke to swim underwater, through oil or debris, or in rough seas. It is probably the best stroke for long-range swimming: it allows you to conserve your energy and maintain a reasonable speed.
- *Sidestroke.* It is a good relief stroke because you use only one arm to maintain momentum and buoyancy.
- *Backstroke.* This stroke is also an excellent relief stroke. It relieves the muscles that you use for other strokes. Use it if an underwater explosion is likely.

If you are in an area where surface oil is burning—

- Discard your shoes and buoyant life preserver.

> ✍ **NOTE**
> If you have an uninflated life preserver, keep it.

- Cover your nose, mouth, and eyes and quickly go underwater.

- Swim underwater as far as possible before surfacing to breathe.
- Before surfacing to breathe and while still underwater, use your hands to push burning fluid away from the area where you wish to surface. Once an area is clear of burning liquid, you can surface and take a few breaths. Try to face downwind before inhaling.
- Submerge feet first and continue as above until clear of the flames.

If you are in oil-covered water that is free of fire, hold your head high to keep the oil out of your eyes. Attach your life preserver to your wrist and then use it as a raft.

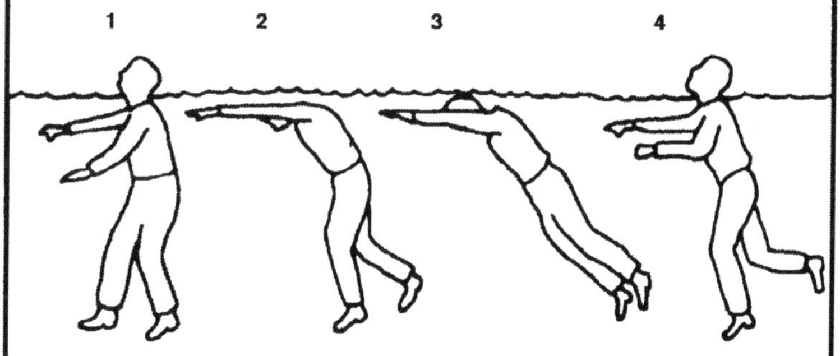

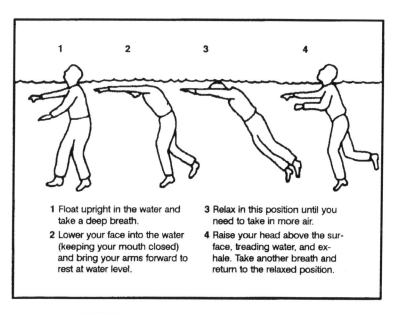

1 Float upright in the water and take a deep breath.

2 Lower your face into the water (keeping your mouth closed) and bring your arms forward to rest at water level.

3 Relax in this position until you need to take in more air.

4 Raise your head above the surface, treading water, and exhale. Take another breath and return to the relaxed position.

FIGURE 5-2: FLOATING POSITION

If you have a life preserver, you can stay afloat for an indefinite period. In this case, use the "HELP" body position: Heat Escaping Lessening Posture (HELP). Remain still and assume the fetal position to help you retain body heat. You lose about 50 percent of your body heat through your head. Therefore, keep your head out of the water. Other areas of high heat loss are the neck, the sides, and the groin. Figure 5-3 illustrates the HELP position.

FIGURE 5-3: HELP Position

If you are in a raft —

- Check the physical condition of all on board. Give first aid if necessary. Take seasickness pills if available. The best way to take these pills is to place them under the tongue and let them dissolve. There are also suppositories or injections against seasickness. Vomiting, whether from seasickness or other causes, increases the danger of dehydration.
- Try to salvage all floating equipment—rations; canteens, thermos jugs, and other containers; clothing; seat cushions; parachutes; and anything else that will be useful to you. Secure the salvaged items in or to your raft. Make sure the items have no sharp edges that can puncture the raft.
- If there are other rafts, lash the rafts together so they are about 7.5 meters apart. Be ready to draw them closer together if you see or hear an aircraft. It is easier for an aircrew to spot rafts that are close together rather than scattered.
- Remember, rescue at sea is a cooperative effort. Use all available visual or electronic signaling devices to signal and make contact with rescuers. For example, raise a flag or reflecting material on an oar as high as possible to attract attention.
- Locate the emergency radio and get it into operation. Operating instructions are on it. Use the emergency transceiver only when friendly aircraft are likely to be in the area.

- Have other signaling devices ready for instant use. If you are in enemy territory, avoid using a signaling device that will alert the enemy. However, if your situation is desperate, you may have to signal the enemy for rescue if you are to survive.
- Check the raft for inflation, leaks, and points of possible chafing. Make sure the main buoyancy chambers are firm (well rounded) but not overly tight (Figure 5-4). Check inflation regularly. Air expands with heat; therefore, on hot days, release some air and add air when the weather cools.
- Decontaminate the raft of all fuel. Petroleum will weaken its surfaces and break down its glued joints.
- Throw out the sea anchor, or improvise a drag from the raft's case, bailing bucket, or a roll of clothing. A sea anchor helps you stay close to your ditching site, making it easier for searchers to find you if you have relayed your location. Without a sea anchor, your raft may drift over 160 kilometers in a day, making it much harder to find you. You can adjust the sea anchor to act as a drag to slow down the rate of travel with the current, or as a means to travel with the current. You make this adjustment by opening or closing the sea anchor's apex. When open, the sea anchor (Figure 5-5) acts as a drag that keeps you in the general area. When closed, it forms a pocket for the current to strike and propels the raft in the current's direction.

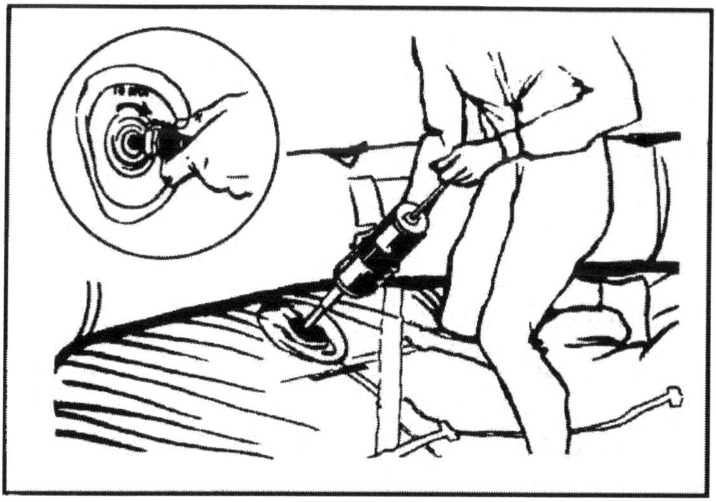

FIGURE 5-4: INFLATING RAFT

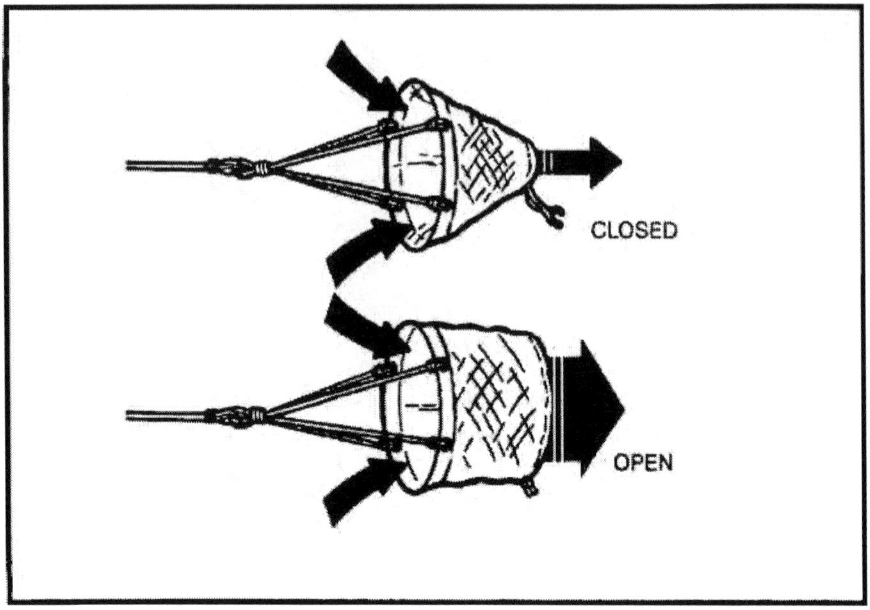

FIGURE 5-5: SEA ANCHOR

Additionally, adjust the sea anchor so that when the raft is on the wave's crest, the sea anchor is in the wave's trough (Figure 5-6).

- Wrap the sea anchor rope with cloth to prevent its chafing the raft. The anchor also helps to keep the raft headed into the wind and waves.
- In stormy water, rig the spray and windshield at once. In a 20-manraft, keep the canopy erected at all times. Keep your raft as dry as possible. Keep it properly balanced. All personnel should stay seated, the heaviest one in the center.
- Calmly consider all aspects of your situation and determine what you and your companions must do to survive. Inventory all equipment, food, and water. Waterproof items that salt water may affect. These include compasses, watches, sextant, matches, and lighters. Ration food and water.
- Assign a duty position to each person: for example, water collector, food collector, lookout, radio operator, signaler, and water bailers. *Note: Lookout duty should not exceed 2 hours. Keep in mind and remind others that cooperation is one of the keys to survival.*

- Keep a log. Record the navigator's last fix, the time of ditching, the names and physical condition of personnel, and the ration schedule. Also record the winds, weather, direction of swells, times of sunrise and sunset, and other navigational data.

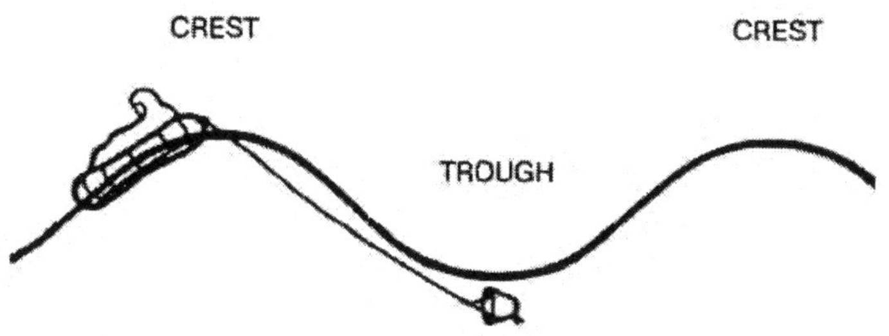

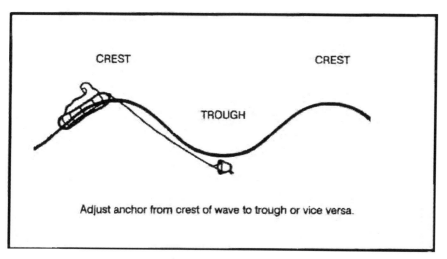

FIGURE 5-6: DEPLOYMENT OF THE SEA ANCHOR

- If you are down in unfriendly waters, take special security measures to avoid detection. Do not travel in the daytime. Throw out the sea anchor and wait for nightfall before paddling or hoisting sail. Keep low in the raft; stay covered with the blue side of the camouflage cloth up. Be sure a passing ship or aircraft is friendly or neutral before trying to attract its attention. If the enemy detects you and you are close to capture, destroy the log book, radio, navigation equipment, maps, signaling equipment, and firearms. Jump overboard and submerge if the enemy starts strafing.
- Decide whether to stay in position or to travel. Ask yourself, "How much information was signaled before the accident? Is your position known to rescuers? Do you know it yourself? Is the weather favorable for a search? Are other ships or aircraft likely to pass your present position? How many days supply of food and water do you have?"

COLD WEATHER CONSIDERATIONS

If you are in a cold climate—

- Put on an antiexposure suit. If unavailable, put on any extra clothing available. Keep clothes loose and comfortable.
- Take care not to snag the raft with shoes or sharp objects. Keep the repair kit where you can readily reach it.
- Rig a windbreak, spray shield, and canopy.
- Try to keep the floor of the raft dry. Cover it with canvas or cloth for insulation.
- Huddle with others to keep warm, moving enough to keep the blood circulating. Spread an extra tarpaulin, sail, or parachute over the group.
- Give extra rations, if available, to men suffering from exposure to cold.

The greatest problem you face when submerged in cold water is death due to hypothermia. When you are immersed in cold water, hypothermia occurs rapidly due to the decreased insulating quality of wet clothing and the result of water displacing the layer of still air that normally surrounds the body. The rate of heat exchange in water is about 25 times greater than it is in air of the same temperature. Table 5-1 lists life expectancy times for immersion in water.

Your best protection against the effects of cold water is to get into the life raft, stay dry, and insulate your body from the cold surface of the bottom of the raft. If these actions are not possible, wearing an antiexposure suit will extend your life expectancy considerably. Remember, keep your head and neck out of the water and well insulated from the cold water's effects when the temperature is below 19 degrees C. Wearing life preservers increases the predicted survival time as body position in the water increases the chance of survival.

HOT WEATHER CONSIDERATIONS

If you are in a hot climate—

- Rig a sunshade or canopy. Leave enough space for ventilation.
- Cover your skin, where possible, to protect it from sunburn. Use sunburn cream, if available, on all exposed skin. Your

eyelids, the back of your ears, and the skin under your chin sunburn easily.

RAFT PROCEDURES

Most of the rafts in the U.S. Army and Air Force inventories can satisfy the needs for personal protection, mode of travel, and evasion and camouflage.

> ✍ **NOTE**
> Before boarding any raft, remove and tether (attach) your life preserver to yourself or the raft. Ensure there are no other metallic or sharp objects on your clothing or equipment that could damage the raft. After boarding the raft, don your life preserver again.

One-Man Raft

The one-man raft has a main cell inflation. If the CO2 bottle should malfunction or if the raft develops a leak, you can inflate it by mouth.

Water Temperature	Time
21.0–15.5 degrees C (70–60 degrees F)	12 hours
15.5–10.0 degrees C (60–50 degrees F)	6 hours
10.0–4.5 degrees C (50–40 degrees F)	1 hour
4.5 degrees C (40 degrees F) and below	less than 1 hour
Note: Wearing an antiexposure suit may increase these times up to a maximum of 24 hours.	

TABLE 5-1: LIFE EXPECTANCY TIMES FOR IMMERSION IN WATER

The spray shield acts as a shelter from the cold, wind, and water. In some cases, this shield serves as insulation. The raft's insulated bottom limits the conduction of cold thereby protecting you from hypothermia (Figure 5-7).

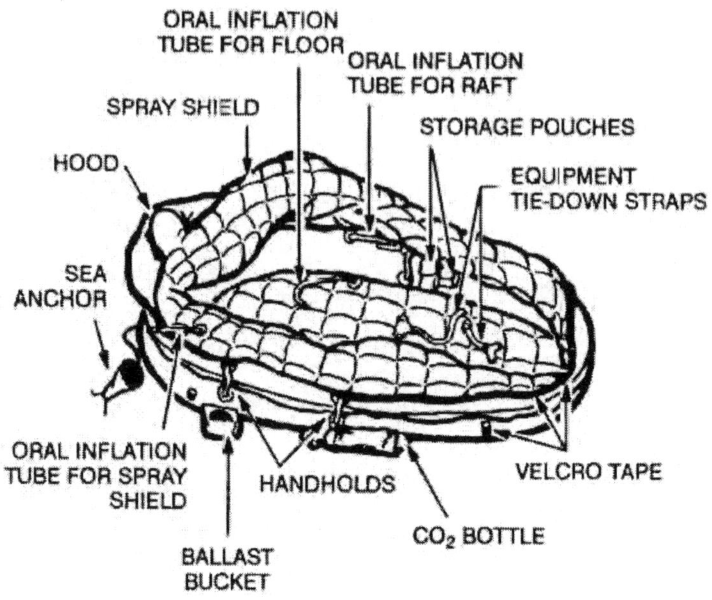

FIGURE 5-7: ONE-MAN RAFT WITH SPRAY SHIELD

You can travel more effectively by inflating or deflating the raft to take advantage of the wind or current. You can use the spray shield as a sail white the ballast buckets serve to increase drag in the water. You may use the sea anchor to control the raft's speed and direction.

There are rafts developed for use in tactical areas that are black. These rafts blend with the sea's background. You can further modify these rafts for evasion by partially deflating them to obtain a lower profile.

A lanyard connects the one-man raft to a parachutist (survivor) landing in the water. You (the survivor) inflate it upon landing. You do not swim to the raft, but pull it to you via the lanyard. The raft may hit the water upside down, but you can right it by approaching the side to which the bottle is attached and flipping the raft over. The spray shield must be in the raft to expose the boarding handles. Follow the steps outlined in the note under raft procedures above when boarding the raft (Figure 5-8).

If you have an arm injury, the best way to board is by turning your back to the small end of the raft, pushing the raft under your buttocks, and lying back. Another way to board the raft is to push down on its small end until one knee is inside and lie forward (Figure 5-9).

In rough seas, it may be easier for you to grasp the small end of the raft and, in a prone position, to kick and pull yourself into the raft. When you are lying face down in the raft, deploy and adjust the sea anchor. To sit upright, you may have to disconnect one side of the seat kit and roll to that side. Then you adjust the spray shield. There are two variations of the one-man raft; the improved model incorporates an inflatable spray shield and floor that provide additional insulation. The spray shield

helps keep you dry and warm in cold oceans and protects you from the sun in the hot climates (Figure 5-10).

FIGURE 5-8: BOARDING THE ONE-MAN RAFT

FIGURE 5-9: BOARDING THE ONE-MAN RAFT (OTHER METHODS)

Seven-Man Raft

Some multiplace aircraft carry the seven-man raft. It is a component of the survival drop kit (Figure 5-10). This raft may inflate upside down and require you to right the raft before boarding. Always work from the bottle side to prevent injury if the raft turns over. Facing into the wind, the wind provides additional help in righting the raft. Use the handles on the inside bottom of the raft for boarding (Figure 5-12).

FIGURE 5-10: ONE-MAN RAFT WITH SPRAY SHIELD INFLATED

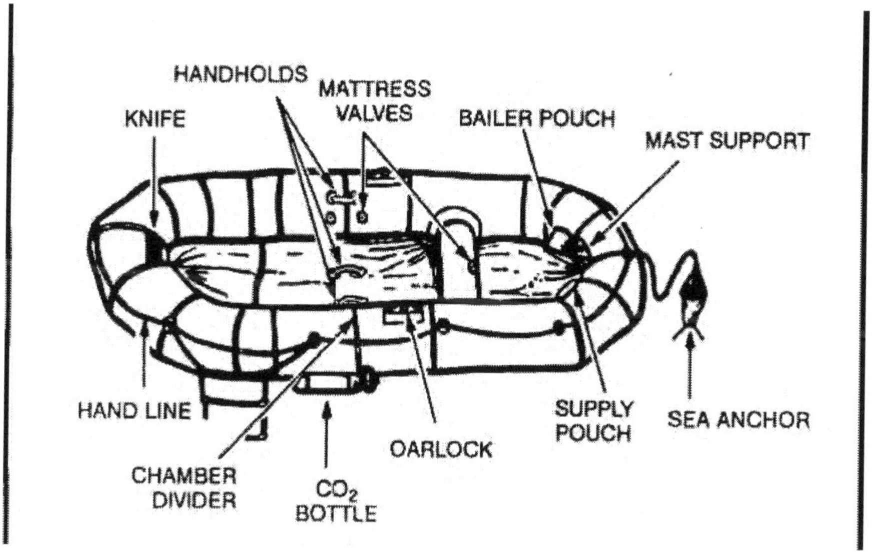

FIGURE 5-11: SEVEN-MAN RAFT

FIGURE 5-12: METHOD OF RIGHTING CRAFT

Use the boarding ramp if someone holds down the raft's opposite side. If you don't have help, again work from the bottle side with the wind at your back to help hold down the raft. Follow the steps outlined in the note under raft procedures above. Then grasp an oarlock and boarding handle, kick your legs to get your body prone on the water, and then kick and pull yourself into the raft. If you are weak or injured, you may partially deflate the raft to make boarding easier (Figure 5-13).

Use the hand pump to keep the buoyancy chambers and cross seat firm. Never overinflate the raft.

Twenty- or Twenty-Five-Man Rafts

You may find 20- or 25-man rafts in multiplace aircraft (Figures 5-14 and 5-15). You will find them in accessible areas of the fuselage or in raft compartments. Some may be automatically deployed from the cockpit, while others may need manual deployment. No matter how the raft lands in the water, it is ready for boarding. A lanyard connects the accessory kit to the raft and you retrieve the kit by hand. You must manually inflate the center chamber with the hand pump. Board the 20- or25-man raft from the aircraft, if possible. If not, board in the following manner:

FIGURE 5-13: METHOD OF BARDING SEVEN-MAN RAFT

- Approach the lower boarding ramp.
- Remove your life preserver and tether it to yourself so that it trails behind you.
- Grasp the boarding handles and kick your legs to get your body into a prone position on the water's surface; then kick and pull until you are inside the raft.

An incompletely inflated raft will make boarding easier. Approach the intersection of the raft and ramp, grasp the upper boarding handle, and swing one leg onto the center of the ramp, as in mounting a horse (Figure 5-16).

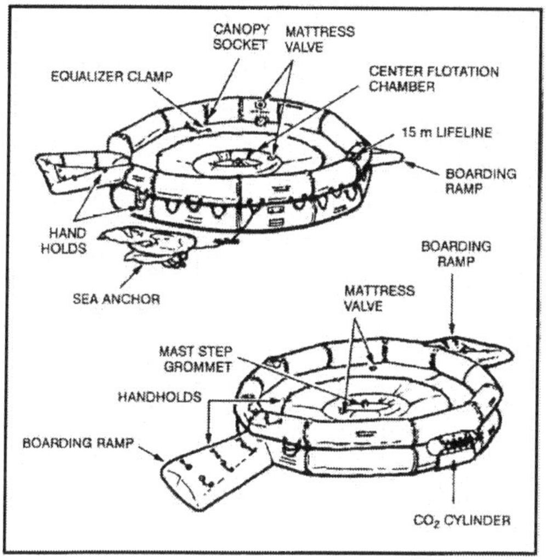

FIGURE 5-14: TWENTY-MAN RAFT

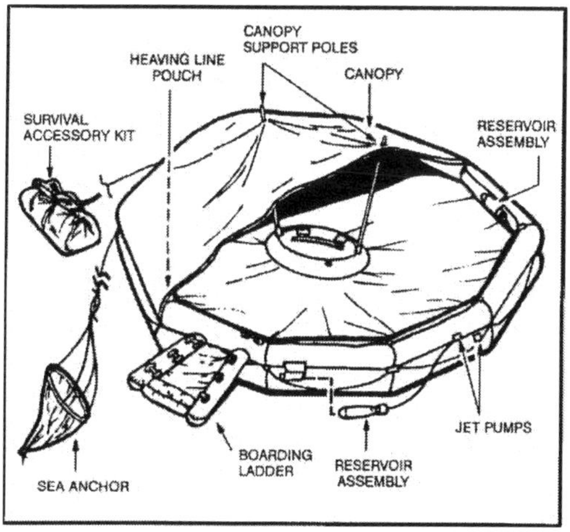

FIGURE 5-15: TWENTY-FIVE-MAN RAFT

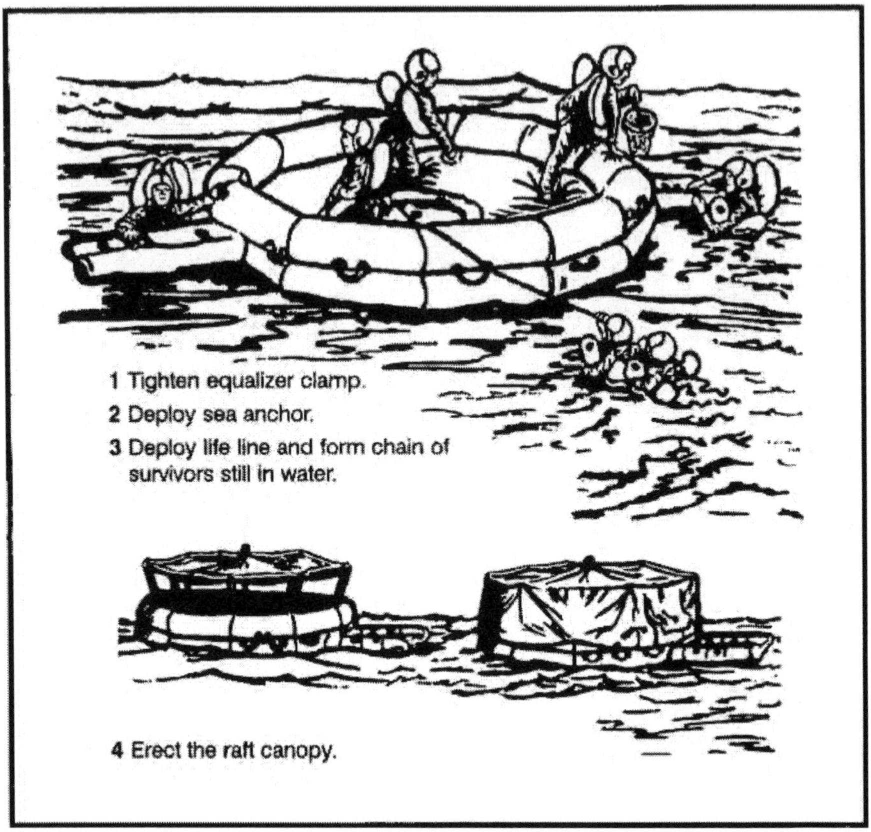

FIGURE 5-16: BOARDING THE 20-MAN RAFT

Immediately tighten the equalizer clamp upon entering the raft to prevent deflating the entire raft in case of a puncture (Figure 5-17).

Use the pump to keep these rafts' chambers and center ring firm. They should be well rounded but not overly tight.

SAILING RAFTS

Rafts do not have keels, therefore, you can't sail them into the wind. However, anyone can sail a raft downwind. You can successfully sail multiplace (except 20- to 25-man) rafts 10 degrees off from the direction of the wind. Do not try to sail the raft unless land is near. If you decide to sail and the wind is blowing toward a desired destination, fully inflate the raft, sit high, take in the sea anchor, rig a sail, and use an oar as a rudder.

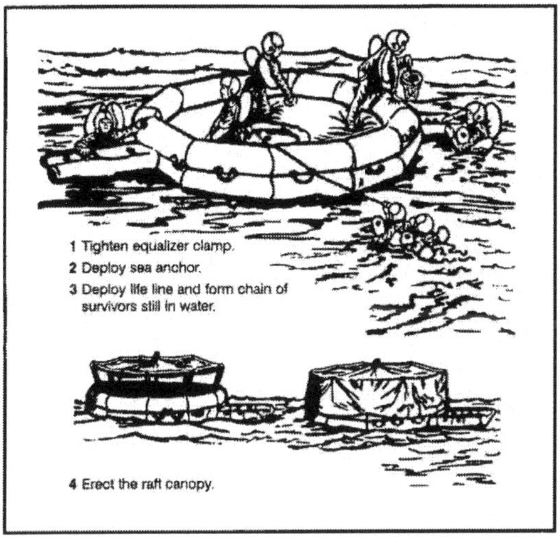

FIGURE 5-17: IMMEDIATE ACTION-MULTIPLACE RAFT

In a multiplace (except 20- to 25-man) raft, erect a square sail in the bow using the oars and their extensions as the mast and crossbar (Figure 5-18). You may use a waterproof tarpaulin or parachute material for the sail.

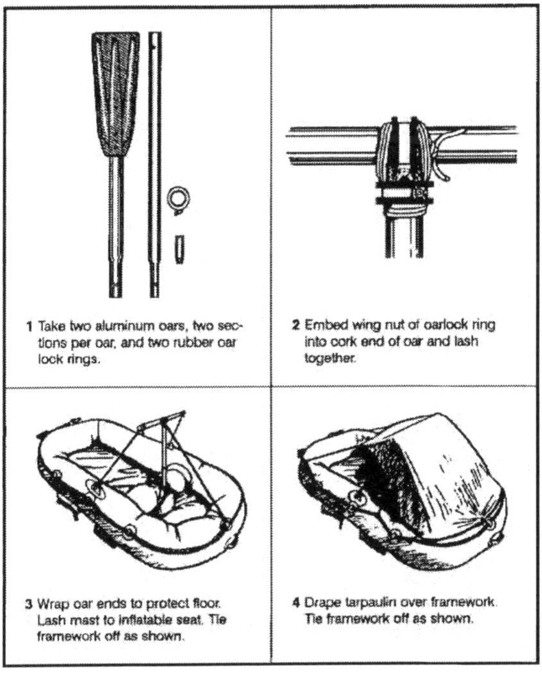

FIGURE 5-18: SAIL CONSTRUCTION

If the raft has no regular mast socket and step, erect the mast by tying it securely to the front cross seat using braces. Pad the bottom of the mast to prevent it from chafing or punching a hole through the floor, whether or not there is a socket. The heel of a shoe, with the toe wedged under the seat, makes a good improvised mast step. Do not secure the corners of the lower edge of the sail. Hold the lines attached to the corners with your hands so that a gust of wind will not rip the sail, break the mast, or capsize the raft.

Take every precaution to prevent the raft from turning over. In rough weather, keep the sea anchor away from the bow. Have the passengers sit low in the raft, with their weight distributed to hold the upwind side down. To prevent falling out, they should also avoid sitting on the sides of the raft or standing up. Avoid sudden movements without warning the other passengers. When the sea anchor is not in use, tie it to the raft and stow it in such a manner that it will hold immediately if the raft capsizes.

WATER

Water is your most important need. With it alone, you can live for ten days or longer, depending on your will to live. When drinking water, moisten your lips, tongue, and throat before swallowing.

Short Water Rations

When you have a limited water supply and you can't replace it by chemical or mechanical means, use the water efficiently. Protect freshwater supplies from seawater contamination. Keep your body well shaded, both from overhead sun and from reflection off the sea surface. Allow ventilation of air; dampen your clothes during the hottest part of the day. Do not exert yourself. Relax and sleep when possible. Fix your daily water ration after considering the amount of water you have, the output of solar stills and desalting kit, and the number and physical condition of your party.

If you don't have water, don't eat. If your water ration is two liters or more per day, eat any part of your ration or any additional food that you may catch, such as birds, fish, shrimp. Anxiety and the life raft's motion may cause nausea. If you eat when nauseated, you may lose your food immediately. If nauseated, rest and relax as much as you can, and take only water.

To reduce your loss of water through perspiration, soak your clothes in the sea and wring them out before putting them on again.

Don't overdo this during hot days when no canopy or sun shield is available. This is a trade-off between cooling and saltwater boils and rashes that will result.

Be careful not to get the bottom of the raft wet.

Watch the clouds and be ready for any chance of showers. Keep the tarpaulin handy for catching water. If it is encrusted with dried salt, wash it in seawater. Normally, a small amount of seawater mixed with rain will hardly be noticeable and will not cause any physical reaction. In rough seas you cannot get uncontaminated fresh water.

At night, secure the tarpaulin like a sunshade, and turn up its edges to collect dew. It is also possible to collect dew along the sides of the raft using a sponge or cloth. When it rains, drink as much as you can hold.

Solar Still

When solar stills are available, read the instructions and set them up immediately. Use as many stills as possible, depending on the number of men in the raft and the amount of sunlight available. Secure solar stills to the raft with care. This type of solar still only works on flat, calm seas.

Desalting Kits

When desalting kits are available in addition to solar stills, use them only for immediate water needs or during long overcast periods when you cannot use solar stills. In any event, keep desalting kits and emergency water stores for periods when you cannot use solar stills or catch rainwater.

Water from Fish

Drink the aqueous fluid found along the spine and in the eyes of large fish. Carefully cut the fish in half to get the fluid along the spine and suck the eye. If you are so short of water that you need to do this, then do not drink any of the other body fluids. These other fluids are rich in protein and fat and will use up more of your reserve water in digestion than they supply.

Sea Ice

In arctic waters, use old sea ice for water. This ice is bluish, has rounded comers, and splinters easily. It is nearly free of salt. New ice is gray, milky, hard, and salty. Water from icebergs is fresh, but icebergs are dangerous to approach. Use them as a source of water only in emergencies.

REMEMBER!
Do not drink seawater.
Do not drink urine.
Do not drink alcohol.
Do not smoke.
Do not eat, unless water is available.

Sleep and rest are the best ways of enduring periods of reduced water and food intake. However, make sure that you have enough shade when napping during the day. If the sea is rough, tie yourself to the raft, close any cover, and ride out the storm as best you can. Relax is the keyword—at least try to relax.

FOOD PROCUREMENT

In the open sea, fish will be the main food source. There are some poisonous and dangerous ocean fish, but, in general, when out of sight of land, fish are safe to eat. Nearer the shore there are fish that are both dangerous and poisonous to eat. There are some fish, such as the red snapper and barracuda that are normally edible but poisonous when taken from the waters of atolls and reefs. Flying fish will even jump into your raft!

Fish

When fishing, do not handle the fishing line with bare hands and never wrap it around your hands or tie it to a life raft. The salt that adheres to it can make it a sharp cutting edge, an edge dangerous both to the raft and your hands. Wear gloves, if they are available, or use a cloth to handle fish and to avoid injury from sharp fins and gill covers.

In warm regions, gut and bleed fish immediately after catching them. Cut fish that you do not eat immediately into thin, narrow strips and hang them to dry. A well-dried fish stays edible for several days. Fish not cleaned and dried may spoil in half a day. Fish with dark meat are very prone to decomposition. If you do not eat them all immediately, do not eat any of the leftovers. Use the leftovers for bait.

Never eat fish that have pale, shiny gills, sunken eyes, flabby skin and flesh, or an unpleasant odor. Good fish show the opposite characteristics. Sea fish have a saltwater or clean fishy odor. Do not confuse eels with sea snakes that have an obviously scaly body and strongly compressed, paddle-shaped tail. Both eels and sea snakes are edible, but you must handle the latter with care because of their poisonous bites. The heart, blood, intestinal wall, and liver of most fish are edible. Cook the intestines. Also edible are the partly digested smaller fish that you may find in the stomachs of large fish. In addition, sea turtles are edible.

Shark meat is a good source of food whether raw, dried, or cooked. Shark meat spoils very rapidly due to the high concentration of urea in the blood, therefore, bleed it immediately and soak it in several changes of water. People prefer some shark species over others. Consider them all edible except the Greenland shark whose flesh contains high quantities of vitamin A. Do not eat the livers, due to high vitamin A content.

Fishing Aids

You can use different materials to make fishing aids as described in the following paragraphs:

- *Fishing line.* Use pieces of tarpaulin or canvas. Unravel the threads and tie them together in short lengths in groups of three or more threads. Shoelaces and parachute suspension line also work well.
- *Fish hooks.* No survivor at sea should be without fishing equipment but if you are, improvise hooks.
- *Fish lures.* You can fashion lures by attaching a double hook to any shiny piece of metal.
- *Grapple.* Use grapples to hook seaweed. You may shake crabs, shrimp, or small fish out of the seaweed. These you may eat or use for bait. You may eat seaweed itself, but only when you have plenty of drinking water. Improvise grapples from wood. Use a heavy piece of wood as the main shaft, and lash three smaller pieces to the shaft as grapples.

- *Bait*. You can use small fish as bait for larger ones. Scoop the small fish up with a net. If you don't have a net, make one from cloth of some type. Hold the net under the water and scoop upward. Use all the guts from birds and fish for bait. When using bait, try to keep it moving in the water to give it the appearance of being alive.

Helpful Fishing Hints

Your fishing should be successful if you remember the following important hints:

- Be extremely careful with fish that have teeth and spines.
- Cut a large fish loose rather than risk capsizing the raft. Try to catch small rather than large fish.
- Do not puncture your raft with hooks or other sharp instruments.
- Do not fish when large sharks are in the area.
- Watch for schools of fish; try to move close to these schools.
- Fish at night using a light. The light attracts fish.
- In the daytime, shade attracts some fish. You may find them under your raft.
- Improvise a spear by tying a knife to an oar blade. This spear can help you catch larger fish, but you must get them into the raft quickly or they will slip off the blade. Also, tie the knife very securely or you may lose it.
- Always take care of your fishing equipment. Dry your fishing lines, clean and sharpen the hooks, and do not allow the hooks to stick into the fishing lines.

Birds

All birds are edible. Eat any birds you can catch. Sometimes birds may land on your raft, but usually they are cautious. You may be able to attract some birds by towing a bright piece of metal behind the raft. This will bring the bird within shooting range, provided you have a firearm.

If a bird lands within your reach, you may be able to catch it. If the birds do not land close enough or land on the other end of the raft, you may be able to catch them with a bird noose. Bait the center of the noose and wait for the bird to land. When the bird's feet are in the center of the noose, pull it tight.

Use all parts of the bird. Use the feathers for insulation, the entrails and feet for bait, and so on. Use your imagination.

Medical Problems Associated With Sea Survival

At sea, you may become seasick, get saltwater sores, or face some of the same medical problems that occur on land, such as dehydration or sunburn. These problems can become critical if left untreated.

Seasickness

Seasickness is the nausea and vomiting caused by the motion of the raft. It can result in—

- Extreme fluid loss and exhaustion.
- Loss of the will to survive.
- Others becoming seasick.
- Attraction of sharks to the raft.
- Unclean conditions.

To treat seasickness—

- Wash both the patient and the raft to remove the sight and odor of vomit.
- Keep the patient from eating food until his nausea is gone.
- Have the patient lie down and rest.
- Give the patient seasickness pills if available. If the patient is unable to take the pills orally, insert them rectally for absorption by the body.

> ✍ **NOTE**
> Some survivors have said that erecting a canopy or using the horizon as a focal point helped overcome seasickness. Others have said that swimming alongside the raft for short periods helped, but extreme care must be taken if swimming.

Saltwater Sores

These sores result from a break in skin exposed to saltwater for an extended period. The sores may form scabs and pus. Do not open or drain. Flush the sores with fresh water, if available, and allow to dry. Apply an antiseptic, if available.

Immersion Rot, Frostbite, and Hypothermia

These problems are similar to those encountered in cold weather environments. Symptoms and treatment are the same as covered in Chapter 3, "Cold Weather Survival."

Blindness/Headache

If flame, smoke, or other contaminants get in the eyes, flush them immediately with salt water, then with fresh water, if available. Apply ointment, if available. Bandage both eyes 18 to 24 hours, or longer if damage is severe. If the glare from the sky and water causes your eyes to become bloodshot and inflamed, bandage them lightly. Try to prevent this problem by wearing sunglasses. Improvise sunglasses if necessary.

Constipation

This condition is a common problem on a raft. Do not take a laxative, as this will cause further dehydration. Exercise as much as possible and drink an adequate amount of water, if available.

Difficult Urination

This problem is not unusual and is due mainly to dehydration. It is best not to treat it, as it could cause further dehydration.

Sunburn

Sunburn is a serious problem in sea survival. Try to prevent sunburn by staying in shade and keeping your head and skin covered. Use cream or Chap Stick from your first aid kit. Remember, reflection from the water also causes sunburn.

SHARKS

Whether you are in the water or in a boat or raft, you may see many types of sea life around you. Some may be more dangerous than others. Generally, sharks are the greatest danger to you. Other animals such as whales, porpoises, and stingrays may look dangerous, but really pose little threat in the open sea.

Of the many hundreds of shark species, only about 20 species are known to attack man. The most dangerous are the great white shark, the hammerhead, the mako, and the tiger shark. Other sharks known to attack man include the gray, blue, lemon, sand, nurse, bull, and oceanic white tip sharks. Consider any shark longer than 1 meter dangerous.

There are sharks in all oceans and seas of the world. While many live and feed in the depths of the sea, others hunt near the surface. The sharks living near the surface are the ones you will most likely see. Their dorsal fins frequently project above the water. Sharks in the tropical and subtropical seas are far more aggressive than those in temperate waters.

All sharks are basically eating machines. Their normal diet is live animals of any type, and they will strike at injured or helpless animals. Sight, smell, or sound may guide them to their prey. Sharks have an acute sense of smell and the smell of blood in the water excites them. They are also very sensitive to any abnormal vibrations in the water. The struggles of a wounded animal or swimmer, underwater explosions, or even a fish struggling on a fish line will attract a shark.

Sharks can bite from almost any position; they do not have to turn on their side to bite. The jaws of some of the larger sharks are so far forward that they can bite floating objects easily without twisting to the side.

Sharks may hunt alone, but most reports of attacks cite more than one shark present. The smaller sharks tend to travel in schools and attack in mass. Whenever one of the sharks finds a victim, the other sharks will quickly join it. Sharks will eat a wounded shark as quickly as their prey.

Sharks feed at all hours of the day and night. Most reported shark contacts and attacks were during daylight, and many of these have been in the late afternoon. Some of the measures that you can take to protect yourself against sharks when you are in the water are—

- *Stay with other swimmers.* A group can maintain a 360-degree watch. A group can either frighten or fight off sharks better than one man.
- *Always watch for sharks.* Keep all your clothing on, to include your shoes. Historically, sharks have attacked the unclothed men in groups first, mainly in the feet. Clothing also protects against abrasions should the shark brush against you.
- *Avoid urinating.* If you must, only do so in small amounts. Let it dissipate between discharges. If you must defecate, do so in

small amounts and throw it as far away from you as possible. Do the same if you must vomit.

If a shark attack is imminent while you are in the water, splash and yell just enough to keep the shark at bay. Sometimes yelling underwater or slapping the water repeatedly will scare the shark away. Conserve your strength for fighting in case the shark attacks.

If attacked, kick and strike the shark. Hit the shark on the gills or eyes if possible. If you hit the shark on the nose, you may injure your hand if it glances off and hits its teeth.

When you are in a raft and see sharks—

- Do not fish. If you have hooked a fish, let it go. Do not clean fish in the water.
- Do not throw garbage overboard.
- Do not let your arms, legs, or equipment hang in the water.
- Keep quiet and do not move around.
- Dispose of all dead as soon as possible. If there are many sharks in the area, conduct the disposal at night.

When you are in a raft and a shark attack is imminent, hit the shark with anything you have, except your hands. You will do more damage to your hands than the shark. If you strike with an oar, be careful not to lose or break it.

DETECTING LAND

You should watch carefully for any signs of land. There are many indicators that land is near.

A fixed cumulus cloud in a clear sky or in a sky where all other clouds are moving often hovers over or slightly downwind from an island.

In the tropics, the reflection of sunlight from shallow lagoons or shelves of coral reefs often causes a greenish tint in the sky.

In the arctic, light-colored reflections on clouds often indicate ice fields or snow-covered land. These reflections are quite different from the dark gray ones caused by open water.

Deep water is dark green or dark blue. Lighter color indicates shallow water, which may mean land is near.

At night, or in fog, mist, or rain, you may detect land by odors and sounds. The musty odor of mangrove swamps and mud flats carry a long way. You hear the roar of surf long before you see the surf. The continued cries of seabirds coming from one direction indicate their roosting place on nearby land.

There usually are more birds near land than over the open sea. The direction from which flocks fly at dawn and to which they fly at dusk may indicate the direction of land. During the day, birds are searching for food and the direction of flight has no significance.

Mirages occur at any latitude, but they are more likely in the tropics, especially during the middle of the day. Be careful not to mistake a mirage for nearby land. A mirage disappears or its appearance and elevation change when viewed from slightly different heights.

You may be able to detect land by the pattern of the waves (refracted) as they approach land (Figure 5-19). By traveling with the waves and parallel to the slightly turbulent area marked "X" on the illustration, you should reach land.

RAFTING OR BEACHING TECHNIQUES

Once you have found land, you must get ashore safely. To raft ashore, you can usually use the one-man raft without danger. However, going ashore in a strong surf is dangerous. Take your time. Select your landing point carefully.

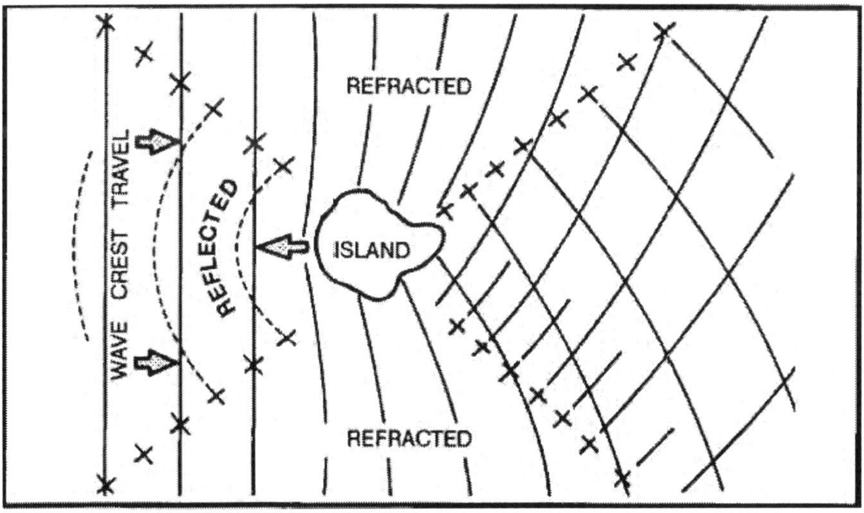

FIGURE 5-19: WAVE PATTERNS ABOUT AN ISLAND

Try not to land when the sun is low and straight in front of you. Try to land on the lee side of an island or on a point of land jutting out into the water. Keep your eyes open for gaps in the surf line, and head for them. Avoid coral reefs and rocky cliffs. There are no coral reefs near the mouths of freshwater streams. Avoid rip currents or strong tidal currents that may carry you far out to sea. Either signal ashore for help or sail around and look for a sloping beach where the surf is gentle.

If you have to go through the surf to reach shore, take down the mast. Keep your clothes and shoes on to avoid severe cuts. Adjust and inflate your life vest. Trail the sea anchor over the stern using as much line as you have. Use the oars or paddles and constantly adjust the sea anchor to keep a strain on the anchor line. These actions will keep the raft pointed toward shore and prevent the sea from throwing the stern around and capsizing you. Use the oars or paddles to help ride in on the seaward side of a large wave.

The surf may be irregular and velocity may vary, so modify your procedure as conditions demand. A good method of getting through the surf is to have half the men sit on one side of the raft, half on the other, facing away from each other. When a heavy sea bears down, half should row (pull) toward the sea until the crest passes; then the other half should row (pull) toward the shore until the next heavy sea comes along.

Against a strong wind and heavy surf, the raft must have all possible speed to pass rapidly through the oncoming crest to avoid being turned broadside or thrown end over end. If possible, avoid meeting a large wave at the moment it breaks.

If in a medium surf with no wind or offshore wind, keep the raft from passing over a wave so rapidly that it drops suddenly after topping the crest. If the raft turns over in the surf, try to grab hold of it and ride it in.

As the raft nears the beach, ride in on the crest of a large wave. Paddle or row hard and ride in to the beach as far as you can. Do not jump out of the raft until it has grounded, then quickly get out and beach it.

If you have a choice, do not land at night. If you have reason to believe that people live on the shore, lay away from the beach, signal, and wait for the inhabitants to come out and bring you in.

If you encounter sea ice, land only on large, stable floes. Avoid icebergs that may capsize and small floes or those obviously disintegrating. Use oars and hands to keep the raft from rubbing on the edge of the ice. Take the raft out of the water and store it well back from the floe's edge. You may be able to use it for shelter. Keep the raft inflated and ready for use. Any floe may break up without warning.

SWIMMING ASHORE

If rafting ashore is not possible and you have to swim, wear your shoes and at least one thickness of clothing. Use the sidestroke or breaststroke to conserve strength.

If the surf is moderate, ride in on the back of a small wave by swimming forward with it. Dive to a shallow depth to end the ride just before the wave breaks.

In high surf, swim toward shore in the trough between waves. When the seaward wave approaches, face it and submerge. After it passes, work toward shore in the next trough. If caught in the undertow of a large wave, push off the bottom or swim to the surface and proceed toward shore as above.

If you must land on a rocky shore, look for a place where the waves rush up onto the rocks. Avoid places where the waves explode with a high, white spray. Swim slowly when making your approach. You will need your strength to hold on to the rocks. You should be fully clothed and wear shoes to reduce injury.

After selecting your landing point, advance behind a large wave into the breakers. Face toward shore and take a sitting position with your feet in front, 60 to 90 centimeters (2 or 3 feet) lower than your head. This position will let your feet absorb the shock when you land or strike submerged boulders or reefs. If you do not reach shore behind the wave you picked, swim with your hands only. As the next wave approaches, take a sitting position with your feet forward. Repeat the procedure until you land.

Water is quieter in the lee of a heavy growth of seaweed. Take advantage of such growth. Do not swim through the seaweed; crawl over the top by grasping the vegetation with overhand movements.

Cross a rocky or coral reef as you would land on a rocky shore. Keep your feet close together and your knees slightly bent in a relaxed sitting posture to cushion the blows against the coral.

PICKUP OR RESCUE

On sighting rescue craft approaching for pickup (boat, ship, conventional aircraft, or helicopter), quickly clear any lines (fishing lines, desalting kit lines) or other gear that could cause entanglement during rescue. Secure all loose items in the raft. Take down canopies and sails to ensure a safer pickup. After securing all items, put on your helmet, if available. Fully inflate your life preserver. Remain in the raft, unless otherwise instructed, and remove all equipment except the preservers. If possible, you will receive help from rescue personnel lowered into the water. Remember, follow all instructions given by the rescue personnel.

- If the helicopter recovery is unassisted, do the following before pickup:
- Secure all the loose equipment in the raft, accessory bag, or in pockets.
- Deploy the sea anchor, stability bags, and accessory bag.
- Partially deflate the raft and fill it with water.
- Unsnap the survival kit container from the parachute harness.
- Grasp the raft handhold and roll out of the raft.
- Allow the recovery device or the cable to ground out on the water's surface.
- Maintain the handhold until the recovery device is in your other hand.
- Mount the recovery device, avoiding entanglement with the raft.
- Signal the hoist operator for pickup.

SEASHORES

Search planes or ships do not always spot a drifting raft or swimmer.

You may have to land along the coast before being rescued. Surviving along the seashore is different from open sea survival. Food and water are more abundant and shelter is obviously easier to locate and construct.

If you are in friendly territory and decide to travel, it is better to move along the coast than to go inland. Do not leave the coast except to avoid obstacles (swamps and cliffs) or unless you find a trail that you know leads to human habitation.

In time of war, remember that the enemy patrols most coastlines. These patrols may cause problems for you if you land on a hostile shore. You will have extremely limited travel options in this situation. Avoid all contact with other humans, and make every effort to cover all tracks you leave on the shore.

SPECIAL HEALTH HAZARDS

Coral, poisonous and aggressive fish, crocodiles, sea urchins, sea biscuits, sponges, anemones, and tides and undertow pose special health hazards.

Coral

Coral, dead or alive, can inflict painful cuts. There are hundreds of water hazards that can cause deep puncture wounds, severe bleeding, and the danger of infection. Clean all coral cuts thoroughly. Do not use iodine to disinfect any coral cuts. Some coral polyps feed on iodine and may grow inside your flesh if you use iodine.

Poisonous Fish

Many reef fish have toxic flesh. For some species, the flesh is always poisonous, for other species, only at certain times of the year. The poisons are present in all parts of the fish, but especially in the liver, intestines, and eggs.

Fish toxins are water soluble—no amount of cooking will neutralize them. They are tasteless, therefore the standard edibility tests are useless. Birds are least susceptible to the poisons. Therefore, do not think that because a bird can eat a fish, it is a safe species for you to eat.

The toxins will produce a numbness of the lips, tongue, toes, and tips of the fingers, severe itching, and a clear reversal of temperature sensations. Cold items appear hot and hot items cold. There will probably also be nausea, vomiting, loss of speech, dizziness, and a paralysis that eventually brings death.

In addition to fish with poisonous flesh, there are those that are dangerous to touch. Many stingrays have a poisonous barb in their tail. There are also species that can deliver an electric shock. Some reef fish, such as stonefish and toadfish, have venomous spines that can cause very painful although seldom fatal injuries. The venom from these spines causes a burning sensation or even an agonizing pain that is out of proportion to the apparent severity of the wound. Jellyfish, while not usually fatal, can inflict a very painful sting if it touches you with its tentacles. See *The Complete U.S. Army Survival Guide to Foraging Skills, Tactics, and Techniques* for details on particularly dangerous fish of the sea and seashore.

Aggressive Fish

You should also avoid some ferocious fish. The bold and inquisitive barracuda has attacked men wearing shiny objects. It may charge lights or shiny objects at night. The sea bass, which can grow to 1.7 meters, is another fish to avoid. The moray eel, which has many sharp teeth and grows to 1.5 meters, can also be aggressive if disturbed.

Sea Snakes

Sea snakes are venomous and sometimes found in mid ocean. They are unlikely to bite unless provoked. Avoid them.

Crocodiles

Crocodiles inhabit tropical saltwater bays and mangrove-bordered estuaries and range up to 65 kilometers into the open sea. Few remain near inhabited areas. You commonly find crocodiles in the remote areas of the East Indies and Southeast Asia. Consider specimens over 1 meter long dangerous, especially females guarding their nests. Crocodile meat is an excellent source of food when available.

Sea Urchins, Sea Biscuits, Sponges, and Anemones

These animals can cause extreme, though seldom fatal, pain. Usually found in tropical shallow water near coral formations, sea urchins resemble small, round porcupines. If stepped on, they slip fine needles of lime or silica into the skin, where they break off and fester. If possible, remove the spines and treat the injury for infection. The other animals mentioned inflict injury similarly.

Tides and Undertow

These are another hazard to contend with. If caught in a large wave's undertow, push off the bottom or swim to the surface and proceed shoreward in a trough between waves. Do not fight against the pull of the undertow. Swim with it or perpendicular to it until it loses strength, then swim for shore.

Food

Obtaining food along a seashore should not present a problem. There are many types of seaweed and other plants and animal life you can easily find and eat.

See *The Complete U.S. Army Survival Guide to Foraging Skills, Tactics, and Techniques,* Chapters 2 and 6 for more information.

Mollusks

Mussels, limpets, clams, sea snails, octopuses, squids, and sea slugs are all edible. Shellfish will usually supply most of the protein eaten by coastal survivors. Avoid the blue-ringed octopus and cone shells (*The Complete U.S. Army Survival Guide to Foraging Skills, Tactics, and Techniques*, Chapter 5). Also beware of "red tides" that make mollusks poisonous. Apply the edibility test on each species before eating.

Worms

Coastal worms are generally edible, but it is better to use them for fish bait. Avoid bristle worms that look like fuzzy caterpillars. Also avoid tubeworms that have sharp-edged tubes. Arrow worms, alias amphioxus, are not true worms. You find them in the sand and are excellent either fresh or dried.

Crabs, Lobsters, and Barnacles

These animals are seldom dangerous to man and are an excellent food source. The pincers of larger crabs or lobsters can crush a man's finger. Many species have spines on their shells, making it preferable to wear gloves when catching them. Barnacles can cause scrapes or cuts and are difficult to detach from their anchor, but the larger species are an excellent food source.

Sea Urchins

These are common and can cause painful injuries when stepped on or touched. They are also a good source of food. Handle them with gloves, and remove all spines.

Sea Cucumbers

This animal is an important food source in the Indo-Pacific regions. Use them whole after evisceration or remove the five muscular strips that run the length of its body. Eat them smoked, pickled, or cooked.

CHAPTER 6:

WATER CROSSINGS

In a survival situation, you may have to cross a water obstacle. It may be in the form of a river, a stream, a lake, a bog, quicksand, quagmire, or muskeg. Even in the desert, flash floods occur, making streams an obstacle. Whatever it is, you need to know how to cross it safely.

RIVERS AND STREAMS

You can apply almost every description to rivers and streams. They maybe shallow or deep, slow or fast moving, narrow or wide. Before you try to cross a river or stream, develop a good plan.

Your first step is to look for a high place from which you can get a good view of the river or stream. From this place, you can look for a place to cross. If there is no high place, climb a tree. Good crossing locations include—

- A level stretch where it breaks into several channels. Two or three narrow channels are usually easier to cross than a wide river.
- A shallow bank or sandbar. If possible, select a point upstream from the bank or sandbar so that the current will carry you to it if you lose your footing.
- A course across the river that leads downstream so that you will cross the current at about a 45-degree angle.

The following areas possess potential hazards; avoid them, if possible:

- *Obstacles on the opposite side of the river that might hinder your travel.* Try to select the spot from which travel will be the safest and easiest.

- *A ledge of rocks that crosses the river.* This often indicates dangerous rapids or canyons.
- *A deep or rapid waterfall or a deep channel.* Never try to ford a stream directly above or even close to such hazards.
- *Rocky places.* You may sustain serious injuries from slipping or falling on rocks. Usually, submerged rocks are very slick, making balance extremely difficult. An occasional rock that breaks the current, however, may help you.
- *An estuary of a river.* An estuary is normally wide, has strong currents, and is subject to tides. These tides can influence some rivers many kilometers from their mouths. Go back upstream to an easier crossing site.
- *Eddies.* An eddy can produce a powerful backward pull downstream of the obstruction causing the eddy and pull you under the surface.

The depth of a fordable river or stream is no deterrent if you can keep your footing. In fact, deep water sometimes runs more slowly and is therefore safer than fast-moving shallow water. You can always dry your clothes later, or if necessary, you can make a raft to carry your clothing and equipment across the river.

You must not try to swim or wade across a stream or river when the water is at very low temperatures. This swim could be fatal. Try to make a raft of some type. Wade across if you can get only your feet wet. Dry them vigorously as soon as you reach the other bank.

RAPIDS

If necessary, you can safely cross a deep, swift river or rapids. To swim across a deep, swift river, swim with the current, never fight it. Try to keep your body horizontal to the water. This will reduce the danger of being pulled under.

In fast, shallow rapids, lie on your back, feet pointing downstream, finning your hands alongside your hips. This action will increase buoyancy and help you steer away from obstacles. Keep your feet up to avoid getting them bruised or caught by rocks.

In deep rapids, lie on your stomach, head downstream, angling toward the shore whenever you can. Watch for obstacles and be careful of backwater eddies and converging currents, as they often contain dangerous swirls. Converging currents occur where new watercourses enter the river or where water has been diverted around large obstacles such as small islands.

To ford a swift, treacherous stream, apply the following steps:

- Remove your pants and shirt to lessen the water's pull on you. Keep your footgear on to protect your feet and ankles from rocks. It will also provide you with firmer footing.
- Tie your pants and other articles to the top of your rucksack or in a bundle, if you have no pack. This way, if you have to release your equipment, all your articles will be together. It is easier to find one large pack than to find several small items.
- Carry your pack well up on your shoulders and be sure you can easily remove it, if necessary. Not being able to get a pack off quickly enough can drag even the strongest swimmers under.
- Find a strong pole about 7.5 centimeters in diameter and 2.1 to 2.4meters long to help you ford the stream. Grasp the pole and plant it firmly on your upstream side to break the current. Plant your feet firmly with each step, and move the pole forward a little downstream from its previous position, but still upstream from you. With your next step, place your foot below the pole. Keep the pole well slanted so that the force of the current keeps the pole against your shoulder (Figure 6-1).
- Cross the stream so that you will cross the downstream current at a 45-degree angle.

Using this method, you can safely cross currents usually too strong for one person to stand against. Do not concern yourself about your pack's weight, as the weight will help rather than hinder you in fording the stream.

If there are other people with you, cross the stream together. Ensure that everyone has prepared their pack and clothing as outlined above. Position the heaviest person on the downstream end of the pole and the lightest on the upstream end. In using this method, the upstream person breaks the current, and those below can move with relative ease in the eddy formed by the upstream person. If the upstream person gets temporarily swept off his feet, the others can hold steady while he regains his footing (Figure 6-2).

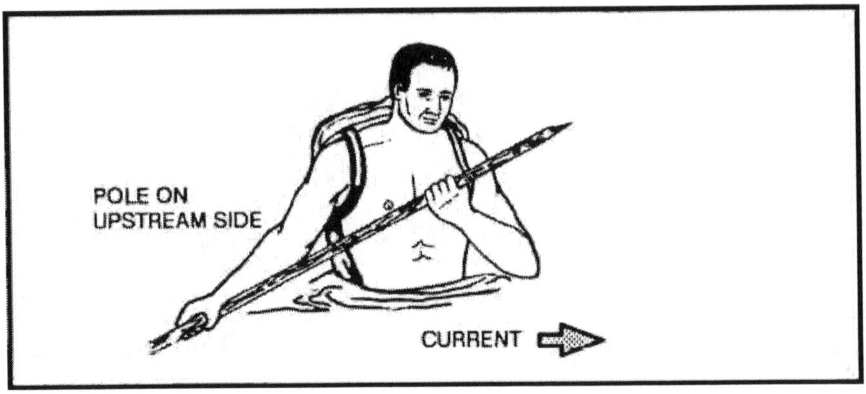

FIGURE 6-1: ONE MAN CROSSING SWIFT STREAM

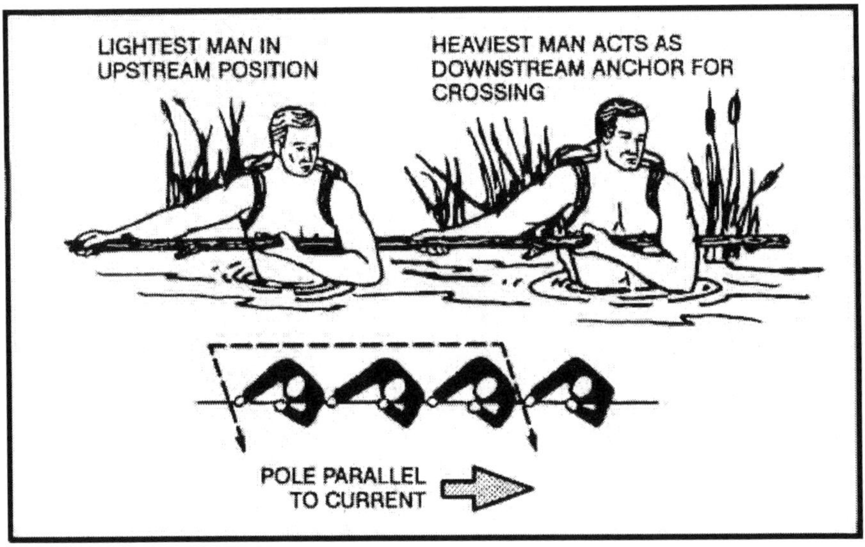

FIGURE 6-2: SEVERAL MEN CROSSING SWIFT STREAM

If you have three or more people and a rope available, you can use the technique shown in Figure 6-3 to cross the stream. The length of the rope must be three times the width of the stream.

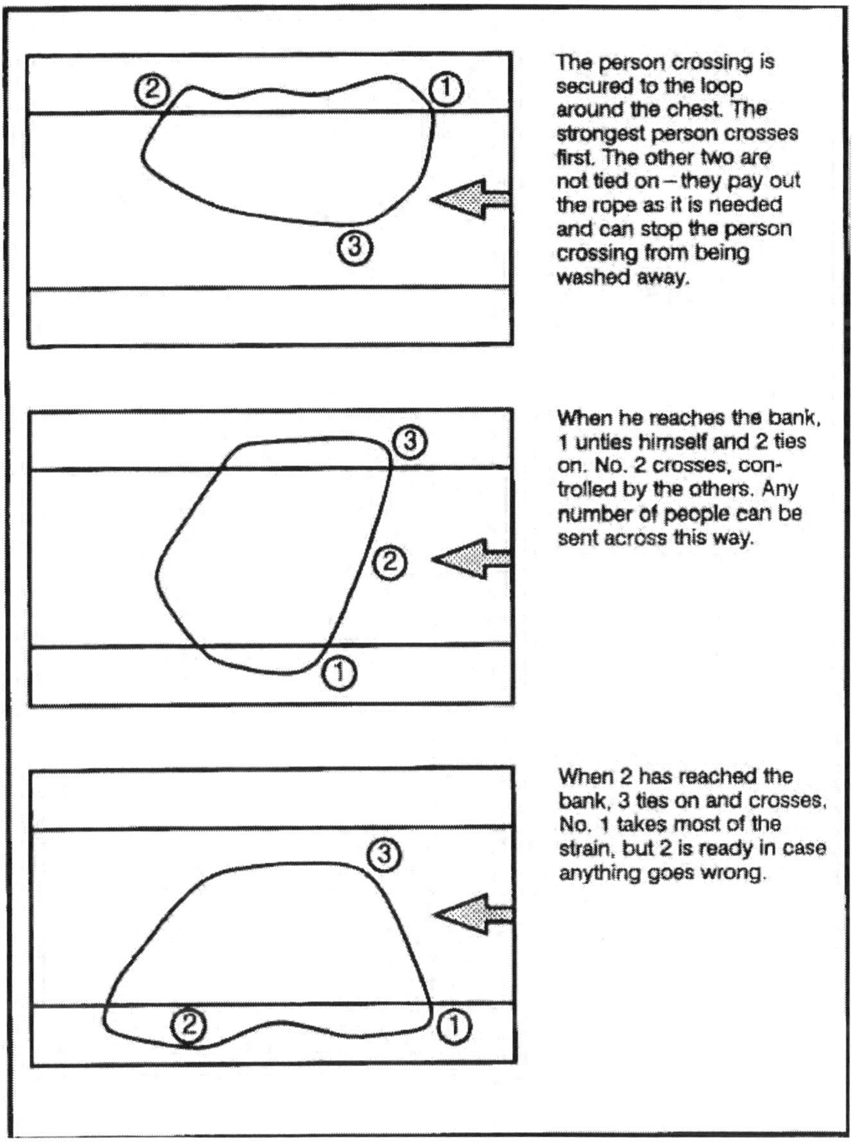

FIGURE 6-3: INDIVIDUALS TIED TOGETHER TO CROSS STREAM

RAFTS

If you have two ponchos, you can construct a brush raft or an Australian poncho raft. With either of these rafts, you can safely float your equipment across a slow-moving stream or river.

Brush Raft

The brush raft, if properly constructed, will support about 115 kilograms. To construct it, use ponchos, fresh green brush, two small saplings, and rope or vine as follows (Figure 6-4):

- Push the hood of each poncho to the inner side and tightly tie off the necks using the drawstrings.
- Attach the ropes or vines at the corner and side grommets of each poncho. Make sure they are long enough to cross to and tie with the others attached at the opposite corner or side.
- Spread one poncho on the ground with the inner side up. Pile fresh, green brush (no thick branches) on the poncho until the brush stack is about 45 centimeters high. Pull the drawstring up through the center of the brush stack.

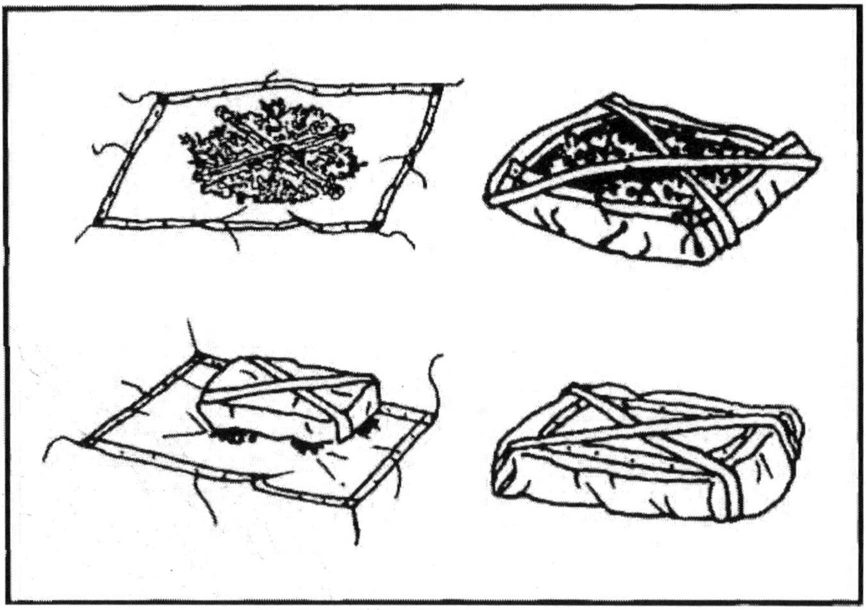

FIGURE 6-4: BRUSH RAFT

- Make an X-frame from two small saplings and place it on top of the brush stack. Tie the X-frame securely in place with the poncho drawstring.
- Pile another 45 centimeters of brush on top of the X-frame, then compress the brush slightly.
- Pull the poncho sides up around the brush and, using the ropes or vines attached to the comer or side grommets, tie them diagonally from corner to corner and from side to side.
- Spread the second poncho, inner side up, next to the brush bundle.
- Roll the brush bundle onto the second poncho so that the tied side is down. Tie the second poncho around the brush bundle in the same manner as you tied the first poncho around the brush.
- Place it in the water with the tied side of the second poncho facing up.

Australian Poncho Raft

If you do not have time to gather brush for a brush raft, you can make an Australian poncho raft. This raft, although more waterproof than the poncho brush raft, will only float about 35 kilograms of equipment. To construct this raft, use two ponchos, two rucksacks, two 1.2-meter poles or branches, and ropes, vines, bootlaces, or comparable material as follows (Figure 6-5):

- Push the hood of each poncho to the inner side and tightly tie off the necks using the drawstrings.
- Spread one poncho on the ground with the inner side up. Place and center the two 1.2-meter poles on the poncho about 45 centimeters apart.
- Place your rucksacks or packs or other equipment between the poles. Also place other items that you want to keep dry between the poles. Snap the poncho sides together.
- Use your buddy's help to complete the raft. Hold the snapped portion of the poncho in the air and roll it tightly down to the equipment. Make sure you roll the full width of the poncho.
- Twist the ends of the roll to form pigtails in opposite directions. Fold the pigtails over the bundle and tie them securely in place using ropes, bootlaces, or vines.
- Spread the second poncho on the ground, inner side up. If you need more buoyancy, place some fresh green brush on this poncho.

- Place the equipment bundle, tied side down, on the center of the second poncho. Wrap the second poncho around the equipment bundle following the same procedure you used for wrapping the equipment in the first poncho.
- Tie ropes, bootlaces, vines, or other binding material around the raft about 30 centimeters from the end of each pigtail. Place and secure weapons on top of the raft.
- Tie one end of a rope to an empty canteen and the other end to the raft. This will help you to tow the raft.

Poncho Donut Raft

Another type of raft is the poncho donut raft. It takes more time to construct than the brush raft or Australian poncho raft, but it is effective. To construct it, use one poncho, small saplings, willow or vines, and rope, bootlaces, or other binding material (Figure 6-6) as follows:

- Make a framework circle by placing several stakes in the ground that roughly outline an inner and outer circle.
- Using young saplings, willow, or vines, construct a donut ring within the circles of stakes.
- Wrap several pieces of cordage around the donut ring about 30 to 60 centimeters apart and tie them securely.

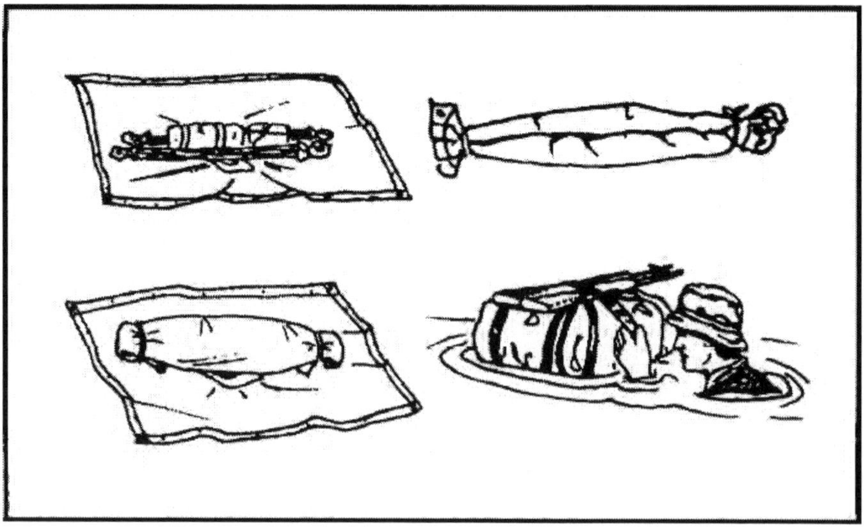

FIGURE 6-5: AUSTRALIAN PONCHO RAFT

- Push the poncho's hood to the inner side and tightly tie off the neck using the drawstring. Place the poncho on the ground, inner side up.
- Place the donut ring on the center of the poncho. Wrap the poncho up and over the donut ring and tie off each grommet on the poncho to the ring.
- Tie one end of a rope to an empty canteen and the other end to the raft. This rope will help you to tow the raft.

When launching any of the above rafts, take care not to puncture or tear it by dragging it on the ground. Before you start to cross the river or stream, let the raft lay on the water a few minutes to ensure that it floats.

If the river is too deep to ford, push the raft in front of you while you are swimming. The design of the above rafts does not allow them to carry a person's full body weight. Use them as a float to get you and your equipment safely across the river or stream.

Be sure to check the water temperature before trying to cross a river or water obstacle. If the water is extremely cold and you are unable to find a shallow fording place in the river, do not try to ford it. Devise other means for crossing. For instance, you might improvise a bridge by felling a tree over the river. Or you might build a raft large enough to carry you and your equipment. For-this, however, you will need an axe, a knife, a rope or vines, and time.

Log Raft

You can make a raft using any dry, dead, standing trees for logs. However, spruce trees found in polar and subpolar regions make the best rafts.

FIGURE 6-6: PONCHO DONUT RAFT

A simple method for making a raft is to use pressure bars lashed securely at each end of the raft to hold the logs together (Figure 6-7).

FLOTATION DEVICES

If the water is warm enough for swimming and you do not have the time or materials to construct one of the poncho-type rafts, you can use various flotation devices to negotiate the water obstacle. Some items you can use for flotation devices are—

- *Trousers*. Knot each trouser leg at the bottom and close the fly. With both hands, grasp the waistband at the sides and swing the trousers in the air to trap air in each leg. Quickly press the sides of the waistband together and hold it underwater so that the air will not escape. You now have water wings to keep you afloat as you cross the body of water. *Note: Wet the trousers before inflating to trap the air better. You may have to reinflate the trousers several times when crossing a large body of water.*
- *Empty containers*. Lash together her empty gas cans, water jugs, ammo cans, boxes, or other items that will trap or hold air. Use them as water wings. Use this type of flotation device only in a slow-moving river or stream.
- *Plastic bags and ponchos*. Fill two or more plastic bags with air and secure them together at the opening. Use your poncho and roll green vegetation tightly inside it so that you have a roll at least 20 centimeters in diameter. Tie the ends of the roll securely. You can wear it around your waist or across one shoulder and under the opposite arm.

FIGURE 6-7: USE OF PRESSURE BARS

- *Logs.* Use a stranded drift log if one is available, or find a log near the water to use as a float. Be sure to test the log before starting to cross. Some tree logs, palm for example, will sink even when the wood is dead. Another method is to tie two logs about 60 centimeters apart. Sit between the logs with your back against one and your legs over the other (Figure 6-8).
- *Cattails.* Gather stalks of cattails and tie them in a bundle 25 centimeters or more in diameter. The many air cells in each stalk cause a stalk to float until it rots. Test the cattail bundle to be sure it will support your weight before trying to cross a body of water.

There are many other flotation devices that you can devise by using some imagination. Just make sure to test the device before trying to use it.

OTHER WATER OBSTACLES

Other water obstacles that you may face are bogs, quagmire, muskeg, or quicksand. Do not try to walk across these. Trying to lift your feet while standing upright will make you sink deeper. Try to bypass these obstacles. If you are unable to bypass them, you may be able to bridge them using logs, branches, or foliage.

A way to cross a bog is to lie face down, with your arms and legs spread. Use a flotation device or form pockets of air in your clothing. Swim or pull your way across moving slowly and trying to keep your body horizontal.

FIGURE 6-8: LOG FLOTATION

In swamps, the areas that have vegetation are usually firm enough to support your weight. However, vegetation will usually not be present in open mud or water areas. If you are an average swimmer, however, you should have no problem swimming, crawling, or pulling your way through miles of bog or swamp.

Quicksand is a mixture of sand and water that forms a shifting mass. It yields easily to pressure and sucks down and engulfs objects resting on its surface. It varies in depth and is usually localized. Quicksand commonly occurs on flat shores, in silt-choked rivers with shifting watercourses, and near the mouths of large rivers. If you are uncertain whether a sandy area is quicksand, toss a small stone on it. The stone will sink in quicksand. Although quicksand has more suction than mud or muck, you can cross it just as you would cross a bog. Lie face down, spread your arms and legs, and move slowly across.

VEGETATION OBSTACLES

Some water areas you must cross may have underwater and floating plants that will make swimming difficult. However, you can swim through relatively dense vegetation if you remain calm and do not thrash about. Stay as near the surface as possible and use the breaststroke with shallow leg and arm motion. Remove the plants around you as you would clothing. When you get tired, float or swim on your back until you have rested enough to continue with the breaststroke.

The mangrove swamp is another type of obstacle that occurs along tropical coastlines. Mangrove trees or shrubs throw out many prop roots that form dense masses. To get through a mangrove swamp, wait for low tide. If you are on the inland side, look for a narrow grove of trees and work your way seaward through these. You can also try to find the bed of a waterway or creek through the trees and follow it to the sea. If you are on the seaward side, work inland along streams or channels. Be on the lookout for crocodiles that you find along channels and in shallow water. If there are any near you, leave the water and scramble over the mangrove roots. While crossing a mangrove swamp, it is possible to gather food from tidal pools or tree roots.

To cross a large swamp area, construct some type of raft.

CHAPTER 7:

SURVIVAL IN NUCLEAR, BIOLOGICAL, AND CHEMICAL ENVIRONMENTS

Nuclear, chemical, and biological weapons have become potential realities on any modern battlefield. Recent experience in Afghanistan, Cambodia, and other areas of conflict has proved the use of chemical and biological weapons (such as mycotoxins). The war fighting doctrine of the NATO and Warsaw Pact nations addresses the use of both nuclear and chemical weapons. The potential use of these weapons intensifies the problems of survival because of the serious dangers posed by either radioactive fallout or contamination produced by persistent biological or chemical agents.

You must use special precautions if you expect to survive in these man-made hazards. If you are subjected to any of the effects of nuclear, chemical, or biological warfare, the survival procedures recommended in this chapter may save your life. This chapter presents some background information on each type of hazard so that you may better understand the true nature of the hazard. Awareness of the hazards, knowledge of this chapter, and application of common sense should keep you alive.

THE NUCLEAR ENVIRONMENT

Prepare yourself to survive in a nuclear environment. Know how to react to a nuclear hazard.

Effects of Nuclear Weapons

The effects of nuclear weapons are classified as either initial or residual. Initial effects occur in the immediate area of the explosion and are hazardous in the first minute after the explosion. Residual effects can last for days or years and cause death. The principal initial effects are blast and radiation.

Blast

Defined as the brief and rapid movement of air away from the explosion's center and the pressure accompanying this movement. Strong winds accompany the blast. Blast hurls debris and personnel, collapses lungs, ruptures eardrums, collapses structures and positions, and causes immediate death or injury with its crushing effect.

Thermal Radiation

The heat and light radiation a nuclear explosion's fireball emits. Light radiation consists of both visible light and ultraviolet and infrared light. Thermal radiation produces extensive fires, skin burns, and flash blindness.

Nuclear Radiation

Nuclear radiation breaks down into two categories-initial radiation and residual radiation.

Initial nuclear radiation consists of intense gamma rays and neutrons produced during the first minute after the explosion. This radiation causes extensive damage to cells throughout the body. Radiation damage may cause headaches, nausea, vomiting, diarrhea, and even death, depending on the radiation dose received. The major problem in protecting yourself against the initial radiation's effects is that you may have received a lethal or incapacitating dose before taking any protective action. Personnel exposed to lethal amounts of initial radiation may well have been killed or fatally injured by blast or thermal radiation.

Residual radiation consists of all radiation produced after one minute from the explosion. It has more effect on you than initial radiation. A discussion of residual radiation follows.

Types of Nuclear Bursts

There are three types of nuclear bursts—airburst, surface burst, and subsurface burst. The type of burst directly affects your chances of survival. A subsurface burst occurs completely underground or underwater. Its effects remain beneath the surface or in the immediate area where the surface collapses into a crater over the burst's location. Subsurface bursts cause you little or no radioactive hazard unless you enter the immediate area of the crater. No further discussion of this type of burst will take place.

An airburst occurs in the air above its intended target. The airburst provides the maximum radiation effect on the target and is, therefore, most dangerous to you in terms of immediate nuclear effects.

A surface burst occurs on the ground or water surface. Large amounts of fallout result, with serious long-term effects for you. This type of burst is your greatest nuclear hazard.

Nuclear Injuries

Most injuries in the nuclear environment result from the initial nuclear effects of the detonation. These injuries are classed as blast, thermal, or radiation injuries. Further radiation injuries may occur if you do not take proper precautions against fallout. Individuals in the area near a nuclear explosion will probably suffer a combination of all three types of injuries.

Blast Injuries

Blast injuries produced by nuclear weapons are similar to those caused by conventional high-explosive weapons. Blast overpressure can produce collapsed lungs and ruptured internal organs. Projectile wounds occur as the explosion's force hurls debris at you. Large pieces of debris striking you will cause fractured limbs or massive internal injuries. Blast overpressure may throw you long distances, and you will suffer severe injury upon impact with the ground or other objects. Substantial cover and distance from the explosion are the best protection against blast injury. Cover blast injury wounds as soon as possible to prevent the entry of radioactive dust particles.

Thermal Injuries

The heat and light the nuclear fireball emits causes thermal injuries. First-, second-, or third-degree burns may result. Flash blindness also occurs. This blindness may be permanent or temporary depending on the degree of exposure of the eyes. Substantial cover and distance from the explosion can prevent thermal injuries. Clothing will provide significant protection against thermal injuries. Cover as much exposed skin as possible before a nuclear explosion. First aid for thermal injuries is the same as first aid for burns. Cover open burns (second- or third-degree) to prevent the entry of radioactive particles. Wash all burns before covering.

Radiation Injuries

Neutrons, gamma radiation, alpha radiation, and beta radiation cause radiation injuries. Neutrons are high-speed, extremely penetrating particles that actually smash cells within your body. Gamma radiation is similar to X-rays and is also a highly penetrating radiation. During the initial fireball stage of a nuclear detonation, initial gamma radiation and neutrons are the most serious threat. Beta and alpha radiation are radioactive particles normally associated with radioactive dust from fallout.

They are short-range particles and you can easily protect yourself against them if you take precautions. See Bodily Reactions to Radiation, below, for the symptoms of radiation injuries.

Residual Radiation

Residual radiation is all radiation emitted after 1 minute from the instant of the nuclear explosion. Residual radiation consists of induced radiation and fallout.

Induced Radiation

It describes a relatively small, intensely radioactive area directly underneath the nuclear weapon's fireball. The irradiated earth in this area will remain highly radioactive for an extremely long time. You should not travel into an area of induced radiation.

Fallout

Fallout consists of radioactive soil and water particles, as well as weapon fragments. During a surface detonation, or if an airburst's nuclear fireball touches the ground, large amounts of soil and water are vaporized along with the bomb's fragments, and forced upward to altitudes of 25,000 meters or more. When these vaporized contents cool, they can form more than 200 different radioactive products. The vaporized bomb contents condense into tiny radioactive particles that the wind carries and they fall back to earth as radioactive dust. Fallout particles emit alpha, beta, and gamma radiation. Alpha and beta radiation are relatively easy to counteract, and residual gamma radiation is much less intense than the gamma radiation emitted during the first minute after the explosion. Fallout is your most significant radiation hazard, provided you have not received a lethal radiation dose from the initial radiation.

Bodily Reactions to Radiation

The effects of radiation on the human body can be broadly classed as either chronic or acute. Chronic effects are those that occur some years after exposure to radiation. Examples are cancer and genetic defects. Chronic effects are of minor concern insofar as they affect your immediate survival in a radioactive environment. On the other hand, acute effects are of primary importance to your survival. Some acute effects occur within hours after exposure to radiation. These effects result from the radiation's direct physical damage to tissue. Radiation sickness and beta burns are examples of acute effects. Radiation sickness symptoms include nausea, diarrhea, vomiting, fatigue, weakness, and loss of hair. Penetrating beta rays cause radiation burns; the wounds are similar to fire burns.

Recovery Capability

The extent of body damage depends mainly on the part of the body exposed to radiation and how long it was exposed, as well as its ability to recover. The brain and kidneys have little recovery capability. Other parts (skin and bone marrow) have a great ability to recover from damage.

Usually, a dose of 600 centigrams (cgys) to the entire body will result in almost certain death. If only your hands received this same dose, your overall health would not suffer much, although your hands would suffer severe damage.

External and Internal Hazards

An external or an internal hazard can cause body damage. Highly penetrating gamma radiation or the less penetrating beta radiation that causes burns can cause external damage. The entry of alpha or beta radiation-emitting particles into the body can cause internal damage. The external hazard produces overall irradiation and beta burns. The internal hazard results in irradiation of critical organs such as the gastrointestinal tract, thyroid gland, and bone. A very small amount of radioactive material can cause extreme damage to these and other internal organs. The internal hazard can enter the body either through consumption of contaminated water or food or by absorption through cuts or abrasions. Material that enters the body through breathing presents only a minor hazard. You can greatly reduce the internal radiation hazard by using good personal hygiene and carefully decontaminating your food and water.

Symptoms

The symptoms of radiation injuries include nausea, diarrhea, and vomiting. The severity of these symptoms is due to the extreme sensitivity of the gastrointestinal tract to radiation. The severity of the symptoms and the speed of onset after exposure are good indicators of the degree of radiation damage. The gastrointestinal damage can come from either the external or the internal radiation hazard.

Countermeasures Against Penetrating External Radiation

Knowledge of the radiation hazards discussed earlier is extremely important in surviving in a fallout area. It is also critical to know how to protect yourself from the most dangerous form of residual radiation—penetrating external radiation.

The means you can use to protect yourself from penetrating external radiation are time, distance, and shielding. You can reduce the level of radiation and help increase your chance of survival by controlling the duration of exposure. You can also get as far away from the radiation source as possible. Finally you can place some radiation-absorbing or shielding material between you and the radiation.

Time

Time is important to you, as the survivor, in two ways. First, radiation dosages are cumulative. The longer you are exposed to a radioactive source, the greater the dose you will receive. Obviously, spend as little time in a radioactive area as possible. Second, radioactivity decreases or decays over time. This concept is known as radioactive half-life. Thus, a radioactive element decays or loses half of its radioactivity within a certain time. The rule of thumb for radioactivity decay is that it decreases in intensity by a factor of ten for every sevenfold increase in time following the peak radiation level. For example, if a nuclear fallout area had a maximum radiation rate of 200 cgys per hour when fallout is complete, this rate would fall to 20 cgys per hour after 7 hours; it would fall still further to 2 cgys per hour after 49 hours. Even an untrained observer can see that the greatest hazard from fallout occurs immediately after detonation, and that the hazard decreases quickly over a relatively short time. As a survivor, try to avoid fallout areas until the radioactivity decays to safe levels. If you can avoid fallout areas long enough for most of the radioactivity to decay, you enhance your chance of survival.

Distance

Distance provides very effective protection against penetrating gamma radiation because radiation intensity decreases by the square of the distance from the source. For example, if exposed to 1,000 cgys of radiation standing 30 centimeters from the source, at 60 centimeters, you would only receive 250 cgys. Thus, when you double the distance, radiation decreases to $(0.5)^2$ or 0.25 the amount. While this formula is valid for concentrated sources of radiation in small areas, it becomes more complicated for large areas of radiation such as fallout areas.

Shielding

Shielding is the most important method of protection from penetrating radiation. Of the three countermeasures against penetrating radiation, shielding provides the greatest protection and is the easiest to use under survival conditions. Therefore, it is the most desirable method. If shielding is not possible, use the other two methods to the maximum extent practical.

Shielding actually works by absorbing or weakening the penetrating radiation, thereby reducing the amount of radiation reaching your body. The denser the material, the better the shielding effect. Lead, iron, concrete, and water are good examples of shielding materials.

Special Medical Aspects

The presence of fallout material in your area requires slight changes in first aid procedures. You must cover all wounds to prevent contamination and the entry of radioactive particles. You must first wash burns of beta radiation, then treat them as ordinary burns. Take extra measures to prevent infection. Your body will be extremely sensitive to infections due to changes in your blood chemistry. Pay close attention to the prevention of colds or respiratory infections. Rigorously practice personal hygiene to prevent infections. Cover your eyes with improvised goggles to prevent the entry of particles.

Shelter

As stated earlier, the shielding material's effectiveness depends on its thickness and density. An ample thickness of shielding material will reduce the level of radiation to negligible amounts.

The primary reason for finding and building a shelter is to get protection against the high-intensity radiation levels of early gamma fallout as fast as possible. Five minutes to locate the shelter is a good guide. Speed in finding shelter is absolutely essential. Without shelter, the dosage received in the first few hours will exceed that received during the rest of a week in a contaminated area. The dosage received in this first week will exceed the dosage accumulated during the rest of a lifetime spent in the same contaminated area.

Shielding Materials

The thickness required to weaken gamma radiation from fallout is far less than that needed to shield against initial gamma radiation. Fallout radiation has less energy than a nuclear detonation's initial radiation. For fallout radiation, a relatively small amount of shielding material can provide adequate protection. Figure 7-1 gives an idea of the thickness of various materials needed to reduce residual gamma radiation transmission by 50 percent. The principle of half-value layer thickness is useful in understanding the absorption of gamma radiation by various materials. According to this principle, if 5 centimeters of brick reduce the gamma radiation level by one-half, adding another 5 centimeters of brick (another half-value layer) will reduce the intensity by another half, namely, to one-fourth the original amount. Fifteen centimeters will reduce gamma radiation fallout levels to one-eighth its original amount, 20 centimeters to one-sixteenth, and so on. Thus, a shelter protected by 1 meter of dirt would reduce a radiation intensity of 1,000 cgys per hour on the outside to about 0.5 cgy per hour inside the shelter.

Natural Shelters

Terrain that provides natural shielding and easy shelter construction is the ideal location for an emergency shelter. Good examples are ditches, ravines, rocky outcropping, hills, and river banks. In level areas without natural protection, dig a fighting position or slit trench.

Trenches

When digging a trench, work from inside the trench as soon as it is large enough to cover part of your body thereby not exposing all your body to radiation. In open country, try to dig the trench from a prone position, stacking the dirt carefully and evenly around the trench. On level ground, pile the dirt around your body for additional shielding.

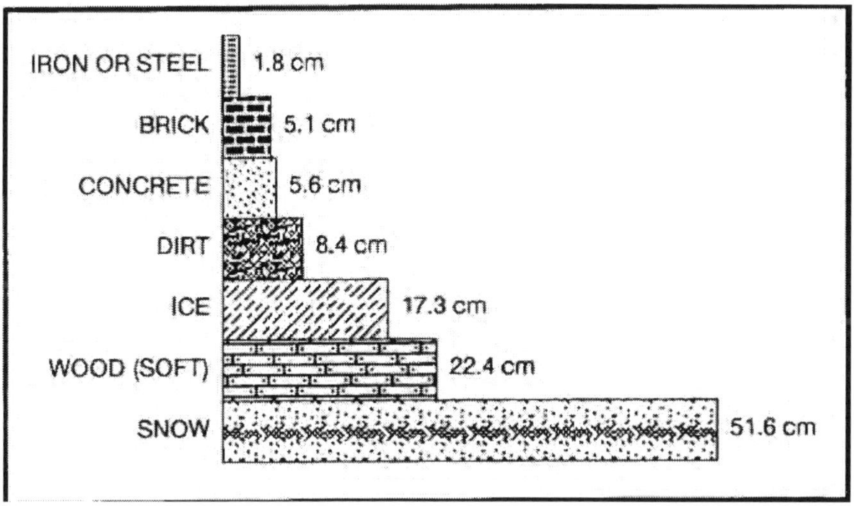

FIGURE 7-1: THICKNESS OF MATERIALS TO REDUCE GAMMA RADIATION

Depending upon soil conditions, shelter construction time will vary from a few minutes to a few hours. If you dig as quickly as possible, you will reduce the dosage you receive.

Other Shelters

While an underground shelter covered by 1 meter or more of earth provides the best protection against fallout radiation, the following unoccupied structures (in order listed) offer the next best protection:

- Caves and tunnels covered by more than 1 meter of earth.
- Storm or storage cellars.
- Culverts.
- Basements or cellars of abandoned buildings.
- Abandoned buildings made of stone or mud.

Roofs

It is not mandatory that you build a roof on your shelter. Build one only if the materials are readily available with only a brief exposure to outside contamination. If building a roof would require extended exposure to penetrating radiation, it would be wiser to leave the shelter roofless. A roof's sole function is to reduce radiation from the fallout source to your body. Unless you use a thick roof, a roof provides very little shielding.

You can construct a simple roof from a poncho anchored down with dirt, rocks, or other refuse from your shelter. You can remove large particles of dirt and debris from the top of the poncho by beating it off from the inside at frequent intervals. This cover will not offer shielding from the radioactive particles deposited on the surface, but it will increase the distance from the fallout source and keep the shelter area from further contamination.

Shelter Site Selection and Preparation

To reduce your exposure time and thereby reduce the dosage received, remember the following factors when selecting and setting up a shelter:

- Where possible, seek a crude, existing shelter that you can improve. If none is available, dig a trench.
- Dig the shelter deep enough to get good protection, then enlarge it as required for comfort.
- Cover the top of the fighting position or trench with any readily available material and a thick layer of earth, if you can do so without leaving the shelter. While a roof and camouflage are both desirable, it is probably safer to do without them than to expose yourself to radiation outside your fighting position.
- While building your shelter, keep all parts of your body covered with clothing to protect it against beta burns.
- Clean the shelter site of any surface deposit using a branch or other object that you can discard. Do this cleaning to remove contaminated materials from the area you will occupy. The cleaned area should extend at least 1.5 meters beyond your shelter's area.
- Decontaminate any materials you bring into the shelter. These materials include grass or foliage that you use as insulation or bedding, and your outer clothing (especially footgear). If the weather permits and you have heavily contaminated outer clothing, you may want to remove it and bury it under a foot of earth at the end of your shelter. You may retrieve it later (after the radioactivity decays) when leaving the shelter. If the clothing is dry, you may decontaminate it by beating or shaking it outside the shelter's entrance to remove the radioactive dust. You may use any body of water, even though contaminated, to rid materials of excess fallout particles. Simply dip

the material into the water and shake it to get rid of the excess water. Do not wring it out, this action will trap the particles.

- If at all possible and without leaving the shelter, wash your body thoroughly with soap and water, even if the water on hand may be contaminated. This washing will remove most of the harmful radioactive particles that are likely to cause beta burns or other damage. If water is not available, wipe your face and any other exposed skin surface to remove contaminated dust and dirt. You may wipe your face with a clean piece of cloth or a handful of uncontaminated dirt. You get this uncontaminated dirt by scraping off the top few inches of soil and using the "clean" dirt.
- Upon completing the shelter, lie down, keep warm, and sleep and rest as much as possible while in the shelter.
- When not resting, keep busy by planning future actions, studying your maps, or making the shelter more comfortable and effective.
- Don't panic if you experience nausea and symptoms of radiation sickness. Your main danger from radiation sickness is infection. There is no first aid for this sickness. Resting, drinking fluids, taking any medicine that prevents vomiting, maintaining your food intake, and preventing additional exposure will help avoid infection and aid recovery. Even small doses of radiation can cause these symptoms which may disappear in a short time.

Exposure Timetable

The following timetable provides you with the information needed to avoid receiving serious dosage and still let you cope with survival problems:

- Complete isolation from 4 to 6 days following delivery of the last weapon.
- A very brief exposure to procure water on the third day is permissible, but exposure should not exceed 30 minutes.
- One exposure of not more than 30 minutes on the seventh day.
- One exposure of not more than 1 hour on the eighth day.
- Exposure of 2 to 4 hours from the ninth day through the twelfth day.
- Normal operation, followed by rest in a protected shelter, from the thirteenth day on.

- In all instances, make your exposures as brief as possible. Consider only mandatory requirements as valid reasons for exposure. Decontaminate at every stop.

The times given above are conservative. If forced to move after the first or second day, you may do so, Make sure that the exposure is no longer than absolutely necessary.

Water Procurement

In a fallout-contaminated area, available water sources may be contaminated. If you wait at least 48 hours before drinking any water to allow for radioactive decay to take place and select the safest possible water source, you will greatly reduce the danger of ingesting harmful amounts of radioactivity.

Although many factors (wind direction, rainfall, sediment) will influence your choice in selecting water sources, consider the following guidelines.

Safest Water Sources

Water from springs, wells, or other underground sources that undergo natural filtration will be your safest source. Any water found in the pipes or containers of abandoned houses or stores will also be free from radioactive particles. This water will be safe to drink, although you will have to take precautions against bacteria in the water.

Snow taken from 15 or more centimeters below the surface during the fallout is also a safe source of water.

Streams and Rivers

Water from streams and rivers will be relatively free from fallout within several days after the last nuclear explosion because of dilution. If at all possible, filter such water before drinking to get rid of radioactive particles. The best filtration method is to dig sediment holes or seepage basins along the side of a water source. The water will seep laterally into the hole through the intervening soil that acts as a filtering agent and removes the contaminated fallout particles that settled on the original body of water. This method can remove up to 99 percent of the radioactivity in water. You must cover the hole in some way in order to prevent further contamination. See *The Complete U.S. Army Survival Guide to Foraging Skills, Tactics, and Techniques*, Chapter 1, Illustration 1-7 for examples of water filters.

Standing Water

Water from lakes, pools, ponds, and other standing sources is likely to be heavily contaminated, though most of the heavier, long-lived radioactive isotopes will settle to the bottom. Use the settling technique to purify this water. First, fill a bucket or other deep container three-fourths full with contaminated water. Then take dirt from a depth of 10 or more centimeters below the ground surface and stir it into the water. Use about 2.5 centimeters of dirt for every 10 centimeters of water. Stir the water until you see most dirt particles suspended in the water. Let the mixture settle for at least 6 hours. The settling dirt particles will carry most of the suspended fallout particles to the bottom and cover them. You can then dip out the clear water. Purify this water using a filtration device.

Additional Precautions

As an additional precaution against disease, treat all water with water purification tablets from your survival kit or boil it.

Food Procurement

Although it is a serious problem to obtain edible food in a radiation-contaminated area, it is not impossible to solve. You need to follow a few special procedures in selecting and preparing rations and local foods for use. Since secure packaging protects your combat rations, they will be perfectly safe for use. Supplement your rations with any food you can find on trips outside your shelter. Most processed foods you may find in abandoned buildings are safe for use after decontaminating them. These include canned and packaged foods after removing the containers or wrappers or washing them free of fallout particles. These processed foods also include food stored in any closed container and food stored in protected areas (such as cellars), if you wash them before eating. Wash all food containers or wrappers before handling them to prevent further contamination.

If little or no processed food is available in your area, you may have to supplement your diet with local food sources. Local food sources are animals and plants.

Animals as a Food Source

Assume that all animals, regardless of their habitat or living conditions, were exposed to radiation. The effects of radiation on animals are similar to those on humans. Thus, most of the wild animals living in a fallout area are likely to become sick or die from radiation during the first month after the nuclear explosion. Even though animals may not be free from harmful radioactive materials, you can and must use them in survival conditions as a food source if other foods are not available. With careful preparation and by following several important principles, animals can be safe food sources.

First, do not eat an animal that appears to be sick. It may have developed a bacterial infection as a result of radiation poisoning. Contaminated meat, even if thoroughly cooked, could cause severe illness or death if eaten.

Carefully skin all animals to prevent any radioactive particles on the skin or fur from entering the body. Do not eat meat close to the bones and joints as an animal's skeleton contains over 90 percent of the radioactivity. The remaining animal muscle tissue, however, will be safe to eat. Before cooking it, cut the meat away from the bone, leaving at least a 3-millimeter thickness of meat on the bone. Discard all internal organs (heart, liver, and kidneys) since they tend to concentrate beta and gamma radioactivity.

Cook all meat until it is very well done. To be sure the meat is well done, cut it into less than 13-millimeter-thick pieces before cooking. Such cuts will also reduce cooking time and save fuel.

The extent of contamination in fish and aquatic animals will be much greater than that of land animals. This is also true for water plants, especially in coastal areas. Use aquatic food sources only in conditions of extreme emergency.

All eggs, even if laid during the period of fallout, will be safe to eat. Completely avoid milk from any animals in a fallout area because animals absorb large amounts of radioactivity from the plants they eat.

Plants as a Food Source

Plant contamination occurs by the accumulation of fallout on their outer surfaces or by absorption of radioactive elements through their roots. Your first choice of plant food should be vegetables such as potatoes, turnips, carrots, and other plants whose edible portion grows underground. These are the safest to eat once you scrub them and remove their skins.

Second in order of preference are those plants with edible parts that you can decontaminate by washing and peeling their outer surfaces. Examples are bananas, apples, tomatoes, prickly pears, and other such fruits and vegetables.

Any smooth-skinned vegetable, fruit, or plant that you cannot easily peel or effectively decontaminate by washing will be your third choice of emergency food.

The effectiveness of decontamination by scrubbing is inversely proportional to the roughness of the fruit's surface. Smooth-surfaced fruits have lost 90 percent of their contamination after washing, while washing rough-surfaced plants removes only about 50 percent of the contamination.

You eat rough-surfaced plants (such as lettuce) only as a last resort because you cannot effectively decontaminate them by peeling or washing.

Other difficult foods to decontaminate by washing with water include dried fruits (figs, prunes, peaches, apricots, pears) and soya beans.

In general, you can use any plant food that is ready for harvest if you can effectively decontaminate it. Growing plants, however, can absorb some radioactive materials through their leaves as well as from the soil, especially if rains have occurred during or after the fallout period. Avoid using these plants for food except in an emergency.

BIOLOGICAL ENVIRONMENTS

The use of biological agents is real. Prepare yourself for survival by being proficient in the tasks identified in your Soldier's Manuals of Common Tasks (SMCTs). Know what to do to protect yourself against these agents.

Biological Agents and Effects

Biological agents are microorganisms that can cause disease among personnel, animals, or plants. They can also cause the deterioration of material. These agents fall into two broad categories-pathogens (usually called germs) and toxins. Pathogens are living microorganisms that cause lethal or incapacitating diseases. Bacteria, rickettsiae, fungi, and viruses are included in the pathogens. Toxins are poisons that plants, animals, or microorganisms produce naturally. Possible biological warfare toxins include a variety of neurotoxic (affecting the central nervous system) and cytotoxic (causing cell death) compounds.

Germs

Germs are living organisms. Some nations have used them in the pastas weapons. Only a few germs can start an infection, especially if inhaled into the lungs. Because germs are so small and weigh so little, the wind can spread them over great distances; they can also enter unfiltered or non-airtight places. Buildings and bunkers can trap them thus causing a higher concentration. Germs do not affect the body immediately. They must multiply inside the body and overcome the body's defenses—a process called the incubation period. Incubation periods vary from several hours to several months, depending on the germ. Most germs must live within another living organism (host), such as your body, to survive and grow. Weather conditions such as wind, rain, cold, and sunlight rapidly kill germs.

Some germs can form protective shells, or spores, to allow survival outside the host. Spore-producing agents are a long-term hazard you must neutralize by decontaminating infected areas or personnel. Fortunately, most live agents are not spore-producing. These agents must find a host within roughly a day of their delivery or they die. Germs have three basic routes of entry into your body: through the respiratory tract, through a break in the skin, and through the digestive tract. Symptoms of infection vary according to the disease.

Toxins

Toxins are substances that plants, animals, or germs produce naturally. These toxins are what actually harm man, not bacteria. Botulin, which produces botulism, is an example. Modern science has allowed large-scale production of these toxins without the use of the germ that produces the toxin. Toxins may produce effects similar to those of chemical agents. Toxic victims may not, however, respond to first aid measures used against chemical agents. Toxins enter the body in the same manner as germs. However, some toxins, unlike germs, can penetrate unbroken skin. Symptoms appear almost immediately, since there is no incubation period. Many toxins are extremely lethal, even in very small doses. Symptoms may include any of the following:

- Dizziness.
- Mental confusion.
- Blurred or double vision.
- Numbness or tingling of skin.
- Paralysis.
- Convulsions.
- Rashes or blisters.
- Coughing.

- Fever.
- Aching muscles.
- Tiredness.
- Nausea, vomiting, and/or diarrhea.
- Bleeding from body openings.
- Blood in urine, stool, or saliva.
- Shock.
- Death.

Detection of Biological Agents

Biological agents are, by nature, difficult to detect. You cannot detect them by any of the five physical senses. Often, the first sign of a biological agent will be symptoms of the victims exposed to the agent. Your best chance of detecting biological agents before they can affect you is to recognize their means of delivery. The three main means of delivery are—

- *Bursting-type munitions.* These may be bombs or projectiles whose burst causes very little damage. The burst will produce a small cloud of liquid or powder in the immediate impact area. This cloud will disperse eventually; the rate of dispersion depends on terrain and weather conditions.
- *Spray tanks or generators.* Aircraft or vehicle spray tanks or ground-level aerosol generators produce an aerosol cloud of biological agents.
- *Vectors.* Insects such as mosquitoes, fleas, lice, and ticks deliver pathogens. Large infestations of these insects may indicate the use of biological agents.

Another sign of a possible biological attack is the presence of unusual substances on the ground or on vegetation, or sick-looking plants, crops, or animals.

Influence of Weather and Terrain

Your knowledge of how weather and terrain affect the agents can help you avoid contamination by biological agents. Major weather factors that affect biological agents are sunlight, wind, and precipitation. Aerosol sprays will tend to concentrate in low areas of terrain, similar to early morning mist.

Sunlight contains visible and ultraviolet solar radiation that rapidly kills most germs used as biological agents. However, natural or man-made cover may protect some agents from sunlight. Other man-made mutant strains of germs may be resistant to sunlight.

High wind speeds increase the dispersion of biological agents, dilute their concentration, and dehydrate them. The further downwind the agent travels, the less effective it becomes due to dilution and death of the pathogens. However, the downwind hazard area of the biological agent is significant and you cannot ignore it.

Precipitation in the form of moderate to heavy rain tends to wash biological agents out of the air, reducing downwind hazard areas. However, the agents may still be very effective where they were deposited on the ground.

Protection Against Biological Agents

While you must maintain a healthy respect for biological agents, there is no reason for you to panic. You can reduce your susceptibility to biological agents by maintaining current immunizations, avoiding contaminated areas, and controlling rodents and pests. You must also use proper first aid measures in the treatment of wounds and only safe or properly decontaminated sources of food and water. You must ensure that you get enough sleep to prevent a run-down condition. You must always use proper field sanitation procedures.

Assuming you do not have a protective mask, always try to keep your face covered with some type of cloth to protect yourself against biological agent aerosols. Dust may contain biological agents; wear some type of mask when dust is in the air.

Your uniform and gloves will protect you against bites from vectors (mosquitoes and ticks) that carry diseases. Completely button your clothing and tuck your trousers tightly into your boots. Wear a chemical protective overgarment, if available, as it provides better protection than normal clothing. Covering your skin will also reduce the chance of the agent entering your body through cuts or scratches. Always practice high standards of personal hygiene and sanitation to help prevent the spread of vectors.

Bathe with soap and water whenever possible. Use germicidal soap, if available. Wash your hair and body thoroughly, and clean under your fingernails. Clean teeth, gums, tongue, and the roof of your mouth frequently. Wash your clothing in hot, soapy water if you can. If you cannot wash your clothing, lay it out in an area of bright sunlight and allow the light to kill the microorganisms. After a toxin attack, decontaminate yourself as if for a chemical attack using the M258A2 kit (if available) or by washing with soap and water.

Shelter

You can build expedient shelters under biological contamination conditions using the same techniques described in Part III. However, you must make slight changes to reduce the chance of biological contamination. Do not build your shelter in depressions in the ground. Aerosol sprays tend to concentrate in these depressions. Avoid building your shelter in areas of vegetation, as vegetation provides shade and some degree of protection to biological agents. Avoid using vegetation in constructing your shelter. Place your shelter's entrance at a 90-degree angle to the prevailing winds. Such placement will limit the entry of airborne agents and prevent air stagnation in your shelter. Always keep your shelter clean.

Water Procurement

Water procurement under biological conditions is difficult but not impossible. Whenever possible, try to use water that has been in a sealed container. You can assume that the water inside the sealed container is not contaminated. Wash the water container thoroughly with soap and water or boil it for at least 10 minutes before breaking the seal.

If water in sealed containers is not available, your next choice, only under emergency conditions, is water from springs. Again, boil the water for at least 10 minutes before drinking. Keep the water covered while boiling to prevent contamination by airborne pathogens. Your last choice, only in an extreme emergency, is to use standing water. Vectors and germs can survive easily in stagnant water. Boil this water as long as practical to kill all organisms. Filter this water through a cloth to remove the dead vectors. Use water purification tablets in all cases.

Food Procurement

Food procurement, like water procurement, is not impossible, but you must take special precautions. Your combat rations are sealed, and you can assume they are not contaminated. You can also assume that sealed containers or packages of processed food are safe. To ensure safety, decontaminate all food containers by washing with soap and water or by boiling the container in water for 10 minutes.

You consider supplementing your rations with local plants or animals only in extreme emergencies. No matter what you do to prepare the food, there is no guarantee that cooking will kill all the biological agents. Use local food only in life or death situations. Remember, you can survive for along time without food, especially if the food you eat may kill you!

If you must use local food, select only healthy-looking plants and animals. Do not select known carriers of vectors such as rats or other vermin. Select and prepare plants as you would in radioactive areas. Prepare animals as you do plants. Always use gloves and protective clothing when handling animals or plants. Cook all plant and animal food by boiling only. Boil all food for at least 10 minutes to kill all pathogens. Do not try to fry, bake, or roast local food. There is no guarantee that all infected portions have reached the required temperature to kill all pathogens. Do not eat raw food.

CHEMICAL ENVIRONMENTS

Chemical agent warfare is real. It can create extreme problems in a survival situation, but you can overcome the problems with the proper equipment, knowledge, and training. As a survivor, your first line of defense against chemical agents is your proficiency in individual nuclear, biological, and chemical (NBC) training, to include donning and wearing the protective mask and overgarment, personal decontamination, recognition of chemical agent symptoms, and individual first aid for chemical agent contamination. The SMCTs cover these subjects. If you are not proficient in these skills, you will have little chance of surviving a chemical environment.

Detection of Chemical Agents

The best method for detecting chemical agents is the use of a chemical agent detector. If you have one, use it. However, in a survival situation, you will most likely have to rely solely on the use of all of your physical senses. You must be alert and able to detect any clues indicating the use of chemical warfare. General indicators of the presence of chemical agents are tears, difficult breathing, choking, itching, coughing, and dizziness. With agents that are very hard to detect, you must watch for symptoms in fellow survivors. Your surroundings will provide valuable clues to the presence of chemical agents; for example, dead animals, sick people, or people and animals displaying abnormal behavior.

Your sense of smell may alert you to some chemical agents, but most will be odorless. The odor of newly cut grass or hay may indicate the presence of choking agents. A smell of almonds may indicate blood agents.

Sight will help you detect chemical agents. Most chemical agents in the solid or liquid state have some color. In the vapor state, you can see some chemical agents as a mist or thin fog immediately after the bomb or shell bursts. By observing for symptoms in others and by observing delivery means, you may be able to have some warning of chemical agents. Mustard gas in the liquid state will appear as oily patches on leaves or on buildings.

The sound of enemy munitions will give some clue to the presence of chemical weapons. Muffled shell or bomb detonations are a good indicator.

Irritation in the nose or eyes or on the skin is an urgent warning to protect your body from chemical agents. Additionally, a strange taste in food, water, or cigarettes may serve as a warning that they have been contaminated.

Protection Against Chemical Agents

As a survivor, always use the following general steps, in the order listed, to protect yourself from a chemical attack:

- Use protective equipment.
- Give quick and correct self-aid when contaminated.
- Avoid areas where chemical agents exist.
- Decontaminate your equipment and body as soon as possible.

Your protective mask and overgarment are the key to your survival. Without these, you stand very little chance of survival. You must take care of these items and protect them from damage. You must practice and know correct self-aid procedures before exposure to chemical agents. The detection of chemical agents and the avoidance of contaminated areas is extremely important to your survival. Use whatever detection kits may be available to help in detection. Since you are in a survival situation, avoid contaminated areas at all costs. You can expect no help should you become contaminated. If you do become contaminated, decontaminate yourself as soon as possible using proper procedures.

Shelter

If you find yourself in a contaminated area, try to move out of the area as fast as possible. Travel crosswind or upwind to reduce the time spent in the downwind hazard area. If you cannot leave the area immediately and have to build a shelter, use normal shelter construction techniques, with a few changes. Build the shelter in a clearing, away from all vegetation. Remove all topsoil in the area of the shelter to decontaminate the area. Keep the shelter's entrance closed and oriented at a 90-degreeangle to the prevailing wind. Do not build a fire using contaminated wood—the smoke will be toxic. Use extreme caution when entering your shelter so that you will not bring contamination inside.

Water Procurement

As with biological and nuclear environments, getting water in a chemical environment is difficult. Obviously, water in sealed containers is your best and safest source. You must protect this water as much as possible. Be sure to decontaminate the containers before opening.

If you cannot get water in sealed containers, try to get it from a closed source such as underground water pipes. You may use rainwater or snow if there is no evidence of contamination. Use water from slow-moving streams, if necessary, but always check first for signs of contamination, and always filter the water as described under nuclear conditions. Signs of water source contamination are foreign odors such as garlic, mustard, geranium, or bitter almonds; oily spots on the surface of the water or nearby; and the presence of dead fish or animals. If these signs are present, do not use the water. Always boil or purify the water to prevent bacteriological infection.

Food Procurement

It is extremely difficult to eat while in a contaminated area. You will have to break the seal on your protective mask to eat. If you eat, find an area in which you can safely unmask. The safest source of food is your sealed combat rations. Food in sealed cans or bottles will also be safe. Decontaminate all sealed food containers before opening, otherwise you will contaminate the food.

If you must supplement your combat rations with local plants or animals, do not use plants from contaminated areas or animals that appear to be sick. When handling plants or animals, always use protective gloves and clothing.

Made in the USA
Las Vegas, NV
16 October 2022

57427660R00236